Nucleic Acid Architectures for Therapeutics, Diagnostics, Devices and Materials

Nucleic Acid Architectures for Therapeutics, Diagnostics, Devices and Materials

Special Issue Editor

Kirill A. Afonin

MDPI • Basel • Beijing • Wuhan • Barcelona • Belgrade

MDPI

Special Issue Editor
Kirill A. Afonin
UNC Charlotte
USA

Editorial Office
MDPI
St. Alban-Anlage 66
4052 Basel, Switzerland

This is a reprint of articles from the Special Issue published online in the open access journal *Nanomaterials* (ISSN 2079-4991) from 2018 to 2019 (available at: https://www.mdpi.com/journal/nanomaterials/special_issues/Nucleic_Acid_Nanotechnology)

For citation purposes, cite each article independently as indicated on the article page online and as indicated below:

LastName, A.A.; LastName, B.B.; LastName, C.C. Article Title. *Journal Name* **Year**, *Article Number*, Page Range.

ISBN 978-3-03921-259-0 (Pbk)
ISBN 978-3-03921-260-6 (PDF)

Contents

About the Special Issue Editor

Kirill A. Afonin graduated from Saint Petersburg State University, obtained his doctoral degree from Bowling Green State University, completed postdoctoral training at the University of California Santa Barbara and at the National Cancer Institute, and then started his tenure-track appointment at UNC Charlotte in 2015, where he was granted tenure in 2019.

Preface to "Nucleic Acid Architectures for Therapeutics, Diagnostics, Devices and Materials"

Nucleic acids (RNA and DNA) and their chemical analogs have been utilized as building materials due to their biocompatibility and programmability. RNA, which naturally possesses a wide range of different functions, is now being widely investigated for its role as a responsive biomaterial which dynamically reacts to changes in the surrounding environment. It is now evident that artificially designed self-assembling RNAs that can form programmable nanoparticles and supra-assemblies will play an increasingly important part in a diverse range of applications, such as macromolecular therapies, drug delivery systems, biosensing, tissue engineering, programmable scaffolds for material organization, logic gates, and soft actuators, to name but a few. The current exciting Special Issue comprises research highlights, short communications, research articles, and reviews that all bring together the leading scientists who are exploring a wide range of the fundamental properties of RNA and DNA nanoassemblies suitable for biomedical applications.

Kirill A. Afonin
Special Issue Editor

nanomaterials

MDPI

Editorial

Editorial for the Special Issue on "Nucleic Acid Architectures for Therapeutics, Diagnostics, Devices and Materials"

Justin R. Halman and Kirill A. Afonin *

Nanoscale Science Program, Department of Chemistry, University of North Carolina at Charlotte, Charlotte, NC 28223, USA
* Correspondence: kafonin@uncc.edu; Tel.: +1-704-687-0685; Fax: +1-704-687-0960

Received: 26 June 2019; Accepted: 29 June 2019; Published: 30 June 2019

The use of nucleic acids (RNA and DNA) offers a unique and multifunctional platform for numerous applications including therapeutics, diagnostics, nanodevices, and materials. The ubiquity of nucleic acids and their broad functional involvement in various biological processes place them before other biocompatible materials suitable for controlled bottom-up nanofabrication. Systematic elucidation of RNA and DNA folding, physicochemical properties, and relevant biological activities have fueled an increased involvement of nucleic acid-based nanotechnology in various biomedical challenges.

This special issue assembles 11 original articles (with seven research manuscripts and four reviews) which nicely outline several advances made in the field of RNA and DNA nanotechnology. Unified by the versatile use of the intriguing biopolymers, these manuscripts explore the various facets of nucleic acid nanostructures such as design, production, and characterization of RNA and DNA nanoassemblies [1–4], rational design of functional molecular machines [5–8], immunorecognition of nucleic acid nanoparticles [8], in vivo delivery of therapeutic nucleic acids [9,10], and nucleic acid-based biosensors [11]. We anticipate this special issue to be accessible to a wide audience, as it explores not only the biological aspects of nucleic acid nanodesigns, but also different methodologies of their production, their interactions with other classes of biological molecules, physicochemical characteristics, and possible applications.

Drawing inspiration from naturally-occurring structural and long-range interacting RNA motifs, Chopra et al. describe a novel design that combines the internal ribosome entry site of the hepatitis C virus together with the RNA kissing loop motifs to form hexagonal assemblies which are amenable to functionalization with various aptamers [1]. The aptamers include malachite green, PP7, Spinach, and an aptamer against streptavidin, demonstrating the coordination of proteins and small molecules on RNA scaffolds with structural regularity. The use of RNA in this nanodesign also allowed for co-transcriptional production of functional nanoassemblies.

O'Hara et al. describe an innovative approach, integrating a split aptamer system into internal portions of hexagonal RNA rings [2]. Using a split-Spinach aptamer, the authors demonstrate that the intensity of the fluorescent signal associated with the formation of the complete aptamer is dependent on the formation of the RNA rings, thus allowing researchers to monitor the assembly of nanostructures.

In a different approach, Yourston et al. describe the use of nanostructured DNA assemblies for DNA-templated production and organization silver nanoclusters (AgNCs) with unique optical properties [3]. Their work expands the protocols for synthesis of AgNCs on various DNA templates and identifies some structural requirements for altering the fluorescent properties of the AgNCs.

Design and characterization of nanostructured nucleic acids is of paramount importance in both achieving and modulating their functional goals. The majority of characterization techniques of nucleic acid nanostructures involve the use of electrophoretic mobility shift assays, atomic force microscopy, cryogenic electron microscopy, or dynamic light scattering; each approach has its unique benefits and drawbacks. Oliver et al. propose a new approach to characterizing nucleic acid nanoparticles using small angle X-ray or neutron scattering [4]. Their in-depth review examines the structural and chemical requirements for studying various biomolecules in their natural confirmations. It also highlights the multitude of requirements for solution and sample preparations, and discusses the precise data yielded from these techniques. Finally, it discusses the prospects for enhancing the characterization of nanostructured nucleic acid architectures.

Following design and characterization, advances in the use of DNA and RNA nanostructures in devices and molecular machines is highlighted in this issue. One such device is the DNA nanopore, which offers a unique membrane-bound tool for furthering biophysical research and biosensing. In this issue, Burns and Howorka describe the design and evaluation of a DNA nanopore in various biological milieu [5]. The authors identify parameters for the assembly of a hexameric DNA nanopore using gel electrophoresis and fluorescence spectroscopy, and further confirm its integrity in cell media, as well as its successful integration into lipid membranes.

Beyond the development of biophysical tools, this special issue contains two original manuscripts detailing the use of DNA and RNA nanostructures in the design of molecular logic gates. Using "light-up" malachite green and Broccoli aptamers, Goldsworthy et al. describe the design of robust and precise RNA nanodevices with several Boolean logic functions, including AND, OR, NAND, and NOR capabilities [6]. The binding of the small molecule dyes to their aptamers in the conditional situations causes a fluorescent output based strictly on the input of the designed strands. Using this approach, the authors designed and integrated parallel XOR and AND gates within a single RNA nanoparticle.

Towards that end, Zakrevsky et al. describe the use of logic gates to mimic complex molecular biological phenomena with the potential for use in cells [7]. In their research, the authors computationally design and characterize DNA/RNA hybrid-based logic gates suited for the conditional release of single- and double-stranded nucleic acids, thus displaying the possibility of therapeutic action. Their set of logic gates offers an eclectic approach inspired by molecular beacons and other trigger-responsive multi-stranded assemblies.

The concept of smart-responsive or logic-based nucleic acid nanoassemblies is further expanded in the review by Chandler and Afonin [8]. In their review, several advances in the design and function of stimuli-responsive DNA and RNA nanoconstructs are discussed, including simple trigger/activation designs of shape-switching nanoparticles with controlled immunomodulatory properties. Furthermore, they outline the potential of using these dynamic nucleic acid nanoparticles for the conditional stimulation of an immune response, which has been determined to be dependent on the structure and composition of the given nucleic acid nanoparticle.

As therapeutics, nanostructured nucleic acids hold great potential in the treatment of various maladies. Gwak and Lee describe the use of a cationic amphiphilic co-polymer nanoparticles to efficiently deliver plasmid DNA, as well as a small molecule anti-viral therapy for successful treatment of spinal cord injuries using "suicide gene therapy" [9]. The authors further describe the stability of the nanoparticles over time and demonstrate their use in animal models.

In the same regard, Caffery et al. describe in depth the possibility of treating glioblastomas with a combination of gene therapy and various nanoscale carriers [10]. Their comprehensive review discusses both the various therapeutic nucleic acids as well as the various delivery platforms to deliver these therapeutic cargos. The authors also discuss possible limitations to each of these approaches, as well as the challenges of crossing the blood brain barrier.

Finally, this special issue features a review by Sun et al. on the use of genetically encoded RNA-based molecular sensors (GERMS) [11]. The review provides a well-rounded outline of the various design principles and applications for GERMS including their most common use of intracellular imaging. The authors go on to discuss the outlook for this novel technology and potential milestones towards developing next generation RNA based sensors.

Together, this special issue details several important advancements in the field of nucleic acid nanotechnology by focusing on design and characterization, use in devices and molecular machines, and implementation into therapeutics and sensors. The works enclosed here demonstrate the various possible applications for these burgeoning technologies.

Author Contributions: J.R.H. and K.A.A. co-wrote this Editorial Letter.

Funding: The research reported in this publication was supported by the National Institute of General Medical Sciences of the National Institutes of Health under Award Number R01GM120487. The content is solely the responsibility of the authors and does not necessarily represent the official views of the National Institutes of Health.

Acknowledgments: K.A.A. would like to thank all of the authors for their superb contributions to this Special Issue as well as the reviewers who assisted in evaluating and improving the quality of all submitted manuscripts. We would also like to express our appreciation for Yueyue Zhang and the editorial staff at Nanomaterials, without whom this Special Issue would not be possible.

Conflicts of Interest: The authors declare no conflict of interest.

References

1. Chopra, A.; Sagredo, S.; Grossi, G.; Andersen, E.S.; Simmel, F.C. Out-of-Plane Aptamer Functionalization of RNA Three-Helix Tiles. *Nanomaterials* **2019**, *9*, 507. [CrossRef] [PubMed]
2. O'Hara, J.M.; Marashi, D.; Morton, S.; Jaeger, L.; Grabow, W.W. Optimization of the Split-Spinach Aptamer for Monitoring Nanoparticle Assembly Involving Multiple Contiguous RNAs. *Nanomaterials* **2019**, *9*, 378. [CrossRef] [PubMed]
3. Yourston, L.E.; Lushnikov, A.Y.; Shevchenko, O.A.; Afonin, K.A.; Krasnoslobodtsev, A.V. First Step towards Larger DNA-Based Assemblies of Fluorescent Silver Nanoclusters: Template Design and Detailed Characterization of Optical Properties. *Nanomaterials* **2019**, *9*, 613. [CrossRef] [PubMed]
4. Oliver, R.C.; Rolband, L.A.; Hutchinson-Lundy, A.M.; Afonin, K.A.; Krueger, J.K. Small-Angle Scattering as a Structural Probe for Nucleic Acid Nanoparticles (NANPs) in a Dynamic Solution Environment. *Nanomaterials* **2019**, *9*, 681. [CrossRef] [PubMed]
5. Burns, J.R.; Howorka, S. Structural and Functional Stability of DNA Nanopores in Biological Media. *Nanomaterials* **2019**, *9*, 490. [CrossRef] [PubMed]
6. Goldsworthy, V.; LaForce, G.; Abels, S.; Khisamutdinov, E.F. Fluorogenic RNA Aptamers: A Nano-platform for Fabrication of Simple and Combinatorial Logic Gates. *Nanomaterials* **2018**, *8*, 984. [CrossRef] [PubMed]
7. Zakrevsky, P.; Bindewald, E.; Humbertson, H.; Viard, M.; Dorjsuren, N.; Shapiro, B.A. A Suite of Therapeutically-Inspired Nucleic Acid Logic Systems for Conditional Generation of Single-Stranded and Double-Stranded Oligonucleotides. *Nanomaterials* **2019**, *9*, 615. [CrossRef] [PubMed]
8. Chandler, M.; Afonin, K.A. Smart-Responsive Nucleic Acid Nanoparticles (NANPs) with the Potential to Modulate Immune Behavior. *Nanomaterials* **2019**, *9*, 611. [CrossRef] [PubMed]
9. Gwak, S.-J.; Lee, J.S. Suicide Gene Therapy by Amphiphilic Copolymer Nanocarrier for Spinal Cord Tumor. *Nanomaterials* **2019**, *9*, 573. [CrossRef] [PubMed]
10. Caffery, B.; Lee, J.S.; Alexander-Bryant, A.A. Vectors for Glioblastoma Gene Therapy: Viral & Non-Viral Delivery Strategies. *Nanomaterials* **2019**, *9*, 105.
11. Sun, Z.; Nguyen, T.; McAuliffe, K.; You, M. Intracellular Imaging with Genetically Encoded RNA-Based Molecular Sensors. *Nanomaterials* **2019**, *9*, 233. [CrossRef] [PubMed]

nanomaterials

MDPI

Article

Out-of-Plane Aptamer Functionalization of RNA Three-Helix Tiles

Aradhana Chopra [1], Sandra Sagredo [1], Guido Grossi [2], Ebbe S. Andersen [2] and Friedrich C. Simmel [1,*]

[1] Physik-Department E14, Technische Universität München, 85748 Garching, Germany; aradhanachopra@gmail.com (A.C.); sandra.sagredo@tum.de (S.S.)
[2] Interdisciplinary Nanoscience Center (iNANO), Aarhus University, 8000 Aarhus C, Denmark; ggrossi@inano.au.dk (G.G.); esa@inano.au.dk (E.S.A.)
* Correspondence: simmel@tum.de; Tel.: +49-(0)89-289-11611; Fax: +49-(0)89-289-11612

Received: 16 February 2019; Accepted: 27 March 2019; Published: 2 April 2019

Abstract: Co-transcriptionally folding RNA nanostructures have great potential as biomolecular scaffolds, which can be used to organize small molecules or proteins into spatially ordered assemblies. Here, we develop an RNA tile composed of three parallel RNA double helices, which can associate into small hexagonal assemblies via kissing loop interactions between its two outer helices. The inner RNA helix is modified with an RNA motif found in the internal ribosome entry site (IRES) of the hepatitis C virus (HCV), which provides a 90° bend. This modification is used to functionalize the RNA structures with aptamers pointing perpendicularly away from the tile plane. We demonstrate modifications with the fluorogenic malachite green and Spinach aptamers as well with the protein-binding PP7 and streptavidin aptamers. The modified structures retain the ability to associate into larger assemblies, representing a step towards RNA hybrid nanostructures extending in three dimensions.

Keywords: RNA nanotechnology; aptamers; cotranscriptional folding

1. Introduction

Over the past two decades, RNA has been found to be involved in many essential cellular processes other than the conventional roles it fulfils as mRNA, tRNA, or rRNA. Non-coding RNAs (ncRNAs) and many RNA–protein complexes are involved in regulatory functions at the transcriptional and translational levels, and have roles in scaffolding [1], genome-editing, RNA interference, clustered regularly interspaced repeats (CRISPR), and chromatin remodeling [2]. Many naturally occurring ncRNAs fold and assemble into complex 3D architectures via a plethora of secondary and tertiary interactions, and also via association with a wide range of RNA-binding proteins. More recently, the exceptional folding capability and modularity of biological RNAs have inspired the emergence of RNA nanotechnology, which aims at the construction and assembly of artificial nanostructures made from RNA [3]. Compared to DNA, RNA offers a variety of interesting features as a material for nanotechnology: RNA nanostructures can draw from a diverse variety of naturally occurring tertiary motifs [4–6], they can be enzymatically generated in large amounts via transcription, and they can be genetically encoded and expressed in cells [7,8].

Most naturally occurring RNAs are single-stranded and contain self-complementary sequences that facilitate intramolecular folding into distinct secondary structures. In addition, rigid structural motifs consisting of canonical or noncanonical base-pairing, kissing interactions, and stacking of helices play a significant role in RNA folding, resulting in complex 3D structures exhibiting helices, loops, junctions, bulges, stems, hairpins, and pseudoknots [4–6]. Seeking inspiration from the design principles found in nature and employing naturally occurring RNA motifs, several assembly strategies have been developed for RNA nano-construction. They include RNA architectonics [9,10], self-assembly of RNA/DNA hybrids [11,12] and single-strand RNA assembly [13]. RNA architectonics is based on the modular character of RNA, which allows 3D RNA motifs to be organized in alternative combinations in order to create different RNA nano-architectures. Large libraries of thermodynamically stable modular RNAs, which include both structural and functional motifs, have been identified and characterized from natural RNA molecules. Furthermore, such motifs have been used as suitable "parts" for designing self-assembling RNA units (or tectoRNAs) [10]. For instance, in this context, in silico methods have been developed to screen for natural RNA motifs capable of self-assembling into closed ring structures [14]. The DNA/RNA hybrid strategy leverages the properties of both RNA and DNA. It has been used for producing large nucleic acid structures based on the DNA origami technique, where a large RNA scaffold is folded with a number of complementary DNA/RNA staples. Single strand RNA assembly relies on a number of RNA strands that are unstructured by themselves but when mixed together assemble into a structure [12].

A technique dubbed cotranscriptional ssRNA origami has been developed, in which a single RNA strand folds into a predefined RNA tile that further assembles into hexagonal and rectilinear lattices while the RNA is produced by the RNA polymerase [13]. It employs a variety of RNA tertiary motifs to mediate the intra- and inter-tile interactions. Improved unimolecular DNA and RNA folding strategies have been developed based on minimizing the knotting complexity by employing parallel crossovers to avoid kinetic trapping during folding. This has resulted in RNA nanostructures of a variety of shapes and up to 6000 nt in length [15].

Rationally designed RNA nanostructures and devices have great potential for applications in synthetic biology, metabolic engineering, and nanomedicine [3,16–19]. In particular, the three-way junction from pRNA (a component of the phage phi29 packaging motor) has been found to be an extremely stable motif that can be used as the basis of multifunctional RNA nanoparticles for therapeutic applications [19]. More recently, RNA tiles designed by the cotranscriptional ssRNA origami technique have been used in combination with fluorogenic RNA aptamers to function as nanoscale aptamer-based Förster resonance energy transfer (FRET) sensors [20]. Additionally, programmed folding of RNA nanostructures of different shapes has also been performed in vivo [21]. Furthermore, hybrid multicomponent RNA-protein nanostructures have been characterized in vivo and in cell-free gene expression systems [22].

In the present work, we construct a novel three-helix "antiparallel even" RNA tile (3H-AE) based on the cotranscriptional ssRNA origami approach. In our design, the outer two helices of the 3H-AE tile provide a rigid RNA scaffold that can interact with other tiles via kissing loop (KL) interactions. The central RNA double-helical extensions, however, are conceived as modular plug-in modules, which can be modified at will without interfering with two-dimensional tile assembly. Specifically, we modified the plug-in helices with subdomain IIa of the internal ribosomal entry site (IRES) of the Hepatitis C virus (HCV) genome [23], which allowed attachment of arbitrary RNA modules perpendicularly protruding either above or below the plane defined by the 3H-AE RNA tile. As examples, we positioned the two fluorogenic malachite green (MG) [24] and Spinach [25,26] aptamers above and below the tile plane, respectively, which allowed us to monitor functional assembly of the tile structures via fluorescence spectroscopy. In order to demonstrate the modularity of the design, the top (malachite green) aptamer was also replaced with other aptamers such as the PP7 aptamer for the viral coat protein PCP [27,28] and an RNA aptamer for streptavidin [22]. We finally show that the streptavidin aptamer module can be used to immobilize the RNA tiles on streptavidin coated surfaces, while presenting a second function via another aptamer, indicating the potential for functionalizing surfaces with unmodified proteins or other ligands in controlled orientations.

Our results demonstrate the modularity of the RNA origami tile approach and represent a step towards multifunctional RNA assemblies extending in three dimensions. The most salient feature of the three-helix tile structure introduced in this work is the modular extension of the middle helix with aptamer functions. This allows for connecting the tiles in 2D while presenting separate binding modalities on the two sides of the 2D assembly. Furthermore, it is conceivable to create 3D lattices from such structures by also polymerizing along the z-direction.

2. Materials and Methods

2.1. Design of Three-Helix RNA Tiles (3H-AE)

3H-AE and its subsequent modifications were designed by the cotranscriptional ssRNA origami method that has been previously described in detail [13,29]. Briefly, using the 3D modeling programs Swiss-PdbViewer [30] and UCSF Chimera [31], three standard A-form RNA double helices (#1-3, Figure 1a) were positioned over one another and rotated to create an optimal spacing for an anti-parallel even (AE) double crossover (DX). An internal 180° KL (HIV-1 DIS, PDB ID: 2B8R) [32] was placed between the crossovers on helix number 1 (top). UUCG tetraloops (extracted from PDB ID: 1F7Y) [33] were positioned at the ends and in between crossovers on helix number 2 (middle). In addition, 120° KLs (RNA i/ii inverse loop, PDB ID: 2BJ2) [34] were positioned at the ends of double helices numbers 1 and 3 to allow formation of hexagonal lattices. The 180° KL forms 6 base pairs between the two loops resulting in a coaxial stack and is in phase with the A-form helix of RNA, whereas the 120° KL forms 7 base pairs between the loops resulting in a continuous, but bent coaxial stack. Modifications in the 3H-AE design included additional RNA motifs such as subdomain IIa (PDB ID: 2PN4) and domain IIa (PDB ID: 1P5M) of the IRES of the HCV genome, which were used as connectors between the RNA tile and the malachite green aptamer (MGA) (PDB ID: 1F1T) [24]. These domains consist of a 90° angle [35,36] which allows for almost perpendicular arrangement of the added RNA motifs to the tile. An additional RNA sequence encoding Spinach aptamer (PDB ID: 4KZD) [26] was connected to 3H-AE with MGA V1 via subdomain IIa to generate the modified structure 3H-AE-MGA-Spinach. Additional variants of 3H-AE-MGA-Spinach without 120° KL were constructed. The MGA was further replaced by either a streptavidin aptamer [22] or an aptamer binding to bacteriophage PP7 coat protein fused with mCherry (PCP-mCherry, PDB ID: 2QUX) [28], respectively.

After the initial modeling, the 3D structures were ligated with a Perl script ("ligate.pl", which was available from [37] and refined using a recursive geometric refinement function in the program *Assemble* [38]. 3D models in this work were rendered in UCSF Chimera. The designed structure was further traced using a Perl script ("trace.pl", which is also available from [37]) and an input was generated that was used for the design of the corresponding RNA sequences in Nupack [39]. The sequences of the 180° KLs, 120° KLs, tetraloops, subdomain IIa, domain IIa, MGA, Spinach aptamer and PP7 aptamer were chosen from the PDB files and added to the respective designs (assuming a stable folding behavior of the aptamers, the corresponding sequences were modularly replaced, keeping the rest of the 3H-AE sequence constant). Additionally, some of the base pairs in the dovetail seam were constrained to be strongly stacking G-C pairs in an attempt to immobilize it to a static position. Sequences were further constrained to contain at least one G-U wobble pair per every eight continuous base pairs in order to avoid secondary structures in the RNA-encoding DNA template and simplify its synthesis. All of the remaining positions were designed by Nupack. The 5'-end of each sequence was constrained to begin with GGG, an optimal leader sequence for T7 RNA polymerase. In addition, 2D blueprints of the final structures resulting from this process are shown in Supplementary Figure S1.

Primers were generated specific to DNA sequence generated by Nupack and their melting temperature was calculated using the NEB Tm calculator [40].

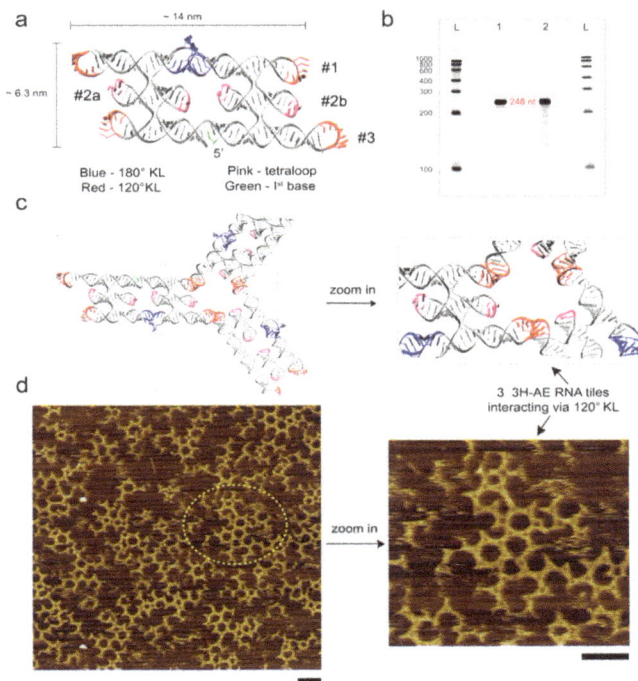

Figure 1. Design and characterization of the 3H-AE RNA tile. (**a**) schematic representation of designed 3H-AE RNA tile. Structural motifs are color coded as described in the legend; (**b**) denaturing PAGE gel showing the correct length of the RNA tile. L: LowRange RiboRuler, 1-2: RNA tile transcribed from DNA template amplified from Phusion High-Fidelity Master mix with HF or GC, buffer, respectively. (**c**) 3H-AE RNA tiles can interact with each other via 120° kissing loop (KL) motifs; (**d**) AFM images showing correct assembly and interaction of the RNA tiles. Tile assemblies were prepared by snap-cooling followed by incubation on mica at 37 °C (cf. Materials and Methods). The region enclosed by the dashed circle is further zoomed in to show the interactions of the 120° KL. Scale bars: 50 nm.

2.2. Preparation of RNA Tiles

Genetic templates for all RNA tiles were amplified from "custom dsDNA gBlocks" from Integrated DNA Technologies (IDT) using the polymerase chain reaction (PCR). RNA tiles were then prepared using in vitro transcription from these templates as described in detail in the Supplementary Methods.

2.3. Characterization of RNA Tiles and Tile Assemblies

The formation of the RNA tiles was characterized using gel electrophoresis and atomic force microscopy (AFM). The transcription of RNA tiles modified with fluorogenic aptamers was followed using fluorescence spectroscopy. RNA tiles containing streptavidin aptamers were further investigated using streptavidin-coated microbeads. Detailed experimental procedures are found in the Supplementary Methods.

3. Results and Discussion

3.1. Design and Folding of the 3H-AE RNA Tile

We designed a novel three-helix "antiparallel even" (3H-AE) RNA tile as described in detail in the Methods section above. In contrast to previously described RNA tiles [13], it was constructed from *three* RNA double helices placed over one another connected via double crossovers, and converted into a cotranscriptionally folding continuous ssRNA using one 180° KLs, four 120° KLs and four tetraloops (Figure 1a). The central (inner) helices (#2a and #2b) of the 3H-AE tile were modularly functionalized with various aptamers as discussed in the following sections.

Custom dsDNA segments containing the sequence for the 3H-AE RNA tile and its subsequent modifications were ordered from a gene synthesis supplier and PCR amplified using a Phusion High-Fidelity PCR Master mix (with HF or GC buffer). The purity and amplification of the samples was checked using agarose gel electrophoresis (Supplementary Figure S2). Transcription and folding of RNA structures (3H-AE RNA tile, 248 nt) were verified via a denaturing PAGE. The transcribed RNA was observed at the correct length as shown in Figure 1b. As indicated in Figure 1c, the 3H-AE RNA tiles were modified with 120° KLs to enable formation of hexagonal assemblies from multiple interacting structures. Correct folding of the RNA tiles and their assembly into super-structures were assessed via atomic force microscopy (Figure 1d). In addition, 120° KL interactions between individual tiles resulted in the assembly of hexagonal lattices on the length scale of a few tens of nanometers. Larger, more disordered molecular networks were formed with hundreds of nanometers in size (Supplementary Figure S3). The dimensions of the RNA tiles observed via AFM were in accordance with the design of Figure 1a.

3.2. Modification of the 3H-AE RNA Tile with Aptamers

We next designed modified versions of the 3H-AE RNA tile with one or more fluorescent RNA aptamers. The first version included the replacement of one of the two interior tetraloops (sitting on one of the central RNA helices of the three-helix structure, (#2a) by two additional RNA motifs—subdomain IIa of the internal ribosomal entry site (IRES) of the HCV genome (PDB ID: 2PN4) and the malachite green aptamer (MGA) (Figure 2a). Subdomain IIa is an L-shaped structural motif that provides a 90° bend of the double helix center axis before and after the motif. The length of helix #2a and subdomain IIa together facilitated the attachment of an additional RNA domain protruding perpendicularly from the plane defined by the 3H-AE tile. As an example, for this attachment strategy, we elongated subdomain IIa with the fluorogenic MG aptamer (3H-AE with MGA V1, cf. Figure 2a).

This modification resulted in an increase in the number of nucleotides in the tile from 248 to 297 nt, which was verified by denaturing PAGE (Figure 2b). The presence of the fluorescent aptamer on the RNA tile further enabled real-time monitoring of the production of RNA tiles via in vitro transcription (Figure 2c). The stability of the signal over more than 20 h suggests negligible degradation of the RNA structure under our reaction conditions. AFM observation of small hexagonal assemblies by the MGA-functionalized tiles showed their correct folding and also demonstrated that the 120° KL interactions remained intact despite the modifications (Figure 2d).

We also created a variation of 3H-AE-MGA V1 by replacing the connecting motif subdomain IIa (PDB ID: 2PN4) by the slightly larger HCV IRES domain IIa (PDB ID: 1P5M), which includes a few additional unpaired bases. As a result, the length of the tile was increased by six bases from 297 nt to 303 nt (3H-AE-MGA V2), and the angle of MGA with respect to the tile was expected to tilt slightly (Supplementary Figure S4). Characterization by fluorescence spectroscopy and AFM showed very similar behaviors for both connection motifs (Supplementary Figure S4a,d).

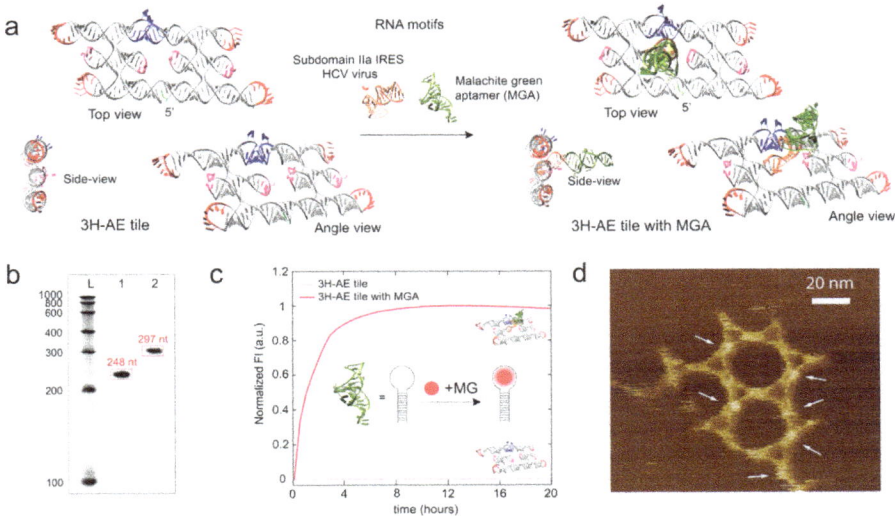

Figure 2. Modification of the 3H-AE RNA tile with the malachite green aptamer (MGA). (**a**) schematic representation showing different views of 3H-AE RNA tiles with and without the addition of MGA (green) connected via subdomain IIa IRES of HCV virus (orange); (**b**) denaturing PAGE gel showing the length of the unmodified (no aptamer) and modified RNA tiles (with MGA). L: LowRange RiboRuler, 1: 3H-AE RNA tile, 2: 3H-AE RNA tile with MGA; (**c**) real-time fluorescence of MG recorded during transcription of MGA modified RNA tiles (fluorescence normalized to maximum value). The transcription reaction typically ceases after 3–4 h due to activity loss of the RNAP; (**d**) AFM image showing formation of a hexagonal mini-lattice with elevated features resulting from the MGA modifications (note that not all of the modifications are visible equally well, probably depending on the orientation of the extensions with respect to the AFM scanning direction). The sample was prepared by snap-cooling and incubation on mica as described in the Materials and Methods.

3.3. Double Functionalization of the 3H-AE RNA Tile with Two Aptamers

Next, the complexity of the MGA-functionalized 3H-AE tiles was further increased by the addition of another aptamer via an additional subdomain IIa connected to the second internal tetraloop of the tile structure. As shown in Figure 3a,b, we used this approach to attach the fluorogenic Spinach aptamer (PDB ID: 4KZD) pointing perpendicularly away from the tile in the direction opposite to the initial MGA functionalization. Accordingly, the length of the tile's RNA sequence increased to 388 nt, which was confirmed by denaturing PAGE (Figure 3c). AFM of the double-modified RNA tiles showed structures with local, 120° KL-mediated hexagonal order (Figure 3d,e). As expected, larger assemblies were not observed, as the extension of the aptamers in opposite directions did not allow the RNA tiles to lie flat on the mica surface. AFM imaging of the double-modified tiles therefore also turned out to be particularly challenging, in part probably due to the reduced contact area of the structures with the mica surface.

As before, transcription of the double-modified RNA tile was monitored via the fluorescence signal generated by the two aptamers in their respective emission channels (Figure 4), indicating proper folding of the aptamer domains and stability of the structure.

We also designed a modification of 3H-AE-MGA-Spinach-KL, in which the 120° KLs were replaced by tetraloops to avoid inter-tile interactions, keeping all the other features unaltered. The non-interacting tile displayed similar behavior in native gel electrophoresis and fluorescence experiments (Supplementary Figures S5–S7). An enhanced fluorescence value observed for the tetraloop-containing structure (Supplementary Figure S6c) potentially indicates a slightly better folding of the tile, which would be consistent with the fact that tetraloops support the intramolecular folding process. Alternatively, it is possible that the two structures are transcribed at different rates given the different sizes and base compositions of the sequences.

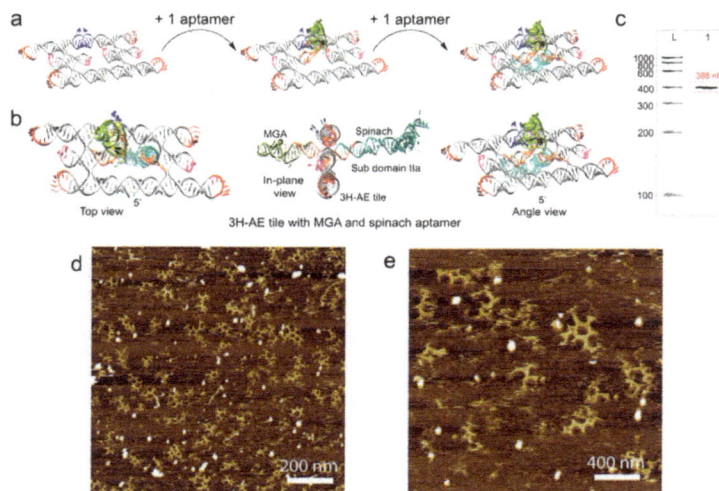

Figure 3. Modification of the MGA functionalized 3H-AE tile with an additional Spinach aptamer (resulting in 3H-AE-MGA-Spinach). (**a**) schematic representation of the addition of MGA (green) and Spinach aptamer (cyan) to the 3H-AE RNA tile connected via subdomain IIa (orange); (**b**) different views of 3H-AE-MGA-Spinach; (**c**) denaturing PAGE gel showing the correct length of 3H-AE-MGA-Spinach. L: LowRange RiboRuler, 1: 3H-AE-MGA-Spinach (388 nt); (**d,e**) AFM images corresponding to different areas of imaging—1 µm × 1 µm and 0.5 µm × 0.5 µm, respectively. Samples prepared by snap-cooling followed by incubation on mica.

For specific applications, the fluorogenic aptamer modifications can be easily replaced with other functional RNA sequences. As a proof of concept, two variants of 3H-AE-MGA-spinach-Tetraloop structure were designed and constructed with the MGA replaced by the PP7 aptamer [28] or a streptavidin aptamer, respectively (Figure 5a and Figure S8). The production of the structures was validated by real-time fluorescence of the (remaining) spinach aptamer and denaturing PAGE (Figure 5b,c). Interactions of the tiles with streptavidin (Figure 5d) and PCP-mCherry fusion proteins (Supplementary Figures S9–S11) were further verified by EMSA.

In order to indicate one potential application for double-functionalized 3H-AE tiles, we utilized the streptavidin aptamer module to specifically arrange RNA tiles on the surface of streptavidin-coated polystyrene microbeads. 3H-AE tiles double-functionalized with the Spinach and streptavidin aptamer are expected to present the Spinach aptamer in an orientation pointing away from the surface of the beads (Figure 6a). Indeed, binding of the RNA tiles to the microparticles is observed only with nanostructures containing the streptavidin aptamer (Figure 6b). While in this experiment the fluorogenic Spinach aptamer was used as the second function on the RNA tile, it is easily conceivable to generate protein-functionalized surfaces or membranes using other aptamer modules in a similar manner. Such surfaces should be of interest for the spatial organization of enzymes or other biochemical functions. As the formation of such functionalized surfaces could be triggered by the transcription of the RNA tiles, this process would also be genetically controllable.

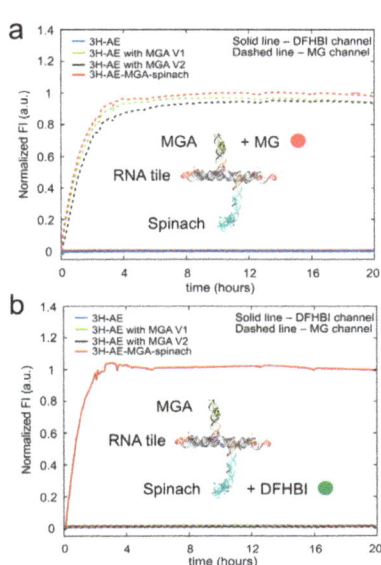

Figure 4. Monitoring transcription of RNA tiles via aptamer fluorescence. MG and the Spinach aptamer fluorophore 3,5-difluoro-4-hydroxybenzylidene imidazolinone (DFHBI) were added to transcription reactions of all four versions of RNA tiles. The structure of the 3H-AE-MG-Spinach-KL RNA tile is shown as an inset to the figures: (**a**) MGA and (**b**) Spinach (DFHBI) fluorescence recorded during transcription of the various RNA tiles constructed (Ex = 615–645 nm, Em = 669–699 nm for MG and Ex = 475–495 nm, Em = 515–545 nm for DFHBI; fluorescence intensities normalized with respect to maximum fluorescence).

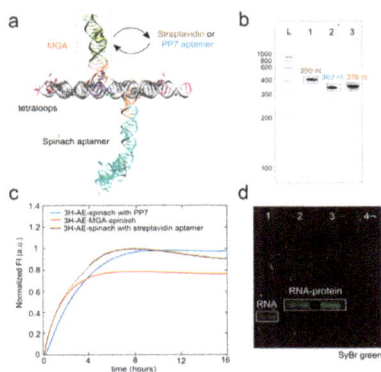

Figure 5. (**a**) schematic representation of 3H-AE-MGA-Spinach RNA tile without 120° KL. The MG aptamer can be modularly replaced by other RNA aptamers as indicated; (**b**) denaturing PAGE gel showing the length of 3H-AE-Spinach with different secondary aptamer modifications. L: LowRange RiboRuler, 1: 3H-AE-Spinach with streptavidin aptamer (390 nt), 2: 3H-AE-Spinach with PP7 aptamer (363 nt), 3: 3H-AE-MGA-Spinach (376 nt); (**c**) real-time fluorescence of DFHBI during transcription of RNA tiles containing different aptamers; (**d**) native agarose gel (2% agarose in 1X tris-borate EDTA (TBE) buffer + 2 mM MgCl$_2$) stained with SyBr Green showing the retardation of RNA tile containing streptavidin aptamer (3H-AE-Spinach with streptavidin aptamer) in presence of streptavidin. 1: RNA tile only, 2: 3H-AE-Spinach with streptavidin aptamer + streptavidin (100 nM), 3: 3H-AE-Spinach with streptavidin aptamer + streptavidin (200 nM), 4: Streptavidin only (100 nM). RNA tiles were prepared by heat denaturation/renaturation as described in the Supporting Materials and Methods.

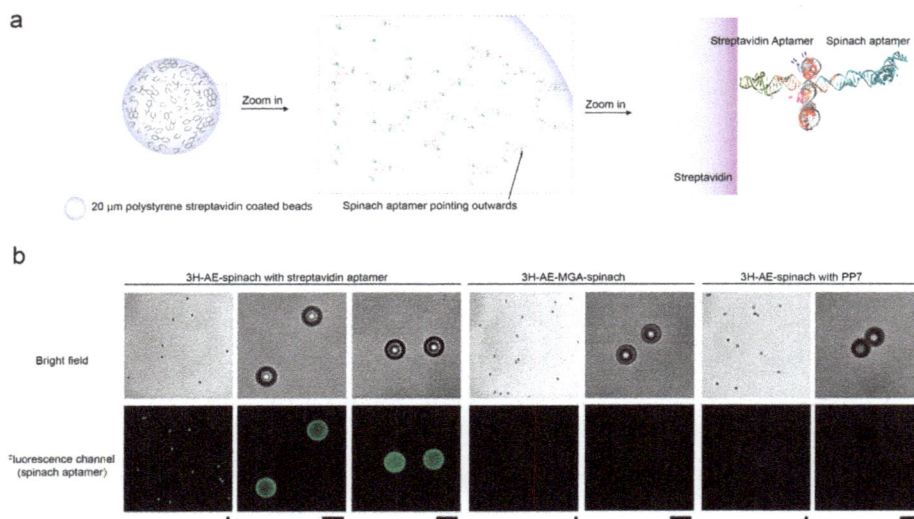

Figure 6. Assembly of double-functionalized 3H-AE tiles on microparticles. (**a**) schematic representation of attachment of 3H-AE-Spinach tiles with streptavidin aptamer on the surface of streptavidin coated beads. The streptavidin aptamer is expected to be anchored on the surface of streptavidin particles and thus the Spinach aptamer points away from it; (**b**) fluorescence images of different RNA tiles interacting with 20 μm streptavidin coated polystyrene beads after washing. Only the RNA tiles carrying Spinach and a streptavidin aptamer lead to a fluorescent signal localized to the beads; scale bars: 20 μm.

4. Conclusions

The results demonstrated here represent initial efforts towards the design and synthesis of multifunctional tile-based RNA nanostructures extending in three dimensions. Three-helix RNA tiles were folded from a single RNA strand, which was composed of a variety of naturally occurring RNA motifs that assist in reaching the desired target shape. The main focus of the present work was put on domain IIa and subdomain IIa, which are RNA motifs found in the HCV IRES. Both these motifs contain a 90° bend that allows positioning of other RNA modules such as fluorogenic RNA aptamers perpendicular to the RNA tile structure. As examples, we created 3H-AE tile structures either modified with a single malachite green aptamer or with MGA and a Spinach aptamer pointing in opposite directions. Importantly, the modified 3H-AE-tiles retained the ability to associate into hexagonal assemblies via kissing loop interactions.

Among the most promising applications of such multi-functionalized RNA tile structures, we envision the creation of artificial ribonucleoprotein complexes, in which RNA structures scaffold the co-localization and arrangement of proteins, e.g., for membrane-less compartmentalization or for the creation of multienzyme structures. As the RNA tiles can be made to polymerize in 2D, it is also conceivable to create extended RNA sheets or "membranes", which are functionalized with aptamers on either side, which could be used to further spatially organize proteins in order to enhance their structural or enzymatic functions. From these, in turn, one could generate RNA/protein covered surfaces (e.g., with catalytic function) via self-assembly of such structures. Importantly, such self-assembly processes could all be controlled via gene expression from a plasmid. From a nanotechnology point of view, another interesting opportunity lies in the possibility to stack several such RNA sheets on top of each other—mediated by the perpendicular RNA extensions, resulting in multilayered RNA nanostructures extending in 3D.

Supplementary Materials: The following are available online at http://www.mdpi.com/2079-4991/9/4/507/s1, Supplementary Methods and Materials, Supplementary Figures S1–S11, Supplementary Tables S1–S7, Supplementary References 1–3.

Author Contributions: Conceptualization, A.C. and F.C.S.; Methodology, A.C., G.G., E.S.A., and S.S.; Software, G.G. and E.S.A.; Investigation, A.C. and S.S.; Writing—Original Draft Preparation, A.C.; Writing—Review and Editing, G.G., E.S.A., and F.C.S; Visualization, A.C. and S.S.; Supervision, F.C.S.; Funding Acquisition, F.C.S. and E.S.A.

Funding: This research was funded by the Deutsche Forschungsgemeinschaft—Sonderforschungsbereich 1032/TPA02; the European Commission FP7—MSCA-ITN EScoDNA, grant No. 317110; the European Research Council, Grant No. 694410, project AEDNA, and Grant No. 683305, project RNA ORIGAMI.

Acknowledgments: A.C., S.S. and F.C.S. gratefully acknowledge financial support by the Deutsche Forschungsgemeinschaft through Sonderforschungsbereich 1032 (TP A2), the European Commission FP7 as part of the MSCA-ITN EScoDNA (No. 317110), and the European Research Council (Grant No. 694410, project AEDNA). G.G. and E.S.A. were supported by European Research Council (Grant No. 683305, project RNA ORIGAMI). The authors would like to thank Ali Aghebat Rafat for help with the AFM and Steffen Sparvath for training A.C. in RNA origami design.

Conflicts of Interest: The authors declare no conflict of interest.

References

1. Ribeiro, D.M.; Zanzoni, A.; Cipriano, A.; Delli Ponti, R.; Spinelli, L.; Ballarino, M.; Bozzoni, I.; Tartaglia, G.G.; Brun, C. Protein complex scaffolding predicted as a prevalent function of long non-coding RNAs. *Nucleic Acids Res.* **2017**, *46*, 917–928. [CrossRef]

2. Mercer, T.R.; Dinger, M.E.; Mattick, J.S. Long non-coding RNAs: Insights into functions. *Nat. Rev. Genet.* **2009**, *10*, 155–159. [CrossRef] [PubMed]

3. Jasinski, D.; Haque, F.; Binzel, D.W.; Guo, P. Advancement of the Emerging Field of RNA Nanotechnology. *ACS Nano* **2017**, *11*, 1142–1164. [CrossRef] [PubMed]

4. Batey, R.T.; Rambo, R.P.; Doudna, J.A. Tertiary Motifs in RNA Structure and Folding. *Angew. Chem. Int. Ed. Engl.* **1999**, *38*, 2326–2343. [CrossRef]

5. Leontis, N.B.; Lescoute, A.; Westhof, E. The building blocks and motifs of RNA architecture. *Curr. Opin. Struct. Biol.* **2006**, *16*, 279–287. [CrossRef] [PubMed]

6. Bindewald, E.; Hayes, R.; Yingling, Y.G.; Kasprzak, W.; Shapiro, B.A. RNAJunction: A database of RNA junctions and kissing loops for three-dimensional structural analysis and nanodesign. *Nucleic Acids Res.* **2007**, *36*, D392–D397. [CrossRef] [PubMed]

7. Delebecque, C.J.; Lindner, A.B.; Silver, P.A.; Aldaye, F.A. Organization of intracellular reactions with rationally designed RNA assemblies. *Science* **2011**, *333*, 470–474. [CrossRef] [PubMed]

8. Sachdeva, G.; Garg, A.; Godding, D.; Way, J.C.; Silver, P.A. In vivo co-localization of enzymes on RNA scaffolds increases metabolic production in a geometrically dependent manner. *Nucleic Acids Res.* **2014**, *42*, 9493–9503. [CrossRef]

9. Jaeger, L.; Chworos, A. The architectonics of programmable RNA and DNA nanostructures. *Curr. Opin. Struct. Biol.* **2006**, *16*, 531–543. [CrossRef]

10. Jaeger, L.; Westhof, E.; Leontis, N.B. TectoRNA: Modular assembly units. *Nucleic Acids Res.* **2001**, *29*, 455–463. [CrossRef]

11. Ko, S.H.; Su, M.; Zhang, C.; Ribbe, A.E.; Jiang, W.; Mao, C. Synergistic self-assembly of RNA and DNA molecules. *Nat. Chem.* **2010**, *2*, 1050. [CrossRef]

12. Afonin, K.A.; Bindewald, E.; Yaghoubian, A.J.; Voss, N.; Jacovetty, E.; Shapiro, B.A.; Jaeger, L. In vitro Assembly of Cubic RNA-Based Scaffolds Designed in silico. *Nat. Nanotechnol.* **2010**, *5*, 676–682. [CrossRef]

13. Geary, C.; Rothemund, P.W.K.; Andersen, E.S. A single-stranded architecture for cotranscriptional folding of RNA nanostructures. *Science* **2014**, *345*, 799–804. [CrossRef] [PubMed]

14. Parlea, L.; Bindewald, E.; Sharan, R.; Bartlett, N.; Moriarty, D.; Oliver, J.; Afonin, K.A.; Shapiro, B.A. Ring Catalog: A resource for designing self-assembling RNA nanostructures. *Methods* **2016**, *103*, 128–137. [CrossRef] [PubMed]

15. Han, D.; Qi, X.; Myhrvold, C.; Wang, B.; Dai, M.; Jiang, S.; Bates, M.; Liu, Y.; An, B.; Zhang, F.; et al. Single-stranded DNA and RNA origami. *Science* **2017**, *358*, eaao2648. [CrossRef]

16. Afonin, K.A.; Grabow, W.W.; Walker, F.M.; Bindewald, E.; Dobrovolskaia, M.A.; Shapiro, B.A.; Jaeger, L. Design and self-assembly of siRNA-functionalized RNA nanoparticles for use in automated nanomedicine. *Nat. Protoc.* **2011**, *6*, 2022. [CrossRef] [PubMed]

17. Afonin, K.A.; Kireeva, M.; Grabow, W.W.; Kashlev, M.; Jaeger, L.; Shapiro, B.A. Co-transcriptional Assembly of Chemically Modified RNA Nanoparticles Functionalized with siRNAs. *Nano Lett.* **2012**, *12*, 5192–5195. [CrossRef] [PubMed]

18. Shu, D.; Khisamutdinov, E.F.; Zhang, L.; Guo, P. Programmable folding of fusion RNA in vivo and in vitro driven by pRNA 3WJ motif of phi29 DNA packaging motor. *Nucleic Acids Res.* **2014**, *42*, e10. [CrossRef]

19. Shu, D.; Shu, Y.; Haque, F.; Abdelmawla, S.; Guo, P. Thermodynamically Stable RNA three-way junctions as platform for constructing multi-functional nanoparticles for delivery of therapeutics. *Nat. Nanotechnol.* **2011**, *6*, 658–667. [CrossRef]

20. Jepsen, M.D.E.; Sparvath, S.M.; Nielsen, T.B.; Langvad, A.H.; Grossi, G.; Gothelf, K.V.; Andersen, E.S. Development of a genetically encodable FRET system using fluorescent RNA aptamers. *Nat. Commun.* **2018**, *9*, 18. [CrossRef]

21. Li, M.; Zheng, M.; Wu, S.; Tian, C.; Liu, D.; Weizmann, Y.; Jiang, W.; Wang, G.; Mao, C. In vivo production of RNA nanostructures via programmed folding of single-stranded RNAs. *Nat. Commun.* **2018**, *9*, 2196. [CrossRef]

22. Schwarz-Schilling, M.; Dupin, A.; Chizzolini, F.; Krishnan, S.; Mansy, S.S.; Simmel, F.C. Optimized Assembly of a Multifunctional RNA-Protein Nanostructure in a Cell-Free Gene Expression System. *Nano Lett.* **2018**, *18*, 2650–2657. [CrossRef]

23. Lukavsky, P.J. Structure and function of HCV IRES domains. *Virus Res.* **2009**, *139*, 166–171. [CrossRef] [PubMed]

24. Baugh, C.; Grate, D.; Wilson, C. 2.8 A crystal structure of the malachite green aptamer. *J. Mol. Biol.* **2000**, *301*, 117–128. [CrossRef]

25. Paige, J.S.; Wu, K.Y.; Jaffrey, S.R. RNA Mimics of Green Fluorescent Protein. *Science* **2011**, *333*, 642. [CrossRef] [PubMed]

26. Huang, H.; Suslov, N.B.; Li, N.S.; Shelke, S.A.; Evans, M.E.; Koldobskaya, Y.; Rice, P.A.; Piccirilli, J.A. A G-quadruplex-containing RNA activates fluorescence in a GFP-like fluorophore. *Nat. Chem. Biol.* **2014**, *10*, 686–691. [CrossRef] [PubMed]

27. Lim, F.; Peabody, D.S. RNA recognition site of PP7 coat protein. *Nucleic Acids Res.* **2002**, *30*, 4138–4144. [CrossRef] [PubMed]

28. Chao, J.A.; Patskovsky, Y.; Almo, S.C.; Singer, R.H. Structural basis for the coevolution of a viral RNA-protein complex. *Nat. Struct. Mol. Biol.* **2008**, *15*, 103–105. [CrossRef]

29. Sparvath, S.L.; Geary, C.W.; Andersen, E.S. Computer-Aided Design of RNA Origami Structures. *Methods Mol. Biol.* **2017**, *1500*, 51–80. [CrossRef]

30. Guex, N.; Peitsch, M.C. SWISS-MODEL and the Swiss-PdbViewer: An environment for comparative protein modeling. *Electrophoresis* **1997**, *18*, 2714–2723. [CrossRef]

31. Pettersen, E.F.; Goddard, T.D.; Huang, C.C.; Couch, G.S.; Greenblatt, D.M.; Meng, E.C.; Ferrin, T.E. UCSF Chimera—A visualization system for exploratory research and analysis. *J. Comput. Chem.* **2004**, *25*, 1605–1612. [CrossRef]

32. Ennifar, E.; Walter, P.; Ehresmann, B.; Ehresmann, C.; Dumas, P. Crystal structures of coaxially stacked kissing complexes of the HIV-1 RNA dimerization initiation site. *Nat. Struct. Biol.* **2001**, *8*, 1064–1068. [CrossRef]

33. Ennifar, E.; Nikulin, A.; Tishchenko, S.; Serganov, A.; Nevskaya, N.; Garber, M.; Ehresmann, B.; Ehresmann, C.; Nikonov, S.; Dumas, P. The crystal structure of UUCG tetraloop. *J. Mol. Biol.* **2000**, *304*, 35–42. [CrossRef] [PubMed]

34. Lee, A.J.; Crothers, D.M. The solution structure of an RNA loop-loop complex: The ColE1 inverted loop sequence. *Structure* **1998**, *6*, 993–1005. [CrossRef]

35. Lukavsky, P.J.; Kim, I.; Otto, G.A.; Puglisi, J.D. Structure of HCV IRES domain II determined by NMR. *Nat. Struct. Biol.* **2003**, *10*, 1033–1038. [CrossRef]

36. Zhao, Q.; Han, Q.; Kissinger, C.R.; Hermann, T.; Thompson, P.A. Structure of hepatitis C virus IRES subdomain IIa. *Acta Crystallogr. D Biol. Crystallogr.* **2008**, *64*, 436–443. [CrossRef] [PubMed]

37. Andersen Lab for Biomolecular Design. Available online: http://bion.au.dk/ (accessed on 27 March 2019).

38. Jossinet, F.; Ludwig, T.E.; Westhof, E. Assemble: An interactive graphical tool to analyze and build RNA architectures at the 2D and 3D levels. *Bioinformatics* **2010**, *26*, 2057–2059. [CrossRef]

39. Zadeh, J.N.; Steenberg, C.D.; Bois, J.S.; Wolfe, B.R.; Pierce, M.B.; Khan, A.R.; Dirks, R.M.; Pierce, N.A. NUPACK: Analysis and design of nucleic acid systems. *J. Comput. Chem.* **2011**, *32*, 170–173. [CrossRef] [PubMed]

40. Tm Calculator v 1.10.2. Available online: http://tmcalculator.neb.com/ (accessed on 27 March 2019).

nanomaterials

MDPI

Article

Optimization of the Split-Spinach Aptamer for Monitoring Nanoparticle Assembly Involving Multiple Contiguous RNAs

Jack M. O'Hara [1], Dylan Marashi [1], Sean Morton [1], Luc Jaeger [2] and Wade W. Grabow [1,*]

[1] Department of Chemistry and Biochemistry, Seattle Pacific University, Seattle, WA 918119-1997, USA;
     oharaj@spu.edu (J.M.O.); marashid@spu.edu (D.M.); mortons@spu.edu (S.M.)
[2] Department of Chemistry and Biochemistry, Biomolecular Science and Engineering Program,
     University of California, Santa Barbara, CA 93106-9510, USA; jaeger@ucsb.edu
*   Correspondence: grabow@spu.edu; Tel.: +1-206-281-2016

Received: 14 February 2019; Accepted: 4 March 2019; Published: 6 March 2019

Abstract: The fact that structural RNA motifs can direct RNAs to fold and self-assemble into predictable pre-defined structures is an attractive quality and driving force for RNA's use in nanotechnology. RNA's recognized diversity concerning cellular and synthetically selected functionalities, however, help explain why it continues to draw attention for new nano-applications. Herein, we report the modification of a bifurcated reporter system based on the previously documented Spinach aptamer/DFHBI fluorophore pair that affords the ability to confirm the assembly of contiguous RNA strands within the context of the previously reported multi-stranded RNA nanoring. Exploration of the sequence space associated with the base pairs flanking the aptamer core demonstrate that fluorescent feedback can be optimized to minimize the fluorescence associated with partially-assembled RNA nanorings. Finally, we demonstrate that the aptamer-integrated nanoring is capable of assembling directly from transcribed DNA in one pot.

Keywords: RNA nanotechnology; RNA self-assembly; light-up aptamer; RNA nanoparticle

1. Introduction

RNA nanotechnology leverages the formation of programmable base pairs and regular three-dimensional folding patterns of structural RNA moieties to construct materials with precise, predefined shapes [1–5]. As a building material, RNA offers several unique benefits such as biocompatibility, the ability to generate or add diverse biological functions, and the potential to generate and assemble nanoparticles directly from DNA transcripts [6–8]. As a consequence, RNA nanoparticles have been used in a variety of applications including the delivery of therapeutics, as stable scaffolds for the attachment of functional moieties, and as molecular signaling devices [9–12]. While much progress has been made in the manufacturing of rationally designed RNA structures, few tools exist to monitor their assembly and/or allow the subsequent tracking of wholly formed nanoparticles. As the design and utilization of nanostructures with increased complexity continues to progress, new methods and systems intended to monitor and verify the assembly of nanoparticles will be required to advance the field of RNA nanotechnology further.

A promising strategy that has been developed to visualize RNA in recent years involves the use of light-up RNA aptamer/fluorophore pairs [13–17]. Several RNA-based aptamer/fluorophore pairs have been developed to allow the monitoring of any RNA transcript [18–21]. Fluorescent-based, label-free RNA tracking methods are thought to offer distinct advantages over other investigative strategies because they can be integrated non-intrusively into virtually any RNA of interest in a variety of different contexts [22–27]. The advent RNA aptamer/fluorophore pairs with tunable wavelengths and the development of user-friendly toolkits continues to provide greater accessibility and inspire new applications [28–30]. Multiple RNA light-up aptamers have been bifurcated in order to enable the monitoring of more than one RNA transcript [18,31–33]. Split-aptamer systems rely on the fact that the functional aptamer forms only when both non-functional halves combine in the presence of a small molecular fluorophore. Given the dynamic nature of aptamer assembly, such fluorogenic systems have opened up new applications that include high-throughput assays, controlled reporting of assembly and processing, the development of logic gates and molecular computation, and more [23,33–35]. While split-aptamers offer the ability to monitor the assembly of two unique RNA strands (or three if formed on a scaffold-strand), most RNA nanoparticles are composed of several unique strands of RNA. Thus the ability to confirm the assembly of multiple RNAs is an important requirement where the assembly of more complex nanoparticles is concerned. With this understanding in mind we set out to design a modified bifurcated platform that provides the capability to monitor the assembly of a nanoparticle comprised of more than two RNA strands.

Given its unique structure and demonstrated ability to be functionalized with a variety of RNA-based functional groups, we chose to integrate the split-Spinach aptamer into the previously reported RNA nanoring [36,37]. The nanoring/split-aptamer reporter system represents a significant expansion of previous uses of the light-up aptamer/fluorophore pairs which rely on the direct interactions at the secondary structure level alone. The goal of our system was to be able to detect tertiary contacts formed by RNAs not directly coupled to the split-aptamer. In this regard, our design focused on finding an optimal thermodynamic balance between split-aptamer assembly and nanoparticle assembly where the formation of the functional aptamer depended more on the assembly of the whole nanoring so that maximum fluorescence occurred only in the presence of the whole nanoparticle. Herein, we report a split-Spinach aptamer system with the ability to monitor the assembly of six strands of RNA in a single nanoparticle. We demonstrate that the integrated light-up aptamer exhibits significant sensitivity for fully-assembled over partially-assembled nanoring nanoparticles. In doing so, we believe this to be the first system developed with the ability to detect adjacent, long-range tertiary interactions as opposed to base pairing directly mediated by the aptamer itself. Finally, we discuss the particular design constraints associated with our system in order to suggest general considerations that could be applied to the development of future multi-stranded reporter systems.

2. Materials and Methods

2.1. Design and Synthesis of Split-Spinach Aptamer and Fluorophore

The previously published Spinach aptamer (PBD ID: 4TS2) [33,38] was modeled into the RNA nanoring [37] using the Swiss PDB-Viewer [39]. Placement of the aptamer inside the nanoring provided an initial estimate regarding stem and linker-strand lengths as well as optimal orientation of the aptamer. The split-Spinach strands were fused to two of the opposing nanoring struts (Figure 1A). Individual RNA strands were rationally-designed and evaluated for unintended secondary folding patterns prior to their synthesis and experimentation [40,41]. Sequences associated with helical regions of the nanoring struts were optimized to avoid secondary structures that would interfere with nanoring loops and/or the core of the Spinach aptamer—both sequence regions which could not be altered. DNA sequences, corresponding to the RNA sequences of interest, were designed by adding a T7 polymerase promoter site sequence (TTCTAATACGACTCACTATA) to the 5′ end of each RNA. Corresponding DNA templates and primers were purchased from Integrated DNA technologies (IDT, San Diego, CA, USA), amplified using *taq* DNA polymerase (Thermo Fisher Scientific, Waltham, MA, USA) via polymerase chain reaction (PCR), and isolated by a DNA purification kit (Epoch Life Sciences, Missouri, MO, USA). Transcription of amplified DNA was accomplished using T7 RNA polymerase (Thermo Fisher Scientific, Waltham, MA, USA) in vitro. The resulting transcripts were purified by polyacrylamide gel electrophoresis (PAGE) [8–10% polyacrylamide, 8 M urea, $1 \times (89$ mM, pH 8.2) Tris Borate (TB)]. Excised gel fragments containing RNA were placed in Crush and Soak buffer (200 mM NaCl, 10 mM Tris pH 7.5, 1 mM filtrated Na_2EDTA pH 8, water), shook overnight at 5 °C, and the RNA was isolated the next day via ethanol precipitation. The fluorophore, 3,5-difluoro-4-hydroxybenzylidene imidazolinone (DFHBI), was synthesized as previously reported according the protocol of the Paige research group [21]. A complete list of RNA sequences used in the study can be found in the Supporting Information.

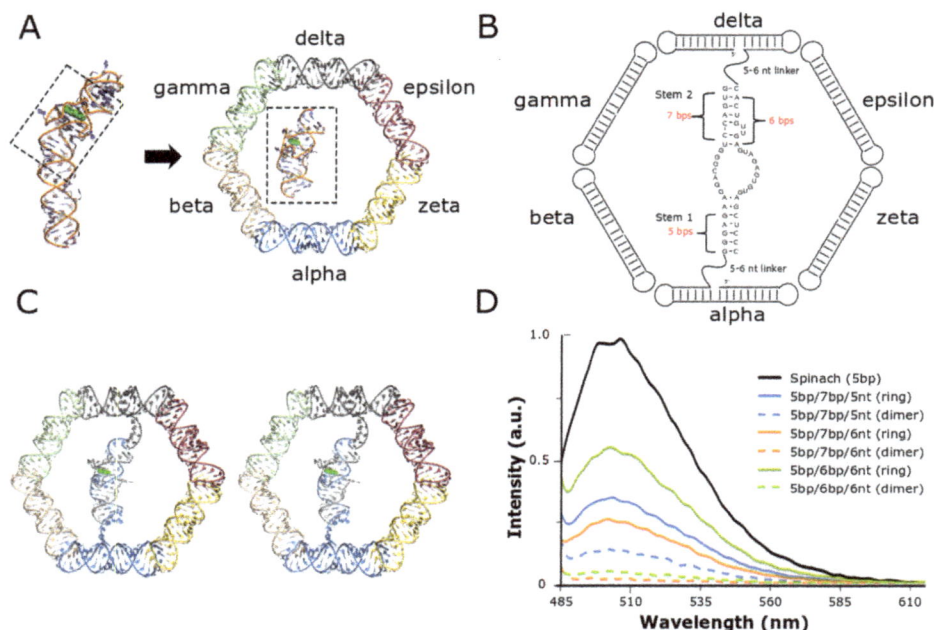

Figure 1. Design and integration of the split-Spinach aptamer and RNA nanoring. (**A**) The central portion of the Spinach aptamer crystallized by Huang et al. (PDB ID: 3IVK) [42] was placed in the interior of the previously reported RNA nanoring and grafted onto two of the nanoring's opposing helical struts. Based on initial placement, the Spinach aptamer was modeled to contain two short stems and single-stranded linkers with variable lengths. (**B**) 2D diagram resulting from initial modification and modeling of the split-Spinach aptamer into the RNA nanoring. (**C**) Stereoview of split-aptamer integrated into RNA nanoring. (**D**) The variable stem and single-stranded linker lengths were tested via fluorescent spectroscopy in the presence of the light-up chromophore DFHBI. The combination containing 5- and 6-bp stems and 6-nt linkers (5bp/6p/6nt) showed the highest response and was therefore chosen as the initial base model for further refinement.

2.2. Monitoring Nanoring Assembly

Assembly of the split-aptamer integrated nanoring was evaluated by native PAGE (40 mM HEPES, pH 8.2 buffer and 2 mM $Mg(OAc)_2$) and fluorescent spectroscopy. RNAs were assembled by combining equimolar concentrations of RNA strands (at a concentration of 500 mM unless noted otherwise) and the snap cool process (2 min at 95 °C and 3 min on ice). After snap cooling, an association buffer was added to achieve a final concentration of 40 mM HEPES (pH 8.2), 1 mM $Mg(OAc)_2$, and 50 mM KCl. This mixture was incubated at 37 °C for 20 min and evaluated by fluorescence spectroscopy with an LS 55 luminescence spectrometer (PerkinElmer, Waltham, MA, USA). DFHBI was added (either before or after incubation) to final concentration of 1 mM. Samples were loaded into a 15 uL quartz cuvette (Starna Cells, Inc., Atascadero, CA, USA) and excited at 469 nm. Emission was recorded at 509 nm. Assembly products were also analyzed by a gel shift assay. Products were loaded into a 7%

polyacrylamide gel of 1× HEPES (40 mM HEPES) buffer and 1 mM Mg(OAc)$_2$. Gels were run at 6 W for 3−4 h at 4 °C. Gels were stained with Sybr Gold (Invitrogen, Carlsbad, CA, USA) and imaged using a FluoroChemQ gel imager (Protein Simple, San Jose, CA, USA).

2.3. Co-Transcriptional Assembly

Amplified DNA (0.35 µM of each individual strand) for the RNA ring pieces and/or aptamer were added to 5× co-transcription buffer (DTT (100 mM), NTPs (25 mM each), IPP (0.1 u/µL), RNasin (40 u/µL), and T7 RNA polymerase (20-120U)). The amount of T7 RNA polymerase was normalized to the total amount of DNA in each reaction mixture. The total volume of reaction mixtures was 20 µL. In a typical experiment, reaction mixtures were incubated at 37 °C for 60 min. After incubation, 3 µL of DNase I (1 u/µL) was added to each reaction mixture and then incubated for an additional 15 min at 37 °C. Aliquots of each reaction mixture were evaluated by fluorescence (17 µL) and/or by gel electrophoresis (3 µL) as described above. All fluorescence signals were normalized to the 5 base pair (bp) full-length Spinach aptamer reported by Huang et al which was used as a control in each individual experiment conducted [42].

3. Results & Discussion

3.1. Initial Design of the Split-Spinach/Nanoring System

The overall structural architecture of the RNA nanoring provided a unique opportunity to place a functional split-aptamer in the interior of the ring (Figure 1). The ability to rationally design and integrate the split-Spinach/fluorophore system into the nanoring was made possible because of the previously reported crystal structure of the full-length Spinach aptamer (PBD ID: 4TS2)—which was essential for evaluating its potential placement within the interior of the six-membered RNA nanoring in silico (Figure 1A) [36,42]. The two main stems flanking, and responsible for stabilizing, the fluorophore binding pocket were both shortened to approximately the same length in order to allow the fully-formed aptamer to sit comfortably in the middle of the interior region of the nanoring (Figure 1A). Previous reports revealed that one of the closing stems responsible for stabilizing the aptamer's core—formed between the 5′ and 3′ ends of the full-length RNA strand and referred to as stem 1—could be reduced to five base pairs (bp) without compromising the binding and fluorescence of the fluorophore [42]. In order to convert the full-length aptamer into a bifurcated system we eliminated the terminal hairpin loop (which was subsequently replaced with a loop from the class I ligase ribozyme to create a binding site for the crystallization chaperone Fab BL3-6). This second closing stem (stem 2), on the opposite side of the aptamer core and which also functions to stabilize the formation of the aptamer's binding pocket, was shortened to seven bps for initial testing (Figure 1B). Visual inspection of the model built using the Swiss PDB-Viewer [39] revealed that the nanoring could readily accommodate a 5-bp stem adjacent to the two uracil bulge near the binding pocket. Each strand of the minimized split-aptamer was tethered to one of the opposing helical struts of the nanoring via flexible single-stranded linkers. Our three-dimensional model based on the previously reported structures of the Spinach aptamer and the RNA nanoring indicated that linker strands of five to six nucleotides were needed to span the gap between the aptamer and nanoring struts (Figure 1B). Finally, realizing that the orientation of the linker strand exiting the nanoring depends on its nucleotide position within the nanoring stem, modeling revealed that grafting the linker strands on the sixth nucleotide from 5′ end of each of two struts of the nanoring directed the formation of the aptamer toward the interior of ring.

Using the visually-constructed three-dimensional model as our guide, we tested a small set of nanoring/split-aptamer systems with variable stem and linker lengths. We evaluated the fluorescence of the ring strands containing the split-aptamer sequences in the presence and absence of the peripheral helical struts responsible for complete nanoring formation. Our results showed that the nanoring/aptamer system possessing stem lengths of five base pairs on one side of the aptamer (stem 1) and six base pairs on the other side (stem 2)—in conjunction with linkers of six nucleotides (referred to as 5bp/6bp/6nt)—produced the greatest difference between ring and split-aptamer only assemblies (Figure 1D). In both cases where a longer stem was used, the fluorescence associated with the ring was lower. This data suggests that the longer stem lengths may have been sterically hindered within the interior of the ring. This is partially corroborated by the fact that 5bp/7bp/5nt exhibited higher fluorescence as a dimer than 5bp/6bp/6nt as expected given the longer stem but was lower when placed in the context of the ring.

3.2. Optimization and Assessment of Split-Spinach Variants

In the ideal case, the functional nanoring/split-aptamer would possess the ability to bind DFHBI and fluoresce only after all six strands of the nanoring were present and able to assemble into the complete nanoring. We theorized that if we wanted to rely on the split-aptamer to identify the assembly of contiguous RNA strands not directly connected to the aptamer core then we had to destabilize the split-aptamer's propensity to assemble and create a functional aptamer on its own. Building off of our initial results, we set out to improve upon the 5bp/6bp/6nt version of the split-aptamer to provide maximum sensitivity for fully-assembled nanorings over partially-assembled ones. We hypothesized that just the right degree of destabilization in the aptamer stems could allow the formation of the nanoring to play a greater factor in promoting the formation of a functional split-aptamer which would in turn function to minimize the fluorescent signal induced by incomplete or partial assemblies. With this goal in mind, we created variants of the split-Spinach aptamer with different base pairs in the two stems surrounding the binding pocket—seeking to find a split-aptamer system that abided by a Goldilocks-like principle: just stable enough, but not too stable. We targeted the three base pairs formed at the 5'/3' interfaces of both respective aptamer halves as prime locations to alter stem stabilities without compromising the aptamer core (Figure 2A). We theorized that they were far away from the aptamer core that their alteration could affect aptamer core stability while having a minimal effect on or interference with DFHBI binding. Base pairs were intentionally mutated and/or deleted at these positions with the goal of destabilizing the aptamer in order to prevent its functional formation in the absence of the supporting ring struts. In each case, the various split-aptamer variants were evaluated by their fluorescence intensities—normalized to the fluorescent intensity of the full-length Spinach aptamer as a control.

As a means of judging aptamer sensitivities, assemblies involving all ring strands were compared to those consisting of just the two ring strands possessing each half of the split-aptamer (alpha and delta strands) in the absence of the nanoring's remaining supporting struts (beta, gamma, epsilon, and zeta strands respectively) (Figure 2). In order to compare and assess the sensitivity of the different variant combinations, we calculated the ratio of fluorescence of the rings to their corresponding dimers without the supporting struts. We hypothesized that substituting stronger interacting base pairs with weaker ones (i.e., replacing GC bps with AU or GU bps) or disrupting base pairs by removing nucleotides in these two regions would destabilize the aptamer and thereby provide greater split-aptamer sensitivity over stems with increased stabilities. Generally, this hypothesis held true. In all cases where the overall stability of the split-aptamer was increased by replacing an AU bp with a GC bp the resulting variants showed higher fluorescent signals in the context of fully-assembled nanoparticles—with one combination showing nearly the same intensity as the single full-length Spinach control. The sensitivity of these stabilized systems however was generally lower than their destabilized counterparts (Figure 2B). This was due to the fact that the fluorescent signals from the dimers also increased (and in greater proportion than that of the fully formed ring systems). In most cases where the stems were destabilized, the split-aptamer showed markedly lower signals for alpha and delta strands in the absence of the four ring struts as desired. In the most extreme instances (e.g., when a nucleotide was removed from the 3' end of each strand to remove a base pair on each side of the aptamer) the fluorescence signal associated with both the assembled ring and dimer were significantly diminished. In other situations however, the dimer signal decreased without affecting the signal from the fully-assembled ring (e.g., in cases where a GC in stem 1 was removed or replaced with an AU or GU). It is also worth noting that attempts to invert all three base pairs in stem 1 produced lower sensitivities and/or lower overall fluorescence signal in all but one case. In a few instances, attempts to destabilize stem 2 actually increased the split-aptamer's fluorescence intensity associated with both the dimers and the assembled rings. For example, the introduction of a G at position 51 of the delta strand (changing an AU bp to a GU bp) showed increased fluorescence over the initial 5bp/6bp/6nt model (Figure 2B). This instance suggests that stem stability alone is not the only factor responsible for the split-aptamer's performance within the context of the nanoring. Other aspects such as folding dynamics and the secondary structure of each strand is also thought to be important.

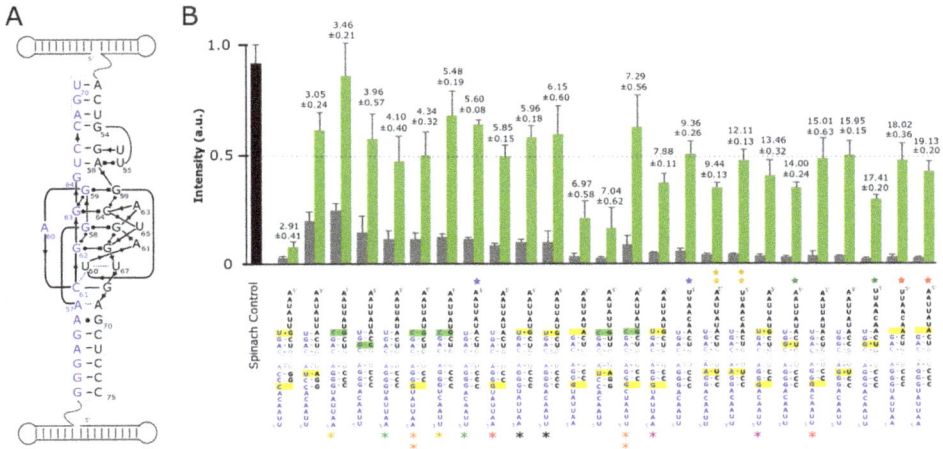

Figure 2. Optimization of split-Spinach as reporter for RNA nanoring assembly. (**A**) 2D diagram of split-Spinach aptamer core (adapted from Ouellet Front. Chem. 2016) [16] (**B**) Fluorescent data of variant split-Spinach sequences tested normalized to unimolecular Spinach control (black bar). Green bars indicate fluorescence associated with the fully-assembled nanoring while grey bars represent fluorescence associated with the incubation of the alpha and delta strands containing the split aptamer alone. Variants are ranked according to their respective sensitivities (i.e., ratio of fluorescence in ring/fluorescence of dimer shown above the green and grey bars). Split-Spinach variants that differ from each other in the linker sequences are identified by colored asterisks (alpha strand) or stars (delta strand). Pairs that differ by only a single linker are identified by the same color. The number of asterisks or stars indicate the identity of the particular linker strand used. Data is based on a minimum of three trials. The error bars represent the standard deviation associated with each collection of measurements.

In addition to varying the composition of base pairs around the aptamer's core we also explored the way in which different linker sequences could affect split-Spinach function. The linker sequences make up an important design element because they represent the only truly unconstrained sequence space associated with each split-aptamer strand (of course, in terms of the whole system the nucleotide sequences associated with the helical struts could also be altered by covariation of base pairs they are thought to have little to no influence on the aptamer system as a whole). We postulated that the primary influence linker sequences could have on the reporter system's performance was through its effect on the individual strand's secondary structure. We reasoned that linkers which were able to fold in a way that partially sequestered the portions of the split-aptamer involved in forming the stabilizing helixes could provide further sensitivity with regard to distinguishing between the presence and absence of fully-assembled nanorings. For example, in the absence of the ring the partially blocked strands would have a more difficult time forming a functional aptamer than when they are brought in close proximity due to the assembly of the nanoring. In terms of their overall design, we intentionally relied predominately on uracils (for flexibility) and adenines (to avoid unintended pairings with the multiple guanine's associated with the aptamer core). We evaluated the folding of candidate sequences in silico using Mfold in order to find the potential secondary structures associated with each unique sequence (see Supporting Information) [41]. Generally, we found that additional secondary structure resulted in increased sensitivity. For example the addition of a GC bp to stem 2 produced the highest fluorescent signal—on par with the full the length aptamer but it lacked sensitivity because the fluorescent signal of the dimer was also the highest out of any of the constructs tested. By introducing a linker sequence that predicted a more robust secondary structure, the overall sensitivity increased by about 60% (from 3.46 to 5.48) (See supporting information, Figure S2). In other cases, strands with similar secondary structures exhibited different sensitivities—suggesting that other factors like individual folding pathways also provide subtle effects. It is clear however that the stability of the stems surrounding the aptamer core had the largest influence on a combination's overall performance.

After evaluating stem stabilities and linker contributions we identified several sequence combinations that provided at least a 10-fold increase in signal between the split-aptamer in the presence of all the ring components compared to split-aptamer strands incubated by themselves with the highest ratio reaching nearly a 20-fold fluorescence gain. Following these results we evaluated the ability for the split-aptamer system to distinguish between fully- and partially-assembled nanorings. Partially-assembled rings (composed of five out of the six nanoring struts) were expected to produce a higher fluorescent signal than the split-aptamer dimers alone because, like the fully-assembled nanorings, they provide a physical conduit for the split-aptamer strands to be brought together. Previous evaluation of the nanoring however showed that partially-assembled rings are not as thermodynamically stable as fully-assembled rings and so we postulated that the split-aptamer system should be able to show some level of discrimination between the two [37] In order to evaluate the utility of split-aptamer/nanoring system further, we selected the seven most sensitive combinations (by comparing fully-assembled and split-aptamer dimers only) for further assessment in the context of partially-formed nanorings (Figure 3).

Evaluation of the partially-assembled rings revealed a number of interesting insights (Figure 3). In the first case, the strength of the fluorescence signal of the partially-assembled rings was found to be influenced by the precise identity of the missing strut. Absence of the same strut did not universally increase fluorescence across different combinations. We found that each particular split-aptamer combination had its own characteristic profile with regards to the missing strut (Figure S3). For example, for some combinations the absence of the epsilon strut produced routinely produced the highest signal while for others the highest signal involved the absence of the gamma, or zeta strut. Given the variances observed between the different split-aptamer systems we chose to compare the fluorescent response associated with the highest partially-assembled nanoring signal within each variant system—providing a worst-case scenario in terms of background signal against the fully-assembled nanoring. Secondly, while the partially-assembled constructs generally produced higher fluorescence signals than the dimer alone, we identified two split-aptamer variants that showed virtually no overall increase in fluorescent signal for partially-assembled rings compared to the dimer. In these two cases, the sensitivity of fully- over partially-assembled remained approximately 15-fold more sensitive for the fully-assembled nanoring over the partially-assembled one (Figure 3C).

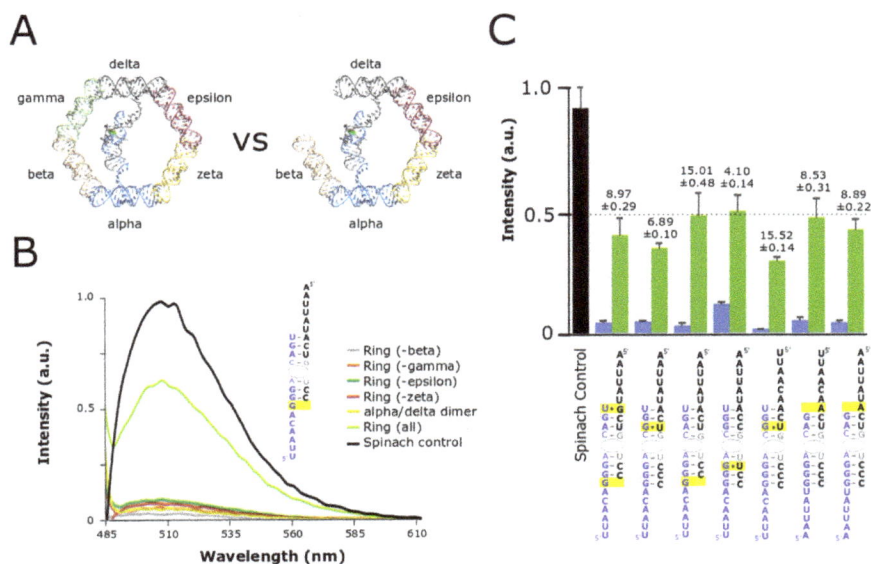

Figure 3. Comparison of fully- and partially-assembled nanorings. (**A**) In order to function as an effective tool for verifying and monitoring nanoring assembly the system needed to be able distinguish between whole and incomplete or partially-assembled nanorings. (**B**) Fluorescent data comparing fully-assembled to nanorings missing one of the four supporting struts (beta, gamma, epsilon, or zeta). (**C**) Summary of average fluorescent values for whole nanoring assemblies and partial assemblies compared to Spinach control where the green bar represents the complete nanoring and the blue bar represents the partially-assembled nanoring. The ratio of fluorescence of ring to partially-assembled (ring missing one strut) is shown above each plot. Data is based on a minimum of three trials. The error bars represent the standard deviation associated with each collection of measurements.

3.3. Co-Transcriptional Assembly

A distinct advantage of RNA-based nanoparticles relates to their ability to self-assemble isothermally directly from their transcription via DNA templates [6,18,43]. This ability, paired with the rise of new RNA aptamer/fluorophore pairs has opened the door for a variety of in vitro applications involving the visualization of RNA transcripts [14,15,17,18,23,44]. The RNA nanoring, in particular, offers an attractive scaffolding platform for further functionalization and the development of high-throughput, automated medicine [6,45]. In order to evaluate the robustness of the split-aptamer/nanoring system we looked at its sensitivity with respect to fully- and partially-assembled nanorings directly resulting from the transcription of the individual DNA templates. Initial experiments evaluated the split-aptamer's performance from equimolar mixtures of its composite RNA strands. In these cases RNA assembly was carefully controlled via a snap-cooling protocol to ensure proper folding of individual components (see Material and Methods). In the case of co-transcriptional assembly, equimolar concentrations of DNA templates were added to a transcription mixture along with T7 RNA polymerase. The resulting transcripts were left to self-assemble in the mixture during the course of the experiment.

We evaluated the self-assembly of the split-aptamer modified RNAs by fluorescence spectroscopy and by gel electrophoresis (Figure 4). Native PAGE gels (40 mM HEPES, pH 8.2 buffer and 2 mM Mg(OAc)$_2$) were used to compare the resulting RNA assemblies of the transcription mixture with an RNA ladder prepared via snap-cooled assembly of the corresponding nanoring. Native PAGE reveals that the fully-assembled nanorings constitute the primary assembly product (Figure 4A). The fluorescent signal associated with dimers, partially-assembled, and fully-assembled nanorings supports this same outcome. While the signaling ratio between fully-assembled and partially-assembled rings was on the whole lower, the fully-assembled rings achieved between 2- and nearly 7-fold increase in fluorescence over the partially-assembled rings. Fully-assembled nanorings have been shown to be more chemically and thermodynamically stable than partially-assembled ones [37]. Given that the partially-assembled nanorings would have much shorter half-lives and be less prone to form, we postulate that sensitivity of the fully-assembled nanoring over the partially-assembled ones could be increased in certain environments. More impressively (considering fluorescence is triggered by the folding and assembly of six independent strands versus a single transcript) the fluorescent signal of three of the ring systems produced signal levels that remained at over 50% of the single-transcript Spinach aptamer. This is particularly remarkable given the fact that maximum fluorescence is achieved only by the production and assembly of six individual strands compared to the transcription and intramolecular folding of the single control strand. Because the fluorescent output of the split-aptamer remains quite strong compared to the full-length Spinach control, it shows promise as a reporter for nanoparticle formation from isothermal assembly of transcription products. Collectively, we believe that these results demonstrate the system's potential for use in high-throughput assembly and other applications.

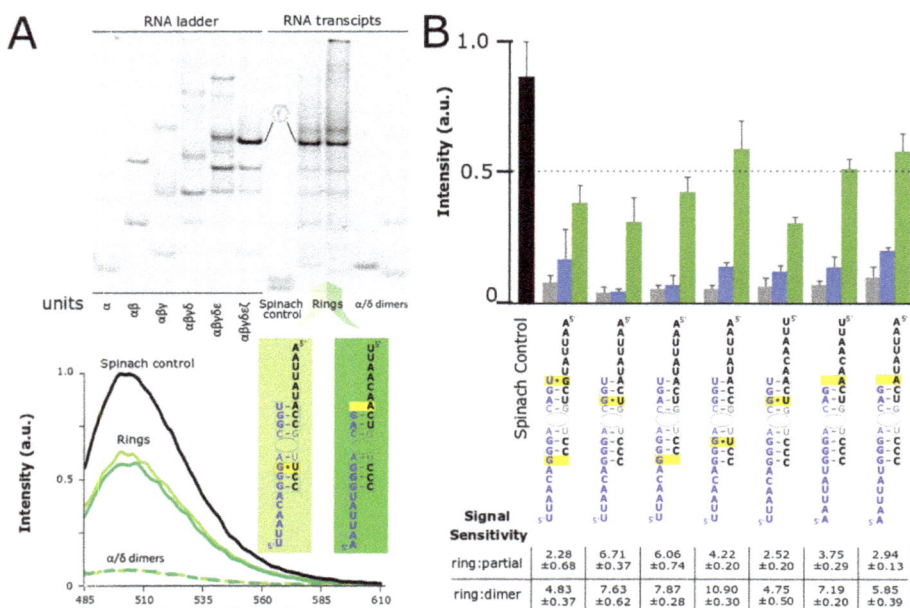

Figure 4. Co-transcriptional assembly of nanorings monitored by native PAGE and fluorescent spectroscopy. (**A**) Native PAGE gel (1× HEPES and 2 mM Mg^{2+}) was used to reveal the composition of nanoring assemblies formed in transcription mix as compared to an RNA ladder that was snapped cooled (top). Fluorescent values associated with the nanoring products were normalized to the Spinach control also formed via transcription (bottom). (**B**) Average fluorescent data for top split-Spinach variants assembled during transcription with ratio of fluorescent signal of fully-assembled ring (green bar) to partially-assembled ring (blue bar) and fully-assembled ring to dimer (grey bar). Data is based on a minimum of three trials. The error bars represent the standard deviation associated with each collection of measurements.

Nanomaterials **2019**, *9*, 378

4. Conclusions

The design and characterization of an aptamer/fluorophore integrated RNA nanoparticle system demonstrates the ability to identify perihperal tertiary interactions between assembled RNA strands of the six-membered RNA nanoring. While many of the particular design constraints discussed above pertain to the specific system at hand, our work reveals a few general principles relating to the design of future split-aptamer systems. During the course of our design we observed that stem stability and performance of any particular sequence (characterized by the various linker designs) require careful testing. This suggests that full-scale automation of the design process remains elusive and that the design of new sytems will undoubtedly contain their own particular parameters which will also require their own experimental optimization. Simply put, the overall performance of variants is and remains context dependent and very difficult to predict. In this regard, even though in vitro studies provide a baseline for assessing the behavior of the a self-assembling nanoparticle/split-aptamer/fluorophore system, evaluation in increasingly complex environments remain necessary for validation and further refinement. Modeling and design provide an essential starting point but due to myriad of parameters, experimental refinement remains absolutely necessity. Our work also shows that the destablization of stems surrounding the aptamer core generally provide increased sensitivity between fully- and partially assembled nanoparticles. The same is likely to be true for future developments with different nanoparticles—where the goal or purpose of nanoparticle assembly functions, at least in part, to restore and/or compensate the intentionally diminished stabilities of the altered stems.

The split-aptamer/nanoparticle system demonstrates the ability, not only to monitor the assembly of multiple strands, but the ability to monitor the assembly of RNA strands not directly linked to the aptamer units. To our knowledge this is the first system developed that demonstrates the ability to detect the formation of tertiary interactions on contiguous strands of RNA. We demonstrate further that fluorescent signaling associated with fully-assembled nanorings can be maximized over the partially assembled split-aptamer system associated with two or five strand mixtures. Given its ability to assemble isothermally from DNA templates, we propose that the co-transcriptional assembly of the split-aptamer/nanoring system provides a robust platform for the further development of a variety of automated and/or high-throughput applications. The nanoring's four other struts are readily functionalizable and because the nanoring's helical struts can be increased a full helical turn to accommodate a larger interior space, it is possible that more complex aptamer systems may be able to be incorporated (such as the addition of secondary modular aptamer domain). Finally, fluorescent reporting of fully-formed nanorings provides instant feedback for the previously proposed automated assembly of therapeutic RNA nanoparticles [45].

Supplementary Materials: The following are available online at http://www.mdpi.com/2079-4991/9/3/378/s1, Figure S1: 2D diagram of aptamer integrated nanoring, Table S1: List of sequences tested, Figure S2: Predicted secondary structures. Figure S3: Fluorescent profile of partially-assembled nanorings.

Author Contributions: The authors of the study all participated and contributed substantially to the research article and their individual contributions are as follows: Conceptualization, L.J. and W.W.G.; Methodology, W.W.G.; Data Acquisition, J.M.O., D.M. and S.M.; Formal Analysis, J.M.O. and W.W.G.; Data Curation, J.M.O., D.M. and S.M.; Writing-Original Draft Preparation, J.M.O. and W.W.G.; Writing-Review & Editing, J.M.O., L.J., and W.W.G.; Visualization, J.M.O. and W.W.G.; Funding Acquisition, W.W.G.

Funding: This research was funded by MJ Murdock Charitable Trust; grant number 2014278:MNL:2/26/2015 and The APC was funded in part by Seattle Pacific University's Faculty Research Grant.

Conflicts of Interest: The authors declare no conflict of interest.

References

1. Jaeger, L.; Westhof, E.; Leontis, N.B. TectoRNA: Modular assembly units for the construction of RNA nano-objects. *Nucleic Acids Res.* **2001**, *29*, 455–463. [CrossRef] [PubMed]
2. Geary, C.; Chworos, A.; Verzemnieks, E.; Voss, N.R.; Jaeger, L. Composing RNA nanostructures from a syntax of RNA structural modules. *Nano Lett.* **2017**, *17*, 7095–7101. [CrossRef] [PubMed]

3. Chworos, A.; Severcan, I.; Koyfman, A.Y.; Weinkam, P.; Oroudjev, E.; Hansma, H.G.; Jaeger, L. Building programmable jigsaw puzzles with RNA. *Science* **2004**, *306*, 2068–2072. [CrossRef] [PubMed]

4. Shu, D.; Moll, W.D.; Deng, Z.; Mao, C.; Guo, P. Bottom-up assembly of RNA arrays and superstructures as potential parts in nanotechnology. *Nano Lett.* **2004**, *4*, 1717–1723. [CrossRef] [PubMed]

5. Grabow, W.W.; Jaeger, L. RNA self-assembly and RNA nanotechnology. *Acc. Chem. Res.* **2014**, *47*, 1871–1880. [CrossRef] [PubMed]

6. Afonin, K.A.; Viard, M.; Koyfman, A.Y.; Martins, A.N.; Kasprzak, W.K.; Panigaj, M.; Desai, R.; Santhanam, A.; Grabow, W.W.; Jaeger, L.; et al. Multifunctional RNA nanoparticles. *Nano Lett.* **2014**, *16*, 1097–1110. [CrossRef] [PubMed]

7. Guo, S.; Tschammer, N.; Mohammed, S.; Guo, P. Specific delivery of therapeutic RNAs to cancer cells via the dimerization mechanism of phi29 motor pRNA. *Hum. Gene Ther.* **2005**, *16*, 1097–1109. [CrossRef] [PubMed]

8. Haque, F.; Pi, F.; Zhao, Z.; Gu, S.; Hu, H.; Yu, H.; Guo, P. RNA versatility, flexibility, and thermostability for practice in RNA nanotechnology and biomedical applications. *Wiley Interdiscip. Rev. RNA* **2018**, *9*, e1452. [CrossRef] [PubMed]

9. Ishikawa, J.; Furuta, H.; Ikawa, Y. RNA tectonics (tectoRNA) for RNA nanostructure design and its application in synthetic biology. *Wiley Interdiscip. Rev. RNA* **2013**, *4*, 651–664. [CrossRef] [PubMed]

10. Grabow, W.; Jaeger, L. RNA modularity for synthetic biology. *F1000Prime Rep.* **2013**, *5*, 46. Available online: http://www.pubmedcentral.nih.gov/articlerender.fcgi?artid=3816761&tool=pmcentrez&rendertype=abstract (accessed on 3 March 2019). [CrossRef] [PubMed]

11. Binzel, D.W.; Shu, Y.; Li, H.; Sun, M.; Zhang, Q.; Shu, D.; Guo, B.; Guo, P. Specific Delivery of MiRNA for High Efficient Inhibition of Prostate Cancer by RNA Nanotechnology. *Mol. Ther.* **2016**, *24*, 1267–1277. [CrossRef] [PubMed]

12. Guo, P. RNA nanotechnology: Engineering, assembly and applications in detection, gene delivery and therapy. *J. Nanosci. Nanotechnol.* **2005**, *5*, 1964–1982. [CrossRef] [PubMed]

13. Bouhedda, F.; Autour, A.; Ryckelynck, M. Light-up RNA aptamers and their cognate fluorogens: From their development to their applications. *Int. J. Mol. Sci.* **2018**, *19*, 44. [CrossRef] [PubMed]

14. Chakraborty, K.; Veetil, A.T.; Jaffrey, S.R.; Krishnan, Y. Nucleic Acid–Based Nanodevices in Biological Imaging. *Annu. Rev. Biochem.* **2016**, *85*, 349–373. [CrossRef] [PubMed]

15. Dolgosheina, E.V.; Unrau, P.J. Fluorophore-binding RNA aptamers and their applications. *Wiley Interdiscip. Rev. RNA* **2016**, *7*, 843–851. [CrossRef] [PubMed]

16. Ouellet, J. RNA Fluorescence with Light-Up Aptamers. *Front. Chem.* **2016**, *4*, 29. [CrossRef] [PubMed]

17. Jaffrey, S.R. RNA-Based Fluorescent Biosensors for Detecting Metabolites in vitro and in Living Cells. *Adv. Pharmacol.* **2018**, *82*, 187–203. [CrossRef] [PubMed]

18. Alam, K.K.; Tawiah, K.D.; Lichte, M.F.; Porciani, D.; Burke, D.H. A Fluorescent Split Aptamer for Visualizing RNA-RNA Assembly in Vivo. *ACS Synth. Biol.* **2017**, *6*, 1710–1721. [CrossRef] [PubMed]

19. Baugh, C.; Grate, D.; Wilson, C. 2.8 Å crystal structure of the malachite green aptamer. *J. Mol. Biol.* **2000**, *301*, 117–128. [CrossRef] [PubMed]

20. Dolgosheina, E.V.; Jeng, S.C.; Panchapakesan, S.S.; Cojocaru, R.; Chen, P.S.; Wilson, P.D.; Hawkins, N.; Wiggins, P.A.; Unrau, P.J. RNA Mango Aptamer-Fluorophore: A Bright, High-Affinity Complex for RNA Labeling and Tracking. *ACS Chem. Biol.* **2014**, *9*, 2412–2420. [CrossRef] [PubMed]

21. Paige, J.S.; Wu, K.Y.; Jaffrey, S.R. RNA mimics of green fluorescent protein. *Science* **2011**, *333*, 642–646. [CrossRef] [PubMed]

22. Pothoulakis, G.; Ceroni, F.; Reeve, B.; Ellis, T. The spinach RNA aptamer as a characterization tool for synthetic biology. *ACS Synth. Biol.* **2014**, *3*, 182–187. [CrossRef] [PubMed]

23. Benenson, Y. RNA-based computation in live cells. *Curr. Opin. Biotechnol.* **2009**, *20*, 471–478. [CrossRef] [PubMed]

24. Kellenberger, C.A.; Wilson, S.C.; Sales-Lee, J.; Hammond, M.C. RNA-Based Fluorescent Biosensors for Live Cell Imaging of Second Messengers Cyclic di-GMP and Cyclic AMP-GMP. *J. Am. Chem. Soc.* **2013**, *135*, 35. [CrossRef] [PubMed]

25. Su, Y.; Hickey, S.F.; Keyser, S.G.L.; Hammond, M.C. In Vitro and in Vivo Enzyme Activity Screening via RNA-Based Fluorescent Biosensors for S-Adenosyl-l-homocysteine (SAH). *J. Am. Chem. Soc.* **2016**, *138*, 7040–7047. [CrossRef] [PubMed]

26. Sunbul, M.; Arora, A.; Jäschke, A. Visualizing RNA in Live Bacterial Cells Using Fluorophore- and Quencher-Binding Aptamers. In *RNA Detection*; Humana Press: New York, NY, USA, 2018; pp. 289–304. [CrossRef]

27. Yerramilli, V.S.; Kim, K.H. Labeling RNAs in Live Cells Using Malachite Green Aptamer Scaffolds as Fluorescent Probes. *ACS Synth. Biol.* **2018**, *7*, 758–766. [CrossRef] [PubMed]

28. Nakano, S.; Nakata, E.; Morii, T. Facile conversion of RNA aptamers to modular fluorescent sensors with tunable detection wavelengths. *Bioorg. Med. Chem. Lett.* **2011**, *21*, 4503–4506. [CrossRef] [PubMed]

29. Jasinski, D.L.; Khisamutdinov, E.F.; Lyubchenko, Y.L.; Guo, P. Physicochemically tunable polyfunctionalized RNA square architecture with fluorogenic and ribozymatic properties. *ACS Nano* **2014**, *8*, 7620–7629. [CrossRef] [PubMed]

30. Chandler, M.; Lyalina, T.; Halman, J.; Rackley, L.; Lee, L.; Dang, D.; Ke, W.; Sajja, S.; Woods, S.; Acharya, S. Broccoli Fluorets: Split Aptamers as a User-Friendly Fluorescent Toolkit for Dynamic RNA Nanotechnology. *Molecules* **2018**, *23*, 3178. [CrossRef] [PubMed]

31. Kikuchi, N.; Kolpashchikov, D.M. A universal split spinach aptamer (USSA) for nucleic acid analysis and DNA computation. *Chem. Commun.* **2017**, *53*, 4977–4980. [CrossRef] [PubMed]

32. Kolpashchikov, D.M. Binary Malachite Green Aptamer for Fluorescent Detection of Nucleic Acids. *J. Am. Chem. Soc.* **2005**, *127*, 12442–12443. [CrossRef] [PubMed]

33. Rogers, T.A.; Andrews, G.E.; Jaeger, L.; Grabow, W.W. Fluorescent monitoring of RNA assembly and processing using the split-spinach aptamer. *ACS Synth. Biol.* **2015**, *4*, 162–166. [CrossRef] [PubMed]

34. Autour, A.; Westhof, E.; Ryckelynck, M. iSpinach: A fluorogenic RNA aptamer optimized for in vitro applications. *Nucleic Acids Res.* **2016**, *44*, 2491–2500. [CrossRef] [PubMed]

35. Goldsworthy, V.; LaForce, G.; Abels, S.; Khisamutdinov, E. Fluorogenic RNA Aptamers: A Nano-platform for Fabrication of Simple and Combinatorial Logic Gates. *Nanomaterials* **2018**, *8*, 984. [CrossRef] [PubMed]

36. Yingling, Y.G.; Shapiro, B.A. Computational design of an RNA hexagonal nanoring and an RNA nanotube. *Nano Lett.* **2007**, *7*, 2328–2334. [CrossRef] [PubMed]

37. Grabow, W.W.; Zakrevsky, P.; Afonin, K.A.; Chworos, A.; Shapiro, B.A.; Jaeger, L. Self-assembling RNA nanorings based on RNAI/II inverse kissing complexes. *Nano Lett.* **2011**, *11*, 878–887. [CrossRef] [PubMed]

38. Warner, K.D.; Chen, M.C.; Song, W.; Strack, R.L.; Thorn, A.; Jaffrey, S.R.; Ferré-D'Amaré, A.R. Structural basis for activity of highly efficient RNA mimics of green fluorescent protein. *Nat. Struct. Mol. Biol.* **2014**, *21*, 658. [CrossRef] [PubMed]

39. Guex, N.; Diamend, A.; Peitsch, M.C.; Schwede, T. *DeepView—Swiss-PdbViewer*; Swiss Institute of Bioinformatics: Geneva, Switzerland, 2017.

40. Zadeh, J.N.; Steenberg, C.D.; Bois, J.S.; Wolfe, B.R.; Pierce, M.B.; Khan, A.R.; Dirks, R.M.; Pierce, N.A. NUPACK: Analysis and design of nucleic acid systems. *J. Comput. Chem.* **2011**, *30*, 170–173. [CrossRef] [PubMed]

41. Zuker, M. Mfold web server for nucleic acid folding and hybridization prediction. *Nucleic Acids Res.* **2003**, *31*, 3406–3415. [CrossRef] [PubMed]

42. Huang, H.; Suslov, N.B.; Li, N.S.; Shelke, S.A.; Evans, M.E.; Koldobskaya, Y.; Rice, P.A.; Piccirilli, J.A. A G-quadruplex-containing RNA activates fluorescence in a GFP-like fluorophore. *Nat. Chem. Biol.* **2014**, *10*, 686–691. [CrossRef] [PubMed]

43. Torelli, E.; Kozyra, J.W.; Gu, J.Y.; Stimming, U.; Piantanida, L.; Voïtchovsky, K.; Krasnogor, N. Isothermal folding of a light-up bio-orthogonal RNA origami nanoribbon. *Nature* **2018**, *8*, 6989. [CrossRef] [PubMed]

44. Qiu, M.; Khisamutdinov, E.; Zhao, Z.; Pan, C.; Choi, J.W.; Leontis, N.B.; Guo, P. RNA nanotechnology for computer design and in vivo computation. *Philos. Trans. A Math. Phys. Eng. Sci.* **2013**, *371*, 20130310. [CrossRef] [PubMed]

45. Afonin, K.A.; Grabow, W.W.; Walker, F.M.; Bindewald, E.; Dobrovolskaia, M.A.; Shapiro, B.A.; Jaeger, L. Design and self-assembly of siRNA-functionalized RNA nanoparticles for use in automated nanomedicine. *Nat. Protoc.* **2011**, *6*, 2022–2034. Available online: http://www.pubmedcentral.nih.gov/articlerender.fcgi?artid=3498981&tool=pmcentrez&rendertype=abstract (accessed on 3 March 2019). [CrossRef] [PubMed]

![nanomaterials logo] *nanomaterials*

MDPI

Article

First Step Towards Larger DNA-Based Assemblies of Fluorescent Silver Nanoclusters: Template Design and Detailed Characterization of Optical Properties

Liam E. Yourston [1], Alexander Y. Lushnikov [2], Oleg A. Shevchenko [3], Kirill A. Afonin [3] and Alexey V. Krasnoslobodtsev [1,2,*]

[1] Department of Physics, University of Nebraska Omaha, Omaha, NE 68182, USA; lyourston@unomaha.edu
[2] Nanoimaging Core Facility at the University of Nebraska Medical Center, Omaha, NE 68198, USA; alushnikov@unmc.edu
[3] Nanoscale Science Program, Department of Chemistry, University of North Carolina at Charlotte, Charlotte, NC 28223, USA; oshevche@uncc.edu (O.A.S.); kafonin@uncc.edu (K.A.A.)
* Correspondence: akrasnos@unomaha.edu or a.krasnoslobodtsev@unmc.edu; Tel.: +1-402-554-3723

Received: 19 March 2019; Accepted: 12 April 2019; Published: 13 April 2019

Abstract: Besides being a passive carrier of genetic information, DNA can also serve as an architecture template for the synthesis of novel fluorescent nanomaterials that are arranged in a highly organized network of functional entities such as fluorescent silver nanoclusters (AgNCs). Only a few atoms in size, the properties of AgNCs can be tuned using a variety of templating DNA sequences, overhangs, and neighboring duplex regions. In this study, we explore the properties of AgNCs manufactured on a short DNA sequence—an individual element designed for a construction of a larger DNA-based functional assembly. The effects of close proximity of the double-stranded DNA, the directionality of templating single-stranded sequence, and conformational heterogeneity of the template are presented. We observe differences between designs containing the same AgNC templating sequence—twelve consecutive cytosines, $(dC)_{12}$. AgNCs synthesized on a single "basic" templating element, $(dC)_{12}$, emit in "red". The addition of double-stranded DNA core, required for the larger assemblies, changes optical properties of the silver nanoclusters by adding a new population of clusters emitting in "green". A new population of "blue" emitting clusters forms only when ssDNA templating sequence is placed on the 5′ end of the double-stranded core. We also compare properties of silver nanoclusters, which were incorporated into a dimeric structure—a first step towards a larger assembly.

Keywords: silver nanoclusters; fluorescence; i-motif DNA; cytosine rich sequences

1. Introduction

The field of nucleic acid (NA) nanotechnology has brought various designs of materials and devices created with facilitation of nucleic acids both DNA and RNA [1–4]. The progress already made in the field makes nucleic acid molecules very promising candidates for fabrication of complex shapes and functional structures at the nanoscale. The NA nanotechnology toolbox utilizes not only traditional DNA and RNA bases but also their chemical analogs and backbone modifications, for example peptide nucleic acid (PNA) [5]. Additional flexibility is added by the architectural elements of NA that employ tertiary interactions, such as kissing loops [6,7], loop receptor interactions [8], paranemic motifs [9], G-quadruplexes [10], and i-motifs [11], just to name a few. All these combined properties of NA provide a remarkable control over the versatile molecular structures that they can be programmed to assemble. NA nanotechnology opens large prospects for practical applications of novel materials design, diagnostics, and nanomedicine.

Besides being used as architectural elements for construction of complex nanostructures, NA can be utilized as scaffolds for hosting various functionalities including moieties with unique optical, electronic, and magnetic properties [12]. One example is the use of DNA template to arrange gold nanoparticles in a plasmonic linear array [12]. Silver nanoclusters (AgNCs) represent a family of such novel functionalities. AgNCs cannot exist on their own—they have to be stabilized with capping agents that (1) limit their size, (2) stabilize nanoclusters, and (3) modify their electronic properties. DNA oligonucleotides have been widely utilized for making AgNCs (reviewed in [13]). DNA protects the metallic core of the clusters and also determines the size and geometry of the AgNCs, which in turn dictates the optical properties and stability of the nanoclusters. The resultant silver nanoclusters are only a few atoms in size. Unique optical properties of AgNCs originate from their size. In this "few atoms" size regime, the continuous density of states breaks up into discrete energy levels in the nanoclusters. With discrete states, the nanoclusters resemble molecular-like behavior with strong fluorescence observed in the visible-near IR range of the spectrum [14]. Additionally, AgNCs are more resistant to photobleaching than widely used organic dyes, quantum dots, or fluorescent proteins, which makes these structures suitable for a plethora of practical applications, including optoelectronics and nanophotonics [15]. It is especially exciting because AgNCs can be naturally integrated with NA assemblies [16] and NA assemblies can be tracked with AgNCs formation [17]. Importantly, NA template sequence can be varied to produce fluorescent silver clusters of distinct optical properties [14].

While promising, AgNCs exhibit complex optical behavior including various fluorescence peaks, effects of templating and non-templating NA sequences [18–21]. Interactions between silver and NA bases might produce long-lived dark electronic states that can be used to enhance fluorescence detection [22]. The complexity of AgNC optical properties is also manifested by the possibility of direct and indirect excitation: Directly in the visible spectral range or indirectly via UV-excitation of DNA bases [23]. The structure and nature of optically active electronic states are poorly understood [24,25]. Therefore, a detailed characterization of these effects is needed to establish how AgNC function in larger assemblies. Recent efforts indicate that large scale photonic applications might benefit from arranging AgNC in DNA-templated arrays [26] and spatial control of nanocluster positioning [27].

In this article, we report a detailed characterization of C_{12}-based templates designed for creation of larger NA assemblies. The design requires the presence of a double-stranded DNA region in close proximity to templating single-stranded sequences. Such proximity of dsDNA has a significant effect on optical properties of silver nanoclusters. We also observe the effect of directionality as optical properties of AgNCs depend strongly on whether C_{12}-templating sequence is positioned on the 3′ or 5′ end of the double-stranded core. Multiple contiguous cytosines in the C_{12} templating sequence trigger the formation of non-canonical structures in the presence of silver atoms, and this modulates the fluorescence of AgNC to a large extent. Atomic Force Microscopy imaging provides an unambiguous proof of conformational heterogeneity of DNA templates. We correlate such conformational differences with optical heterogeneity of AgNCs. A detailed comparison of emissive properties of AgNCs between UV excitation and visible excitation also points to the complex heterogeneous nature of generated AgNCs. The designed constructs presented here can be further programmed into NA nanostructures and networks bearing multiple functionalities. We demonstrate a simple assembly of two nanoclusters into a linear construct, which is the first step towards larger patterned NA-based functional assemblies. Practical applications of such networks are discussed.

2. Materials and Methods

2.1. Materials

All DNA oligonucleotides were purchased from Integrated DNA Technologies, Inc. (Coralville, IA, USA) as desalted products and used without further purification. Oligonucleotide sequences are: $(dC)_{12}$—CCCCCCCCCCCC; ssSD9-OEC_{12}—GAGATGCTAACATGGCTCTAGTCGACGATCCCCCCCCCCCC; ssSD9-BEC_{12}—CCCCCCCCCCCCCTGAGATGCTAACATGGCTCTAGTCGACGATCCCCCCCCCCCC; ssSD9-5′-

OEC$_{12}$: CCCCCCCCCCCCTGAGATGCTAACATGGCTCTAGTCGACGA; ssSD9-DimerC–TCAACATCAGTCTGATAAGCTATTTCGTCGACTAGAGCCATGTTAGCATCTC.

ssSD9-DimerC is complimentary to ssSD9-OEC$_{12}$ and to the following two sequences making "forward" or "reverse" assembly:

ssSD9-DimerC-R-forward—TAGCTTATCAGACTGATGTTGATCCCCCCCCCCCC; ssSD9-DimerC-R-reverse CCCCCCCCCCCCTTAGCTTATCAGACTGATGTTGA.

Sodium borohydride was purchased from TCI (TCI America, Inc., Portland, OR, USA), all other reagents were purchased from Sigma-Aldrich (Sigma-Aldrich, Inc., St. Louis, MO, USA).

2.2. Synthesis of Ag-DNA Nanoclusters

In a typical preparation, DNA and AgNO$_3$ aqueous solutions were mixed at 55 °C and incubated for 25 min at room temperature in the ammonium acetate buffer. Next, NaBH$_4$ aqueous solution was added and samples were placed on ice and stirred vigorously. The final concentrations of the components were C_{DNA} = 10 µM, C_{AgNO3} = 100 µM, C_{NaBH4} = 100 µM, and C_{NH4Ac} = 50 mM. The solution then was incubated in the dark for 24 h at 4 °C.

For manufacturing the assembly, the duplet sequence was utilized. To avoid interference of the neighboring C$_{12}$–C$_{12}$ sequences, nanoclusters were first created on ssSD9-C$_{12}$ or ssR21-C$_{12}$ strands as described above and subsequently annealed onto the larger complimentary ssDNA. For annealing, the mix was subjected to 1 h thermal treatment in water bath at 40 °C, followed by cooling in an ice-water bath. For fluorescence and atomic force microscopy (AFM) measurements, DNA-AgNC samples were used without further purification.

2.3. Atomic Force Microscopy Imaging

AFM imaging of DNA-templated AgNCs was performed on MultiMode AFM Nanoscope IV system (Bruker Instruments, Santa Barbara, CA, USA) in tapping mode. Briefly, 5 µL of the DNA-templated AgNC solution were deposited on amino-propyl-silatrane (APS) modified mica [28] for a total of 2 min [29]. Excess sample was washed with DI water and gently dried under a flow of high purity argon gas and under vacuum overnight. AFM images in air were then recorded with a 1.5 Hz scanning rate using TESPA-300 probes (Bruker AFM Probes, Santa Barbara, CA, USA) The probes have ~320 kHz resonance frequency and a spring constant of about 40 N/m. Images were processed using the "FemtoScan Online" software package (Advanced Technologies Center, Moscow, Russia). Heights of the structures were measured using the "cross section" tool within the "FemtoScan Online" program. Height was measured as the highest point in the cross-section profile with the background subtracted.

2.4. Fluorescence Measurements

The excitation and emission spectra were acquired on a Cary Eclipse Fluorescence Spectrophotometer (Agilent Technologies, Santa Clara, CA, USA). In all the measurements, the concentration of DNA was kept the same at 10 µM. Experiments were carried out at a room temperature of ~22 °C. Fluorescence measurements were carried out in Sub-Micro Fluorometer Cell, model 16.40F-Q-10 (StarnaCells, Inc., Atascadero, CA, USA). Only freshly prepared solutions were used for spectroscopic measurements. The excitation–emission matrix spectra (EEMS) were recorded with 2 nm resolution. Fluorescence spectra were recorded with the emission wavelength range from 270 nm to 800 nm, and the initial excitation wavelength was set to 220 nm and final excitation wavelength was set to 700 nm with an increment of 2 nm. The slits were open to 5 µm and the PMT voltage was set to 700 V. EEMS data were then used for the 2D contour plot using MagicPlot Pro software (Magicplot Systems, LLC, Saint Petersburg, Russia).

3. Results

3.1. Design of AgNC DNA Template

A number of DNA sequences that template and stabilize AgNCs have been studied (reviewed in [13]). Recently, it was proposed that identification of specific genome sequences can be made based on the properties of AgNCs that these sequences template [21]. The most common ones used in AgNC synthesis are cytosine-rich sequences due to the very high affinity of cytosines to silver cations [20,30]. Our primary goal was to characterize the properties of DNA-templated AgNC and explore the possibility of incorporating AgNCs into larger NA networks with functional properties via a bottom-up assembly. We have designed sequences consisting of two parts: one that templates AgNCs formation and another that is used for the further construction of the assembly. The designed sequences have a double-stranded core, schematically illustrated in Figure 1A, with a random 30 bp sequence, which does not form any secondary stable folds or hairpins [31]. AgNC templating sequence has a stretch of twelve consecutive cytosines—$(dC)_{12}$, Figure 1A. This DNA construct can be used simultaneously as a template for AgNC formation and as a building block for making larger assemblies. Two sequences were initially studied: SD9 core modified with $(dC)_{12}$ at only one end and SD9 core modified with $(dC)_{12}$ at both ends. These sequences were named SD9-OEC$_{12}$ and SD9-BEC$_{12}$, respectively. Comparison with $(dC)_{12}$ sequence alone–named C$_{12}$ later in the text–was performed to elucidate the effect of the neighboring duplex region on the properties of AgNCs created with the single-stranded DNA template.

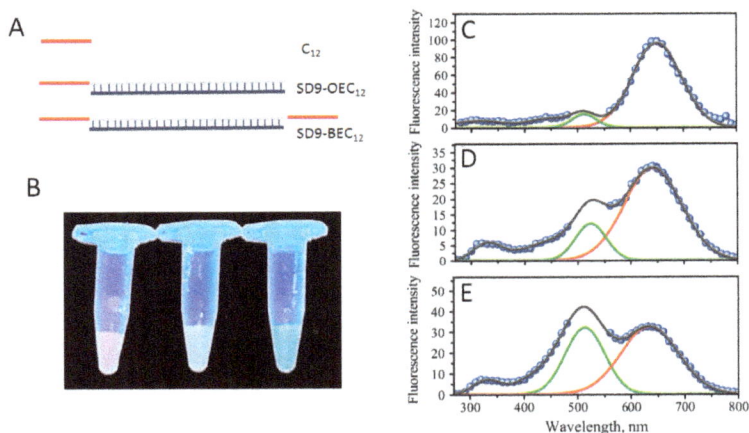

Figure 1. (**A**) Schematic representation of the template design, (**B**) a photograph of fluorescent glowing of solutions containing fluorescent silver nanoclusters (AgNC) templated on C$_{12}$ (red), SD9-OEC$_{12}$ (yellow), and SD9-BEC$_{12}$ (green), (left to right) under the UV excitation on a trans-illuminator. Fluorescence spectra of AgNCs recorded with 254 nm excitation wavelength for the following templates (**C**) C$_{12}$, (**D**) SD9-OEC$_{12}$, and (**E**) SD9-BEC$_{12}$. Gaps in the spectra are due to the removal of the second order scattering. Green and red solid lines are plotted as "guide for the eye" in the positions of major "green" and "red" emission peaks.

3.2. The Formation of Silver Nanoclusters and Their Optical Properties under UV Excitation

The formation of DNA-templated AgNCs is manifested by the changes in solution that are observed after 24 h incubation after the addition of Ag$^+$ and sodium borohydride. Fluorescent glowing of the solutions is evident with UV excitation under a trans-illuminator. Figure 1B shows such glowing effect of AgNC-DNA solutions for three samples corresponding to sequences C$_{12}$, SD9-OEC$_{12}$, and SD9-BEC$_{12}$. Since the templating sequence is the same for all three samples—C$_{12}$

stretch, we expected similar colors. It appears, however, that all three samples have distinct fluorescence colors: Red for C_{12}, yellow for SD9-OEC$_{12}$, and green for SD9-BEC$_{12}$ (Figure 1B). The entire fluorescence spectrum with excitation at 254 nm is shown in Figure 1 for all three sequences C_{12}, SD9-OEC$_{12}$, and SD9-BEC$_{12}$ (Figure 1C–E, respectively).

The spectra have many similar fluorescent peaks. For example, the broad peak in the 300–400 nm part of the spectrum, which is assigned to the fluorescence of DNA bases [32]. The colors of AgNC observed in the trans-illuminator suggested that fluorescence peaks in the visible should be detected. SD9-OEC$_{12}$ and SD9-BEC$_{12}$ samples have a broad fluorescence in the visible region with two major peaks "red" at $\lambda_R = 640$ nm and "green", $\lambda_G = 525$ nm for SD9-OEC$_{12}$, and $\lambda_G = 510$ nm for SD9-BEC$_{12}$, confirming that DNA template sequence has a direct effect on the fluorescent properties of silver nanoclusters. Closer inspection of the fluorescence spectra also revealed two shoulders to the major peaks at $\lambda = 715$ nm (red shoulder) and $\lambda = 445$ nm (green shoulder). Similar peaks are observed for C_{12}–templated AgNC but the red fluorescence at 640 nm dominates the spectrum (Figure 1C). Table 1 shows the relative intensity of "green" to "red" peaks measured as areas of Gaussian fits for the two peaks in the spectrum plotted as fluorescence intensity vs. energy (Supplementary Figure S1). There is a clear trend of dominant red fluorescence in the following order: C_{12} > SD9-OEC$_{12}$ > SD9-BEC$_{12}$. The apparent color observed under UV excitation in the trans-illuminator thus represents a mix of two primary fluorescent peaks with different proportions—rather than a single color.

Table 1. Relative ratios of fluorescence intensities for "green" and "red" emitters obtained from spectra recorded using 254 nm excitation. The numbers were obtained as areas under the Gaussian fits to intensity of fluorescence plotted vs. energy (shown in Supplementary Figure S1).

$F_{MAX, nm}$	C_{12}	SD9-OEC$_{12}$	SD9-BEC$_{12}$
510-525 (green)	13	42	69
640 (red)	87	58	31

Despite the similarities of the templating sequences, C_{12}, for all three designs, we observe large differences in the optical properties of the synthesized AgNCs. These differences go beyond the expected double intensity for SD9-BEC$_{12}$ compared to SD9-OEC$_{12}$. The intensity of "red" emission is similar, while "green" emission is dramatically enhanced for SD9-BEC$_{12}$-templated AgNCs compared to SD9-OEC$_{12}$. The SD9-OEC$_{12}$ and C_{12} also exhibit noticeable differences. The "green" AgNCs have low fluorescence yield when C_{12} template is used. The "green" emission, however, seems to be largely present in SD9-OEC$_{12}$ and even more intense in SD9-BEC$_{12}$. It is tempting to speculate that the formation of "green" emitters is triggered by the double-stranded core SD9. Alternatively, a potential self-assembly of individual strands (Supplementary Figure S2) of SD9-OEC$_{12}$ and SD9-BEC$_{12}$, as predicted by NUPACK [33], may result in reorganization of the C_{12} strands in 3D space, thus changing the AgNC formation.

We have further characterized the properties of AgNCs templated with SD9-BEC$_{12}$, SD9-OEC$_{12}$, and C_{12} by employing fluorescence excitation-emission matrix (EEM) spectroscopy. The excitation/emission relationship of the optical response of the AgNCs can be a little complicated and a better way to represent such responses is through measuring the entire 3D excitation/emission matrix presented as contour maps. EEM has the advantage of showing the relationship between excitation, emission, and their intensities in the wide range of wavelengths, allowing for complete and quantitative characterization of sample's fluorescence profile [25].

Figure 2 shows EEM maps for all three DNA sequences in the range 220–370 nm for the excitation while the entire emission spectrum in the UV and visible spanning 270–800 nm wavelengths was recorded. Some similarities in the behavior of emissions are obvious. For example, emissions of nucleobases spanning ~300 to ~350 nm was excited with UV: 220 to ~230 nm [32]. However, there are major differences observed in the emission of AgNCs in the visible spectral range excited via UV.

Similar to the spectra shown in Figure 1, the red emission dominates C_{12}, while green emission is dominant in the case of SD9-BEC$_{12}$ template.

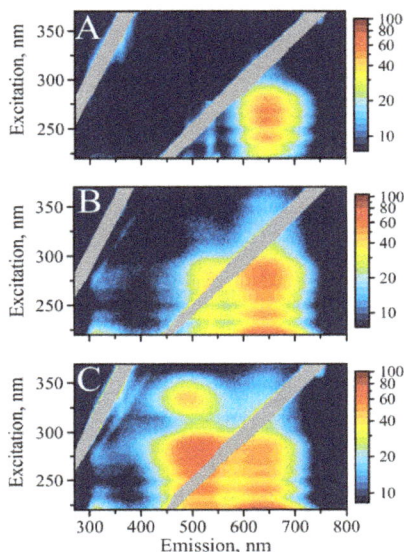

Figure 2. Excitation/emission maps of silver nanoclusters with excitation in UV: (**A**) AgNC templated on C_{12}, (**B**) AgNC templated on SD9-OEC$_{12}$, (**C**) AgNC templated on SD9-BEC$_{12}$.

Close inspection of the emission peaks reveals that C_{12} has the maximum fluorescence at 645 nm, SD9-OEC$_{12}$ at 640 nm, and SD9-BEC$_{12}$ at 635 nm. Interestingly, this trend in the shift to lower wavelength also follows the increase in intensity of green fluorescence in the following order: C_{12}, SD9-OEC$_{12}$, and SD9-BEC$_{12}$. These observations suggest that the "red" emitters synthesized using the three templates are either different species or the same species experiencing distinct surroundings.

While the "green" emission is practically absent in C_{12}, it becomes detectable for SD9-OEC$_{12}$ and is very intense for SD9-BEC$_{12}$. The maximum of green emission for SD9-OEC$_{12}$ is observed at 525 nm while for SD9-BEC$_{12}$ it is observed at 510 nm, a noticeable 15 nm shift. Again, SD9- BEC$_{12}$ template causes the fluorescence maximum to shift to lower wavelengths. Another major difference is an extra excitation band at 335 nm that only appears for the SD9-BEC$_{12}$-templated AgNC and only for the green emission. This fluorescence peak has a maximum at 490 nm ("blue"). This serves as additional evidence of different surroundings of the emitting AgNC. We hypothesize that such differences are related to conformational heterogeneity of the DNA template induced by the presence of double-stranded SD9 core. Our EEM spectroscopic measurements reveal that the color of emission changes dramatically from template to template suggesting that AgNC of the same kind, "red" or "green", experience different surroundings for the three templates used here.

It appears that the emission of AgNCs in the visible part of the spectrum is universally excited via DNA bases, which agrees well with published data [23]. It has been shown before that visible emission of AgNCs can be excited by both direct excitation into the visible band and by UV light into the absorption peak of DNA nucleobases [30,34]. A detailed study of UV excitation of DNA-templated silver nanoclusters suggests optical interactions between the DNA template and AgNC [23]. Excitation of AgNC fluorescence in the area of nucleobase absorption is a valuable feature of a DNA template. As such, optical properties of nucleobases can be beneficial for the excitation of AgNCs. The proximity of AgNC to DNA nucleobases, which not only nucleate and stabilize AgNC but also transfer the energy of the excitation, prompted more detailed studies of the phenomenon [23]. It has been proposed that DNA serves as an antenna in funneling the energy of UV light to fluorescence of AgNC [35], but the mechanism of energy transfer from the UV-excited nucleobases remains unclear. Nevertheless, the presence of coordinated DNA bases is necessary for the excitation of AgNC fluorescence by the UV light. Treating DNA-templated AgNC with DNAse reduces fluorescence intensity with time of treatment (Supplementary Figure S3). These results indicate an intimate relationship between optical properties of AgNC and templating DNA bases. In addition, during treatment "green" color disappears while "red" color still persists, although faintly, suggesting tighter connection of "green" emitters with DNA bases.

3.3. Optical Properties of AgNC under Visible Light Excitation

Next, we measured excitation/emission maps for all three sequences while exciting in the range of visible wavelengths between 370 nm and 700 nm. While EEM with UV excitation reveal properties of the AgNCs that are in close proximity to nucleobases where efficient energy transfer from DNA bases is possible, the visible excitation probes optical properties of the nanoclusters in general, regardless of their association with the bases. Figure 3 shows the maps for C_{12}, SD9-OEC$_{12}$, and SD9-BEC$_{12}$ (A, B, and C respectively).

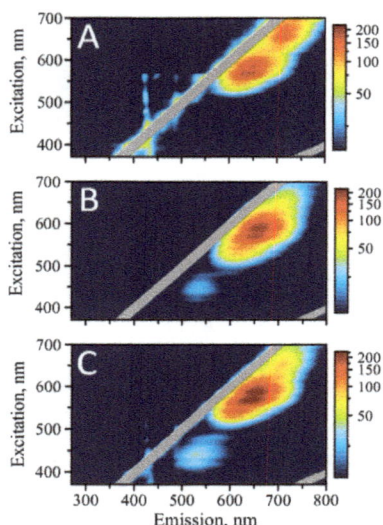

Figure 3. Excitation/emission maps of silver nanoclusters with excitation in the visible: (**A**) AgNC templated on C_{12}, (**B**) AgNC templated on SD9-OEC$_{12}$, (**C**) AgNC templated on SD9-BEC$_{12}$.

It is obvious that there is one peak, which is similar for all three templates—the red emission above 600 nm. Interestingly, this peak appears as a duplet with maxima at 640 nm and 650 nm for all three samples. Another similarity is the emission at ~715 nm. It is well pronounced in C_{12} template but faintly present in both SD9-OEC$_{12}$ and SD9-BEC$_{12}$.

Consistent with the UV excitation, EEM maps for the visible excitation do not show any green fluorescence for the C_{12} template, while it is present for SD9-OEC$_{12}$ and SD9-BEC$_{12}$. We infer that such differences might be associated with the influence of the core SD9, which either stabilizes green emissive species or alters the conformation of C_{12}–stretch in such a way that also allows for the formation of the clusters emitting in green.

Unlike other studies, we find that green and red emissions are not equivalent for the excitation in UV and the visible spectral region. Despite its bright appearance when excited in UV, green emission seems to be relatively silent when excitation is performed in the visible range of wavelengths.

The maximum of emission is also shifted to lower wavelengths when comparing visible vs. UV excitation. The maximum of emission shifts by 10 nm for SD9-OEC$_{12}$, with λ = 535 nm for visible excitation and λ = 525 nm for the excitation in UV. The maximum of emission for SD9-BEC$_{12}$ shifts by more than 20 nm, with λ = 530 nm (for visible excitation) and λ = 510 nm (for UV excitation).

Additionally, red emission has very similar intensity for all three templates, while green clearly dominates the emission for SD9-BEC$_{12}$-templated AgNCs when excited via UV. We hypothesize that SD9-OEC$_{12}$ and SD9-BEC$_{12}$ templates are capable of producing both types of AgNCs, green emitting and red emitting species with population largely shifted towards red species. The coordination with DNA bases in the case of green emitting clusters allows for more efficient energy transfer, larger quantum yield, and, therefore, greater fluorescence intensity of the green emitters when excited via DNA bases in UV.

One interesting feature of the red emission behavior for the AgNCs is that the emission maximum progressively shifts to the red as the excitation wavelength is increased starting from 515 nm all the way to ~700 nm (see Figure 3 and Figure S5). The magnitude of Stokes shift ($\lambda_{EM} - \lambda_{EX}$), however, remains similar at ~70 nm throughout the visible part of the map. This bathochromic shifting in the emission band with the increase of the excitation wavelength resembles the phenomenon observed previously for polar organic fluorophores in "rigid" solutions [36–38]. The phenomenon of fluorescence dependence on the excitation wavelength is known as "red-edge excitation shift" (REES) [39]. This kind of dependence seems to be inconsistent with the Kasha's rule, which states that the fluorescence spectrum should originate from the lowest vibrational level of the first excited singlet state irrespective of the excitation wavelength. Our AgNCs do not seem to obey this rule, exhibiting an obvious shift of emission to the red when excitation wavelength is increased, which is consistent with the REES effect. One possible explanation of such an effect in AgNCs is the "rigid" nature of clusters and their surroundings. The randomization of the local environment is not fast enough to allow for efficient relaxation of the excited state of the cluster. They are forced to emit not from the lowest level of the first excited singlet state but from the higher levels.

It is quite peculiar that no bathochromic shift is observed with the emission bands when excited via DNA bases. Quite smooth spectra with the same maxima are observed regardless of the change in excitation wavelength throughout the UV region. This observation suggests that Kasha's rule is obeyed for the excitation of nanoclusters via UV, providing additional ground for speculation that the two populations of clusters are different in their photophysical properties, which stem from their different surroundings. The differences in optical properties between nanoclusters excited in UV and visible spectral range suggests a large degree of AgNC heterogeneity.

3.4. AFM Topography of DNA-Templated AgNC

We have further investigated the morphology of the DNA-templated AgNC complexes using Atomic Force Microscopy. Figure 4 shows AFM topography images of samples for all three AgNC templating sequences: C$_{12}$, SD9-OEC$_{12}$, and SD9-BEC$_{12}$. The AFM images of SD9-OEC$_{12}$ show that templated silver nanoclusters adopt mostly small elongated shapes (Figure 4B, I) when deposited on mica substrate. The spherical structures were expected based on previously published data for long C-rich strands [40]. We assign these elongated shapes to a single SD9-OEC$_{12}$. In addition to small elongated shapes, the SD9-OEC$_{12}$ forms longer strands (Figure 4B, II), some of which have an angled junction. The total length of the entire SD9-OEC$_{12}$ template is expected to be ~14 nm, if one assumes 0.34 nm base–base distance as in dsDNA. The measured contour length for elongated shapes in the AFM topography image exceeds the expected 14 nm. The strands with the angled junctions morphologies most likely represent adducts of two SD9-OEC$_{12}$ templates joined together. Inset II in Figure 4B shows the length of the arms for angled shape is ~11 nm, which corresponds well to the expected length of 30 bp of the SD9 core. Inset III in Figure 4B shows another representative shape where two molecules form an adduct with a straight junction. This suggests that SD9-OEC$_{12}$ sequences form a variety of shapes in solution as well as adducts between two or more molecules.

Some shapes are tri-point or Y shapes with three strands joined together. Figure 4B inset IV shows one such representative shape, each strand measures ~14 nm indicating that the tri-point is a result of three connected SD9-OEC$_{12}$ monomers.

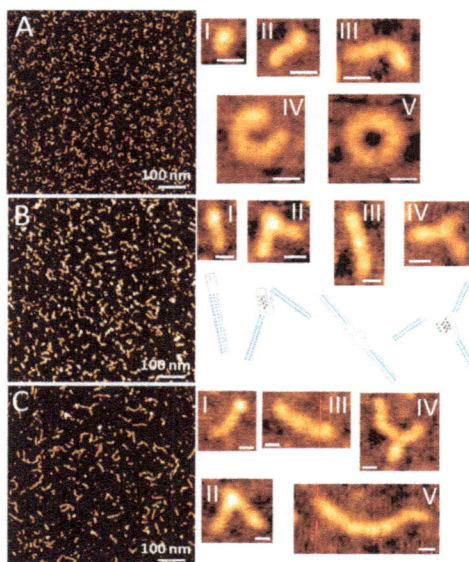

Figure 4. Atomic force microscopy (AFM) topography images of AgNC template with (**A**) C$_{12}$, (**B**) SD9-OEC$_{12}$, (**C**) SD9-BEC$_{12}$. Insets show representative structures of the template sequences. Scale bar in all insets is 10 nm.

The structures of SD9-BEC$_{12}$-templated AgNCs are markedly different. There are still some elongated shapes observed in the AFM topography image shown in Figure 4C, but there are more longer strands in the image. The length of the strands for the SD9-BEC$_{12}$ sample is rather heterogeneous with contour lengths up to 80 nm (Figure 4C, inset V). With extended C$_{12}$ stretches, a single SD9-BEC$_{12}$ molecule should measure ~18 nm assuming 0.34 nm base–base distance. The 80 nm long shape should be comprised of ~5 monomeric SD9-BEC$_{12}$ molecules. Angled and straight dimeric/multimeric strands can also be found in topography images (insets II–IV in Figure 4B).

Longer assemblies are observed in the images of the C$_{12}$ sample as well. Figure 4A shows an AFM image of C$_{12}$-templated AgNCs. Elongated curved structures absolutely dominate the image. The length of the observed shapes exceeds the length of 12 base long oligonucleotide suggesting that the adducts of multiple C$_{12}$ are formed. We further examined the cross-sections of the elongated shapes in the images. Typical AFM images in tapping mode produce the height of double-stranded DNA at around 0.5–0.8 nm [41–43]. While height values for most of the strands measured in our images do fall within this range, some areas show height values that exceed the height of the regular DNA duplex. A cross section in Figure 4B (inset II) for SD9-OEC$_{12}$ indicates h = 0.7 nm in the arms of the dimeric adduct and h = 1.1 nm at the junction. Even visually, these two regions appear very different. The height of the point where two molecules are joined together suggests that the morphology of the DNA structure different from duplex is formed at the junction.

The observed differences in elongated shapes for C$_{12}$, SD9-OEC$_{12}$, and SD9-BEC$_{12}$ are obviously due to the differences in the sequence that allow such distinct shapes to appear. We propose that the elongated strands observed are due to single-stranded (ss) C$_{12}$ stretch forming a non-canonically bonded double-stranded DNA form via Ag+ linkage. Such a possibility has been demonstrated recently [44]. Alternatively, a tetra-stranded structure termed i-motif [45] could also form with the help of silver cations [46]. While connected via Ag$^+$ linkage, double-stranded form is expected to appear as regular dsDNA in the AFM images. The i-motif, on the other hand, is composed of four strands and should appear higher on the image, as we observe at the junction (Figure 4B, inset II).

Both alternative structures differ from ss-DNA and yet serve as AgNC templates in a distinct way. While AFM provides unambiguous evidence for the formation of different shapes, we also observe remarkable differences in fluorescence between nanoclusters templated by C_{12}, SD9-OEC$_{12}$, and SD9-BEC$_{12}$. We hypothesize that the differences in optical properties of silver nanoclusters are likely associated with the extent of non-canonical structure formations such as Ag$^+$-C$_{12}$ duplex or i-motif. Visually, C_{12} templates appear more curved, providing ground for speculation that curved geometry of template is responsible for the red emission of AgNCs in C_{12} while green emitters are predominately formed in linearly shaped templates (SD9-BEC$_{12}$). The results of this section are an additional manifestation of how sensitive the optical properties are towards surroundings, namely the sequence of templating DNA and its secondary structure.

3.5. The First Step Towards Assembly

Dimeric assembly with a dual cluster templating sequence can represent a first step towards a larger DNA-based network. Figure 5 schematically illustrates such a dimeric design. We explored two different variants of the dimeric assembly: (1) "Forward" design, where C_{12} templating stretches are separated by 30 bp double-stranded core sequences—SD9 (Figure 5A), and (2) "reverse" design, with C_{12}-templating sequences brought together by this assembly (Figure 5B). Fluorescence properties of AgNCs templated by the "forward" assembly resemble well the properties of AgNCs formed with SD9-BEC$_{12}$ sequence including the unusual 335/490 nm "blue" peak (Figure 5C,D). The intensities of all the peaks, both in UV-excited emission and visibly-excited emission for the "forward" assembly are exactly the same as for SD9-BEC$_{12}$, confirming the similarity of the structure and conformation of DNA and AgNCs.

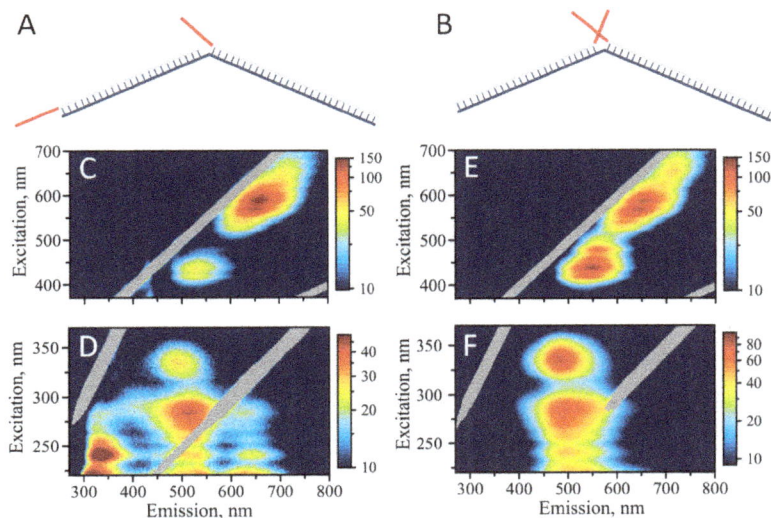

Figure 5. (**A**) Schematic representation of template design for the "forward" assembly, (**B**) schematic representation of template design for the "reverse" assembly. Templating C_{12} sequences are shown in red. Excitation/emission maps of silver nanoclusters: (**C**) AgNC templated on "forward" assembly with excitation in the visible spectral range, (**D**) AgNC templated on "forward" assembly with excitation in the UV, (**E**) AgNC templated on "reverse" assembly with excitation in the visible spectral range, (**F**) AgNC templated on "reverse" assembly with excitation in the UV.

It is quite interesting that the optical properties of the resultant assemblies do not resemble SD9-OEC$_{12}$ but appear similar to SD9-BEC$_{12}$ including the prevalence of green fluorescence and the "blue"—335/490 nm peak. Visible excitation of emission produces a similar map as for SD9-BEC$_{12}$ and SD9-OEC$_{12}$ with the characteristic double peak: 570/650 nm and 590/660 nm. In addition, a small peak at 435/530 nm is observed, similar to both SD9-OEC$_{12}$ and SD9-BEC$_{12}$ templates.

Dramatic changes of the AgNC optical properties are observed when "reverse" design was utilized, Figure 5E,F. The enhancement of the green emission is quite obvious. There are only slight remnants of the red fluorescence while green emission dominates the 2D map when excited in the UV (Figure 5F). The 335/490 nm "blue" peak is now dominant in the excitation/emission map, providing additional clues to the origin of this peak, which are described in the Discussion section. The 2D excitation/emission map with visible excitation for the "reverse" design appears to be different from the one for "forward" design. "Green" peak is intensified and shifted to longer wavelengths in the "reverse" assembly. In addition, "red" fluorescence at ~700 nm is enhanced in the "reverse" design compare to "forward" assembly. Similar to individual templates, the maxima of emission in the assemblies do not coincide when AgNCs are excited via UV and visible, as clearly observed in the maps. "Red" emission is intensified with visible excitation and "green" is shifted to lower wavelengths when excited via UV.

4. Discussion

NA-based tools of nanotechnology open large prospects in designing and controlling the size and shape of nanostructures. DNA can serve as a template for the controlled synthesis of novel nanomaterials that are arranged in a highly organized network of functional entities, exemplified in this work by AgNCs. Beneficial novel properties may result from placing AgNCs into nanostructures with organized patterns. Examples of such beneficial properties are the use of AgNCs as logic optical gates [47], optoelectronics [15], bioimaging, or detection of specific nucleic acid sequences where the signal is amplified due to optical coherence of the signal from AgNCs organized in patterned nanostructures. Additionally, controlled AgNC formation can be used as the means of label-free visualization of NA supra-assemblies and nanomaterials.

One of the questions that we address with this study is, what would be the properties of AgNCs when they are assembled into larger networks using DNA as an architectural element? The effects needed to be probed are the close proximity of the DNA on the optical properties of templated AgNCs such as double-stranded core, loose ends of nucleic acids, and other structural elements used in the field of DNA nanotechnology [48]. Our choice of template sequence was single-stranded cytosine stretch—C$_{12}$, which has been previously shown to produce silver nanoclusters with high yield of fluorescence in the visible range of wavelengths [30]. Additional double-stranded region was placed next to C$_{12}$ mimicking the nanostructure design when larger assembly is created. Double-stranded DNA show no template activity unless modified to contain a mismatched site, an abasic site, a gap, a bulge, or a loop [13]. Silver nanoclusters generated by the double-stranded templates exhibit negligible fluorescence signal. We verified the inability of SD9 core to generate and stabilize any type of clusters by running a control experiment where SD9 core duplex was mixed with AgNC forming components: Ag$^+$ and NaBH$_4$. No detectable fluorescence was observed without the C$_{12}$-templating sequence (Supplementary Figure S4).

While C-rich sequences have high affinity to silver, which makes them very good templating sequences for AgNC synthesis, they also have the ability to form alternative non-canonical DNA structures. For example, a duplex stabilized by silver in a C-Ag-C pairing rather than canonical Watson-Crick [49]. Such structures are quite stable, Ag$^+$ bridges the N3 atoms of the C bases resulting in the C-Ag-C pairing link [50]. Another example of non-canonical form is the i-motif where semiprotonated C-rich oligonucleotides are arranged in a tetra-stranded structure [51]. Typically, the formation of i-motif structures is favored at mild acidic conditions due to protonation of N3 in the cytosine. The entire structure is stabilized via hydrogen bonds between C$^+$–C in the semi-protonated base pairs. However, silver cation can join C strands in a C-Ag$^+$-C bridge even at neutral pH values [46]. The affinity to silver is so specific that an Ag$^+$ detection has been proposed based on i-motif formation due to the presence of Ag [52].

At neutral pH, the formation of alternative structures leading to elongated shapes observed in AFM experiments depends critically on the silver content in the solution. Our data suggest that the elongated strands are linked together via a non-canonical structure with the help of Ag$^+$. We considered both versions of the possible structures: The silver-stabilized duplex between two C$_{12}$ stretches and the i-motif. The duplex between C$_{12}$ strands stabilized by silver is not expected to appear much different from regular DNA duplex in the AFM topography images. An example is round shaped adducts observed in AFM topography images for the C$_{12}$ sequence. While it is clearly an adduct of many molecules of C$_{12}$, the height of the structure appears similar to dsDNA at ~0.7 nm. Contour length analysis of the shape shown in Figure 4A, inset V (L$_C$ = 30 nm) suggests that up to 20 molecules of C$_{12}$ need to join together to form such a shape. On the other hand, the i-motif is structurally different from single-stranded or double-stranded DNA. The four-stranded nature of the i-motif increases the lateral size of the structure, which can be easily identified using AFM. Indeed, AFM allowed us to recognize the features that are larger than DNA duplex. Examples of structures formed with i-motif are dimeric and trimeric adducts with junctions that exceed sizes of ds-DNA (Figure 4B, insets I and II). Figure 4 presents our vision as to how such i-motif junctions can form. It has been shown that both inter-molecular packing and intra-molecular folding of strands into i-motif is possible [53]. Several studies have already started exploring the use of i-motif as a structural element in DNA nanotechnology [11].

The i-motif formation results in a more compact shape than duplex. It has a shorter distance between the bases, 0.31 nm as compared to 0.34 nm in a regular double strand [40,54]. Therefore, i-motif creates a denser population of silver cations holding the i-motif together. The alignment of silver cations and the shape of the resultant nanoclusters templated by the i-motif are expected to be different from those of both single-stranded C$_{12}$ and double-stranded C$_{12}$-Ag-C$_{12}$. Because of such difference, the i-motif structure should also affect optical properties of templated AgNCs. It has been stated previously that i-motif structures primarily produce red emissive AgNCs [55]. Further studies using single nanocluster fluorescence measurements and their correlation with AFM imaging will unambiguously assign the color and photophysics of AgNCs to the template type, specifically i-motif, and we are currently considering these single nanocluster studies.

The measured excitation/emission matrices for C$_{12}$, SD9-OEC$_{12}$, and SD9-BEC$_{12}$ sequences indicate that there are mostly two distinct color clusters—"green" and "red". Previously, the distinct fluorescence peaks have been associated with clusters of different sizes [30]. DNA templates are capable of stabilizing various (Ag)$_N$ clusters where N is the number of silver atoms forming a nanocluster. It is commonly accepted that the DNA templated (Ag)$_N$ nanoclusters are rod like in shape and include both neutral Ag^0 atoms and charged Ag$^+$ [56]. Recent study revealed that a "magic number" of neutral silver atoms is required for distinct "green" and "red" fluorescence to occur. Namely, it has been inferred that four neutral atoms produce green fluorescence and six neutral Ag atoms produce red fluorescence regardless of the number of Ag$^+$ [57]. In general, larger AgNCs emit at longer wavelengths [58]. Therefore, it seems appropriate in our case to assign green fluorescence to AgNCs of smaller cluster size with four Ag^0 and red fluorescence at 640 nm to larger clusters with six Ag^0.

The environment of the clusters composed of nucleobases, although very important, is poorly understood. The tuning of fluorescent properties of AgNCs is often done only empirically. Recent studies have tried to look into a deeper understanding of the role different nucleobases play in optical/photophysical properties of AgNCs [23,59]. When placing AgNC into larger assemblies of nucleic acid structures, several factors may play a role in altering the AgNC fluorescence: (1) Templating sequence, (2) adjacent double-stranded sequence, and (3) the overhang sequence. As all three are contributing factors to fluorescence, changing them provides additional degrees of freedom for tuning the fluorescence properties of AgNCs. A vivid example of the effect of non-templating sequence on optical properties of AgNCs is the presence of overhang strands. G-rich overhangs produce drastic changes in excitation, emission, and quantum yield of fluorescence [18]. Our results demonstrate that simply changing the direction of dimeric assembly from "forward" to "reverse" drastically increases "green" fluorescence when excited via DNA bases with UV. "Red" emitting species are still formed in the "reverse" assembly, as indicated by the excitation in the visible, but the UV excitation channel via DNA bases is completely turned off for the "red" emitting nanoclusters. It has been established that bases are electronically coupled to silver nanoclusters. One manifestation of such coupling is the AgNC emission when excited at nucleobases' UV absorption. The heteroatoms of bases, particularly N7, N3, and N2 change the electronic environment of the cluster. This, in turn, stabilizes particular emissive levels, enhances the fluorescence quantum yield, and increases fluorescence lifetime by several nanoseconds [59]. Our observations point to the complexity of the electronic structure of the nanoclusters and the abundance of possibilities of the de-excitation processes (Supplementary Figure S5).

We hypothesize that the "green" emitting nanoclusters are tightly associated with DNA bases leading to their selective UV excitation. This is contrary to the "red" emitting clusters where association is not as tight, resulting in preferential excitation in the visible. More evidence of tight DNA association with "green" emitting clusters is the apparent lack of red-edge emission shift (REES) for these AgNCs excited in the visible. In general, REES is observed when the process of internal conversion is very slow compared to fluorescence lifetime. REES has been observed for polar fluorophores in viscous media [38]. Relaxation due to solvent molecule reorientation is not fast enough within the fluorescence lifetime, leading to the observation of the emission maximum shifting to the red. No REES is observed for the AgNCs, both "green" and "red" emission, when excited via UV. This indicates that nanoclusters tightly associated with DNA bases experience an environment that provides effective randomization. The "red" emission excited in the visible has a clear REES effect. On the other hand, the maximum of the 550 nm peak in the fluorescence of AgNCs excited in the visible for the "reverse" design appears to be shifting only slightly to the red but not as fast as the emission in the ~600–700 nm region. This situation appears somewhat intermediate between the "rigid" and "flexible" environments, providing some degree of randomization of the surroundings. We hypothesize that this effective randomization originates from tighter association of "green" nanoclusters with DNA bases. While the observed differences between "green" and "red" emitting species indicate a distinct AgNC environment, additional time-resolved detailed spectroscopic measurements will help to more fully describe this phenomenon. One of the most intriguing results of our study is the appearance of the "blue" emitting species, 335/490 nm. These species are excited via UV but emit in the visible spectral range. The magnitude of Stokes shift for these species, $\Delta\lambda \approx 155$ nm, is almost double that for nanoclusters excited in the visible range with $\Delta\lambda \approx 70$ nm, but not nearly as high as the Stokes shift observed for UV excited emission, $\Delta\lambda$ up to 300 nm for "green" or $\Delta\lambda$ up to 420 nm for "red", indicating uniqueness of these hybrid species. The formation of the "blue" emitters is favored whenever C_{12} templating sequence is placed at the 5' end of the double-stranded core. This is the case with SD9-BEC$_{12}$ as well as both dimer assemblies. We have verified the link between 5' template and blue nanocluster formation by placing C_{12} at 5' in the SD9-5'-OEC$_{12}$. There is a clear peak in the blue indicating the formation of the "blue" emitters (Supplementary Figure S6). We, therefore, conclude that C_{12} sequence placed at 5' somehow stimulates the formation of the "blue" emitters, perhaps by adopting favorable conformation.

Overall, our results suggest a high degree of heterogeneity of synthesized AgNCs. Since the resulting AgNCs are so heterogeneous, it might be very difficult to purify a single population of AgNCs, which will exhibit homogeneous optical properties as the environment variation affects individual clusters. Therefore, ensemble-based measurements may not be an appropriate choice of method for studying properties of AgNCs. It is reasonable to suggest that single nanocluster spectroscopic studies should provide a clearer understanding of the AgNC optical properties free of complications originating from the heterogeneous nature of nanoclusters in ensemble.

5. Conclusions

In summary, our studies offer a detailed characterization of AgNC synthesized using C_{12}-based template sequences designed for construction of larger DNA-based assemblies. We have shown the possibility of combining silver nanoclusters into larger assemblies of nucleic acids using a simple design containing a single-stranded template and a double-stranded core. A full excitation/emission data matrix has proven to be an effective tool in exploring the complex optical behavior of silver nanoclusters templated on nucleic acid sequences. Our findings indicate that close proximity of double-stranded region changes preferential formation of "red emitters" with $\lambda_{EM} = 640$ nm observed on C_{12} sequences to the formation of "green emitters" with $\lambda_{EM} = 510$ nm when double-stranded region is present. Further detailed exploration of NA templates, specifically their conformation and the possibility of non-canonical structure formation, will facilitate our understanding and control of AgNC properties. Our results also show that the fluorescence of AgNCs can be modulated by controlling the position of the AgNCs within the dimeric assembly. Placing nanoclusters closer together dramatically enhances the fluorescence of AgNCs, but also changes the electronic structure of the nanoclusters, primarily affecting the excitation/emission of the "green" and the "blue" emitters. Our observations point to the complexity of the electronic structure of the nanoclusters and the abundance of possibilities of the de-excitation processes. This in turn allows for fine-tuning of fluorescent properties of AgNCs. Effective randomization of the local environment for the "green" emitting species, as manifested by the reduction in REES, is primarily observed for the clusters placed in proximity. We envision a construction of large assemblies using nucleic acids as architectural templates with unique tunable fluorescent properties of AgNCs that could be used in novel materials, devices, and diagnostic applications.

Supplementary Materials: The following are available online at http://www.mdpi.com/2079-4991/9/4/613/s1, Figure S1: Fluorescence spectra of AgNCs recorded with 254 nm excitation wavelength and plotted as fluorescence intensity vs. energy (eV), Figure S2: Predicted secondary structures using NUPACK, Figure S3: Fluorescence intensity change of C_{12}-templated AgNCs treated with RQ1 DNase, Figure S4: Full excitation/emission map for double-stranded SD9-core template, Figure S5: Full excitation/emission energy map of AgNCs templated on SD9-OEC$_{12}$ sequence with schematic Jablonski diagrams showing de-excitation pathways for the "red" and the "green" emitters, Figure S6: UV excitation/emission map for SD9-5'-OEC$_{12}$ templated AgNCs.

Author Contributions: Conceptualization, A.V.K.; methodology, L.E.Y., O.A.S., A.Y.L., K.A.A., and A.V.K.; data acquisition, L.E.Y., O.A.S., and A.Y.L.; data analysis, L.E.Y., O.A.S., A.Y.L., K.A.A., and A.V.K.; resources, K.A.A. and A.V.K.; writing—original draft preparation, A.V.K.; writing—review and editing, L.E.Y., O.A.S., A.Y.L., K.A.A., and A.V.K.; project administration, K.A.A. and A.V.K.; funding acquisition, A.V.K.

Funding: The work was partially supported by the following grants: NIH 2P20GM103480 and ACS Petroleum Research Fund to A.V.K.

Acknowledgments: AFM imaging of the DNA-templated AgNC was performed with the use of instruments at the Nanoimaging core facility at the University of Nebraska Medical Center. The Facility is partly supported with funds from the Nebraska Research Initiative (NRI).

Conflicts of Interest: The authors declare no conflict of interest.

References

1. Afonin, K.A.; Kasprzak, W.K.; Bindewald, E.; Kireeva, M.; Viard, M.; Kashlev, M.; Shapiro, B.A. In silico design and enzymatic synthesis of functional RNA nanoparticles. *Accounts Chem. Res.* **2014**, *47*, 1731–1741. [CrossRef]

2. Pinheiro, A.V.; Han, D.; Shih, W.M.; Yan, H. Challenges and opportunities for structural DNA nanotechnology. *Nat. Nanotechnol.* **2011**, *6*, 763–772. [CrossRef] [PubMed]

3. Guo, P. The emerging field of RNA nanotechnology. *Nat. Nanotechnol.* **2010**, *5*, 833–842. [CrossRef] [PubMed]

4. Shukla, G.C.; Haque, F.; Tor, Y.; Wilhelmsson, L.M.; Toulmé, J.-J.; Isambert, H.; Guo, P.; Rossi, J.J.; Tenenbaum, S.A.; Shapiro, B.A. A boost for the emerging field of RNA nanotechnology. *ACS Nano* **2011**, *5*, 3405–3418. [CrossRef] [PubMed]

5. Nielsen, P.E.; Egholm, M.; Berg, R.H.; Buchardt, O. Sequence-selective recognition of DNA by strand displacement with a thymine-substituted polyamide. *Science* **1991**, *254*, 1497. [CrossRef] [PubMed]

6. Grabow, W.W.; Zakrevsky, P.; Afonin, K.A.; Chworos, A.; Shapiro, B.A.; Jaeger, L. Self-assembling RNA nanorings based on RNAI/II inverse kissing complexes. *Nano Lett.* **2011**, *11*, 878–887. [CrossRef]

7. Rackley, L.; Stewart, J.M.; Salotti, J.; Krokhotin, A.; Shah, A.; Halman, J.; Juneja, R.; Smollett, J.; Roark, B.; Viard, M.; et al. RNA Fibers as Optimized Nanoscaffolds for siRNA Coordination and Reduced Immunological Recognition. *Adv. Funct. Mater.* **2018**, *28*, 1805959. [CrossRef]

8. Afonin, K.A.; Lin, Y.P.; Calkins, E.R.; Jaeger, L. Attenuation of loop-receptor interactions with pseudoknot formation. *Nucleic Acids Res.* **2012**, *40*, 2168–2180. [CrossRef] [PubMed]

9. Afonin, K.A.; Cieply, D.J.; Leontis, N.B. Specific RNA self-assembly with minimal paranemic motifs. *J. Am. Chem. Soc.* **2008**, *130*, 93–102. [CrossRef]

10. Collie, G.W.; Parkinson, G.N. The application of DNA and RNA G-quadruplexes to therapeutic medicines. *Chem. Soc. Rev.* **2011**, *40*, 5867–5892. [CrossRef]

11. Dong, Y.; Yang, Z.; Liu, D. DNA Nanotechnology Based on i-Motif Structures. *Accounts Chem. Res.* **2014**, *47*, 1853–1860. [CrossRef] [PubMed]

12. Tan, S.J.; Campolongo, M.J.; Luo, D.; Cheng, W. Building plasmonic nanostructures with DNA. *Nat. Nanotechnol.* **2011**, *6*, 268. [CrossRef]

13. New, S.Y.; Lee, S.T.; Su, X.D. DNA-templated silver nanoclusters: Structural correlation and fluorescence modulation. *Nanoscale* **2016**, *8*, 17729–17746. [CrossRef]

14. Gwinn, E.; Schultz, D.; Copp, S.M.; Swasey, S. DNA-Protected Silver Clusters for Nanophotonics. *Nanomaterials* **2015**, *5*, 180–207. [CrossRef]

15. Lee, T.-H.; Gonzalez, J.I.; Zheng, J.; Dickson, R.M. Single-Molecule Optoelectronics. *Accounts Chem. Res.* **2005**, *38*, 534–541. [CrossRef] [PubMed]

16. Gwinn, E.G.; O'Neill, P.; Guerrero, A.J.; Bouwmeester, D.; Fygenson, D.K. Sequence-Dependent Fluorescence of DNA-Hosted Silver Nanoclusters. *Adv. Mater.* **2008**, *20*, 279–283. [CrossRef]

17. Afonin, K.A.; Schultz, D.; Jaeger, L.; Gwinn, E.; Shapiro, B.A. Silver nanoclusters for RNA nanotechnology: Steps towards visualization and tracking of RNA nanoparticle assemblies. *Methods Mol. Biol.* **2015**, *1297*, 59–66. [CrossRef]

18. Yeh, H.-C.; Sharma, J.; Han, J.J.; Martinez, J.S.; Werner, J.H. A DNA–Silver Nanocluster Probe That Fluoresces upon Hybridization. *Nano Lett.* **2010**, *10*, 3106–3110. [CrossRef]

19. Sylwia, W.; Kiyoshi, M.; Moin, A.; Juewen, L. Towards understanding of poly-guanine activated fluorescent silver nanoclusters. *Nanotechnology* **2014**, *25*, 155501.

20. Schultz, D.; Gwinn, E. Stabilization of fluorescent silver clusters by RNA homopolymers and their DNA analogs: C,G versus A,T(U) dichotomy. *Chem. Commun.* **2011**, *47*, 4715–4717. [CrossRef]

21. Copp, S.M.; Gorovits, A.; Swasey, S.M.; Gudibandi, S.; Bogdanov, P.; Gwinn, E.G. Fluorescence Color by Data-Driven Design of Genomic Silver Clusters. *ACS Nano* **2018**, *12*, 8240–8247. [CrossRef] [PubMed]

22. Fleischer, B.C.; Petty, J.T.; Hsiang, J.-C.; Dickson, R.M. Optically Activated Delayed Fluorescence. *J. Phys. Chem. Lett.* **2017**, *8*, 3536–3543. [CrossRef]

23. O'Neill, P.R.; Gwinn, E.G.; Fygenson, D.K. UV Excitation of DNA Stabilized Ag Cluster Fluorescence via the DNA Bases. *J. Phys. Chem. C* **2011**, *115*, 24061–24066. [CrossRef]

24. Thyrhaug, E.; Bogh, S.A.; Carro-Temboury, M.R.; Madsen, C.S.; Vosch, T.; Zigmantas, D. Ultrafast coherence transfer in DNA-templated silver nanoclusters. *Nat. Commun.* **2017**, *8*, 15577. [CrossRef] [PubMed]

25. Ramsay, H.; Simon, D.; Steele, E.; Hebert, A.; Oleschuk, R.D.; Stamplecoskie, K.G. The power of fluorescence excitation–emission matrix (EEM) spectroscopy in the identification and characterization of complex mixtures of fluorescent silver clusters. *RSC Adv.* **2018**, *8*, 42080–42086. [CrossRef]

26. Copp, S.M.; Schultz, D.E.; Swasey, S.; Gwinn, E.G. Atomically Precise Arrays of Fluorescent Silver Clusters: A Modular Approach for Metal Cluster Photonics on DNA Nanostructures. *ACS Nano* **2015**, *9*, 2303–2310. [CrossRef]

27. O'Neill, P.R.; Young, K.; Schiffels, D.; Fygenson, D.K. Few-Atom Fluorescent Silver Clusters Assemble at Programmed Sites on DNA Nanotubes. *Nano Lett.* **2012**, *12*, 5464–5469. [CrossRef] [PubMed]

28. Shlyakhtenko, L.S.; Gall, A.A.; Filonov, A.; Cerovac, Z.; Lushnikov, A.; Lyubchenko, Y.L. Silatrane-based surface chemistry for immobilization of DNA, protein-DNA complexes and other biological materials. *Ultramicroscopy* **2003**, *97*, 279–287. [CrossRef]

29. Lyubchenko, Y.L.; Gall, A.A.; Shlyakhtenko, L.S. Visualization of DNA and protein-DNA complexes with atomic force microscopy. *Methods Mol. Biol.* **2014**, *1117*, 367–384. [CrossRef]

30. Ritchie, C.M.; Johnsen, K.R.; Kiser, J.R.; Antoku, Y.; Dickson, R.M.; Petty, J.T. Ag Nanocluster Formation Using a Cytosine Oligonucleotide Template. *J. Phys. Chem. C* **2007**, *111*, 175–181. [CrossRef]

31. IDT-DNA oligoanalyzer tool was used to analyze the design of the sequence and modify it to prevent any stable hairpin/fold formation.

32. Onidas, D.; Markovitsi, D.; Marguet, S.; Sharonov, A.; Gustavsson, T. Fluorescence Properties of DNA Nucleosides and Nucleotides: A Refined Steady-State and Femtosecond Investigation. *J. Phys. Chem. B* **2002**, *106*, 11367–11374. [CrossRef]

33. Zadeh, J.N.; Steenberg, C.D.; Bois, J.S.; Wolfe, B.R.; Pierce, M.B.; Khan, A.R.; Dirks, R.M.; Pierce, N.A. NUPACK: Analysis and design of nucleic acid systems. *J. Comput. Chem.* **2011**, *32*, 170–173. [CrossRef]

34. Petty, J.T.; Zheng, J.; Hud, N.V.; Dickson, R.M. DNA-Templated Ag Nanocluster Formation. *J. Am. Chem. Soc.* **2004**, *126*, 5207–5212. [CrossRef] [PubMed]

35. Volkov, I.L.; Reveguk, Z.V.; Serdobintsev, P.Y.; Ramazanov, R.R.; Kononov, A.I. DNA as UV light–harvesting antenna. *Nucleic Acids Res.* **2018**, *46*, 3543–3551. [CrossRef]

36. Galley, W.C.; Purkey, R.M. Role of Heterogeneity of the Solvation Site in Electronic Spectra in Solution. *Proc. Natl. Acad. Sci. USA* **1970**, *67*, 1116–1121. [CrossRef] [PubMed]

37. Mukherjee, S.; Chattopadhyay, A. Wavelength-selective fluorescence as a novel tool to study organization and dynamics in complex biological systems. *J. Fluorescence* **1995**, *5*, 237–246. [CrossRef] [PubMed]

38. Demchenko, A.P. The red-edge effects: 30 years of exploration. *Luminescence* **2002**, *17*, 19–42. [CrossRef]

39. Lakowicz, J.R. *Principles of Fluorescence Spectroscopy*, 3rd ed.; Springer: New York, NY, USA, 2006.

40. Zikich, D.; Liu, K.; Sagiv, L.; Porath, D.; Kotlyar, A. I-motif nanospheres: Unusual self-assembly of long cytosine strands. *Small* **2011**, *7*, 1029–1034. [CrossRef]

41. Hansma, H.G.; Revenko, I.; Kim, K.; Laney, D.E. Atomic Force Microscopy of Long and Short Double-Stranded, Single-Stranded and Triple-Stranded Nucleic Acids. *Nucleic Acids Res.* **1996**, *24*, 713–720. [CrossRef]

42. Moreno-Herrero, F.; Colchero, J.; Baró, A.M. DNA height in scanning force microscopy. *Ultramicroscopy* **2003**, *96*, 167–174. [CrossRef]

43. Guo, Y.; Zhou, X.; Sun, J.; Li, M.; Hu, J. Height measurement of DNA molecules with lift mode AFM. *Chin. Sci. Bull.* **2004**, *49*, 1574–1577. [CrossRef]

44. Ono, A.; Cao, S.; Togashi, H.; Tashiro, M.; Fujimoto, T.; Machinami, T.; Oda, S.; Miyake, Y.; Okamoto, I.; Tanaka, Y. Specific interactions between silver(i) ions and cytosine–cytosine pairs in DNA duplexes. *Chem. Commun.* **2008**, *39*, 4825–4827. [CrossRef]

45. Gehring, K.; Leroy, J.-L.; Guéron, M. A tetrameric DNA structure with protonated cytosine-cytosine base pairs. *Nature* **1993**, *363*, 561. [CrossRef]

46. Day, H.A.; Huguin, C.; Waller, Z.A.E. Silver cations fold i-motif at neutral pH. *Chem. Commun.* **2013**, *49*, 7696–7698. [CrossRef] [PubMed]

47. Lee, T.H.; Dickson, R.M. Discrete two-terminal single nanocluster quantum optoelectronic logic operations at room temperature. *Proc. Natl. Acad. Sci. USA* **2003**, *100*, 3043–3046. [CrossRef]

48. Seeman, N.C. Structural DNA nanotechnology: An overview. *Methods Mol. Biol.* **2005**, *303*, 143–166. [CrossRef]

49. Swasey, S.M.; Rosu, F.; Copp, S.M.; Gabelica, V.; Gwinn, E.G. Parallel Guanine Duplex and Cytosine Duplex DNA with Uninterrupted Spines of AgI-Mediated Base Pairs. *J. Phys. Chem. Lett.* **2018**, *9*, 6605–6610. [CrossRef] [PubMed]

50. Swasey, S.M.; Leal, L.E.; Lopez-Acevedo, O.; Pavlovich, J.; Gwinn, E.G. Silver (I) as DNA glue: Ag+-mediated guanine pairing revealed by removing Watson-Crick constraints. *Sci. Rep.* **2015**, *5*, 10163. [CrossRef]

51. Abou Assi, H.; Garavís, M.; González, C.; Damha, M.J. i-Motif DNA: Structural features and significance to cell biology. *Nucleic Acids Res.* **2018**, *46*, 8038–8056. [CrossRef] [PubMed]

52. Shi, Y.; Sun, H.; Xiang, J.; Yu, L.; Yang, Q.; Li, Q.; Guan, A.; Tang, Y. i-Motif-modulated fluorescence detection of silver(I) with an ultrahigh specificity. *Anal. Chim. Acta* **2015**, *857*, 79–84. [CrossRef]

53. Nonin-Lecomte, S.; Leroy, J.L. Structure of a C-rich strand fragment of the human centromeric satellite III: A pH-dependent intercalation topology11Edited by K. Nagai. *J. Mol. Biol.* **2001**, *309*, 491–506. [CrossRef]

54. Ghodke, H.B.; Krishnan, R.; Vignesh, K.; Kumar, G.V.P.; Narayana, C.; Krishnan, Y. The I-Tetraplex Building Block: Rational Design and Controlled Fabrication of Robust 1D DNA Scaffolds through Non-Watson–Crick Interactions. *Angew. Chem. Int. Ed.* **2007**, *46*, 2646–2649. [CrossRef]

55. Li, T.; He, N.; Wang, J.; Li, S.; Deng, Y.; Wang, Z. Effects of the i-motif DNA loop on the fluorescence of silver nanoclusters. *RSC Adv.* **2016**, *6*, 22839–22844. [CrossRef]

56. Schultz, D.; Gardner, K.; Oemrawsingh, S.S.R.; Markešević, N.; Olsson, K.; Debord, M.; Bouwmeester, D.; Gwinn, E. Evidence for Rod-Shaped DNA-Stabilized Silver Nanocluster Emitters. *Adv. Mater.* **2013**, *25*, 2797–2803. [CrossRef]

57. Copp, S.M.; Schultz, D.; Swasey, S.; Pavlovich, J.; Debord, M.; Chiu, A.; Olsson, K.; Gwinn, E. Magic Numbers in DNA-Stabilized Fluorescent Silver Clusters Lead to Magic Colors. *J. Phys. Chem. Lett.* **2014**, *5*, 959–963. [CrossRef]

58. Schultz, D.; Gwinn, E.G. Silver atom and strand numbers in fluorescent and dark Ag:DNAs. *Chem. Commun.* **2012**, *48*, 5748–5750. [CrossRef]

59. Petty, J.T.; Ganguly, M.; Yunus, A.I.; He, C.; Goodwin, P.M.; Lu, Y.-H.; Dickson, R.M. A DNA-Encapsulated Silver Cluster and the Roles of Its Nucleobase Ligands. *J. Phys. Chem. C* **2018**, *122*, 28382–28392. [CrossRef]

nanomaterials

MDPI

Review

Small-Angle Scattering as a Structural Probe for Nucleic Acid Nanoparticles (NANPs) in a Dynamic Solution Environment

Ryan C. Oliver [1], Lewis A. Rolband [2], Alanna M. Hutchinson-Lundy [2], Kirill A. Afonin [2] and Joanna K. Krueger [2,*]

[1] Neutron Scattering Division, Oak Ridge National Laboratory, Oak Ridge, TN 37830, USA; ryanoliver5683@gmail.com

[2] UNC Charlotte Chemistry Department, Charlotte, NC 28223, USA; lrolband@uncc.edu (L.A.R.); ahutch27@uncc.edu (A.M.H.-L.); kafonin@uncc.edu (K.A.A.)

* Correspondence: Joanna.Krueger@uncc.edu; Tel.: +1-704-687-1642

Received: 29 March 2019; Accepted: 19 April 2019; Published: 2 May 2019

Abstract: Nucleic acid-based technologies are an emerging research focus area for pharmacological and biological studies because they are biocompatible and can be designed to produce a variety of scaffolds at the nanometer scale. The use of nucleic acids (ribonucleic acid (RNA) and/or deoxyribonucleic acid (DNA)) as building materials in programming the assemblies and their further functionalization has recently established a new exciting field of RNA and DNA nanotechnology, which have both already produced a variety of different functional nanostructures and nanodevices. It is evident that the resultant architectures require detailed structural and functional characterization and that a variety of technical approaches must be employed to promote the development of the emerging fields. Small-angle X-ray and neutron scattering (SAS) are structural characterization techniques that are well placed to determine the conformation of nucleic acid nanoparticles (NANPs) under varying solution conditions, thus allowing for the optimization of their design. SAS experiments provide information on the overall shapes and particle dimensions of macromolecules and are ideal for following conformational changes of the molecular ensemble as it behaves in solution. In addition, the inherent differences in the neutron scattering of nucleic acids, lipids, and proteins, as well as the different neutron scattering properties of the isotopes of hydrogen, combined with the ability to uniformly label biological macromolecules with deuterium, allow one to characterize the conformations and relative dispositions of the individual components within an assembly of biomolecules. This article will review the application of SAS methods and provide a summary of their successful utilization in the emerging field of NANP technology to date, as well as share our vision on its use in complementing a broad suite of structural characterization tools with some simulated results that have never been shared before.

Keywords: small-angle X-ray scattering; small-angle neutron scattering; contrast variation; nucleic acid nanoparticle; structural characterization

1. Introduction

Nucleic acid-based nanoparticles (NANPs) [1–13] and other nucleic acid-based nanodevices [14–17] are an emerging research focus area in pharmacological and biological studies. NANPs can be designed and manipulated to produce a variety of different functionalized nanostructured scaffolds; the novel resultant structures require detailed characterization prior to further biomedical transition and in vivo studies [7,18–22]. Conventional characterization techniques include the routinely used analysis of NANPs by native-PAGE (non-denaturing PolyAcrylamide Gel Electrophoresis) [23], dynamic light scattering (DLS) [4], atomic force microscopy (AFM) [24], and more sophisticated methods employing cryogenic-electron microscopy (cryo-EM) [25], nuclear magnetic resonance (NMR) [26], and X-ray crystallography [27] (see Table 1 for a comparison of the advantages and limitations of these structural characterization techniques). However, none of the aforementioned techniques allows for direct visualization of large (>100 kDa) three-dimensional NANPs in solution, and they often require working with high concentrations of NANPs.

Table 1. Comparison of nanoparticle structural characterization techniques.

Technique	Parameters Analyzed/Advantages	Limitations
Solid State/Static Techniques		
Crystallography [28]	High resolution molecular structure Broad Mass range Model building is well-developed	Static crystalline state structure; may not reflect dynamic or flexible structures Sample must form a crystal
Scanning Electron Microscopy (SEM) [28,29]	Particle size, size distributions, shape Sample preparation is relatively simple Structure in native state Allows analysis of hydrated materials without fixation, drying, freezing, or coating	Limited to larger molecules (up to ~200 nm) Highly dependent on electron microscopy (EM) techniques and access to costly equipment Cannot be used on certain biological materials due to degradation caused by the electron beam Low resolution
Transmission Electron Microscopy (TEM) [28]	Particle size, size distributions, shape Produces high resolution images that can provide information about structure and elemental composition High resolution TEM has Å resolution	Harsh chemical treatment of the sample Statistics are highly dependent on technique 2D images Samples need to be dehydrated, collected on metal mesh, and stained Small viewing section of sample
Atomic Force Microscopy (AFM) [30,31]	Provides a three-dimensional surface profile Minimal sample preparation Shown to give true atomic resolution in ultra-high vacuum (UHV) and, more recently, in liquid environments	Can only image a maximum height on the order of 10–20 micrometers and a maximum scanning area of about 150 × 150 micrometers Images can also be affected by hysteresis of the piezoelectric material Possibility of image artifacts Must immobilize the sample onto a substrate
Solution State/Native Techniques		
Static Light Scattering (SLS)/Dynamic Light Scattering (DLS)/ Zeta Potential [28]	Hydrodynamic particle size, size distributions, surface charge Sample volumes are small (μL) Particle size across a broad range (~0.1 nm to ~10 μm) Allows measurements under physiological conditions	Can only measure solid particles, polymers, and proteins dispersed in a solvent or emulsions Light absorption by the dispersant or sample can interfere with detection Concentration dependent Samples need to be homogenous Little shape information; size of particles can be under or over-estimated Dust/traces of agglomerates can interfere with results Cannot distinguish between similarly sized populations without coupling to a separation

Table 1. *Cont.*

Technique	Parameters Analyzed/Advantages	Limitations
Nuclear Magnetic Resonance (NMR) [32]	High resolution structure 3D structure in solution Dynamics can be studied	High sample purity and concentration required Computational simulation is challenging Sample MWs typically limited to below 40–50 kDa Water soluble samples
Small-Angle X-ray Scattering (SAXS) [28,33]	Structure in native state Particle size and shape, size distribution, particle interactions and interatomic distances: some parameters determined with sub Å precision Small sample size (10–25 µL solution; 0.01–10 mg/mL) Highflux synchrotron sources allow for time-resolved, kinetic studies	Low-resolution shape information interpreted from interatomic distance distributions Highest level of structural information requires pure, monodisperse samples
Small-Angle Neutron Scattering (SANS) [28,33]	Amenable to contrast variation Sensitive to fluctuations in the nuclei density of the sample	Experiments require access to user facilities with appropriate neutron source and instrumentation Flux of neutron source is intrinsically low

Small-angle scattering (SAS) is a structural characterization technique that is well placed to determine the conformation of NANPs under varying solution conditions, which will allow for optimization of their design and pipe-line production. SAS experiments yield information on the overall shapes and electron (or nuclear) density distribution within macromolecules in solution (for additional primers on this technique, see [33–37]). In addition, the inherent differences in the neutron scattering of nucleic acids, lipids, and proteins allow for the characterization of the conformations and relative dispositions of the individual components of an assembly of biomolecules using methods of contrast variation or solvent matching. Also, due to the different neutron scattering properties of the isotopes of hydrogen the neutron scattering contrast can be enhanced by labeling one component within a macromolecular assembly with deuterium. An example application would include using contrast variation methods to examine the overall shape of NANPs that have been functionalized with short interfering RNAs (siRNAs), ribozymes, aptamers, proteins, or other small molecules (for a review of functionalized nanoparticles see [38,39]. Conformational changes within the NANP itself as a result of direct or indirect fusion with these therapeutically relevant molecules could then be observed independently. Small-angle neutron scattering (SANS) combined with contrast variation or contrast matching methods would allow for detection of the conformation of each of the individual components within the resultant NANP assembly, as well as the distance between their centers of mass. A structural basis for understanding the resultant functionalized NANPs will be essential to guarantee precise control over the composition and stoichiometry of therapeutic modules for simultaneous delivery into diseased cells and eventually for their successful transitions to in vivo preclinical studies [7,11,18–20,40]. Certainly, the direct visualization of various NANPs and NA multi-stranded assemblies can be, and has been, achieved with AFM [41] and cryo-EM [7], as mentioned previously. However, the resolution of these techniques is currently limited by the size (with smaller NPs < 20 nm being preferred), shape, and composition of the nanoparticles. Also, neither of these techniques addresses the complicated, dynamic environment of the particle in solution. Therefore, to gain additional information about the structure of NANPs and to understand more completely the structure-to-function relationship, thus possibly enhancing its functionality, several techniques must be combined. This article will review the application of small-angle X-ray and neutron scattering methods and provide a summary of their successful utilization in the emerging field of NANP technology to date. Importantly, we share our vision for how it may be used in the near future to add to and complement a broad suite of structural characterization tools with some guidance and the feasibility of the proposed applications by including some simulated results that have never been shared before.

2. Discussion

A model for the structure of nucleic acids was initially proposed by Watson and Crick in 1953. Interpretation of the X-ray diffraction patterns from meticulously prepared 2D deoxyribonucleic acid (DNA) crystals recorded by Rosalind Franklin provided the key to understanding this structure. Since then, X-ray diffraction has played a vital role in further discoveries of numerous types of nucleic acid structures. An excellent review of the applications of various synchrotron-based spectroscopic techniques, including small-angle X-ray scattering (SAXS), has been recently published by Yi Lu and his research group [42]. High-resolution techniques such as X-ray crystallography have the capability to provide atomistic structural views. However, SAS techniques can be applied to molecules in solution and can give insights into systems in which inherent flexibility, which may cause problems for crystallization, is in fact essential for its proper function. SAS was first described for biomolecules (proteins) in the late 1940s [43] and has been widely used for several decades to solve problems requiring an understanding of the nanoscale phenomena. SAS techniques in fact depend on the same physics as the corresponding larger angle scattering methods (X-ray or neutron diffraction), but reveal larger, more global structures due to the inverse relationship between length scale and scattering angle; refer to Glatter and Kratky [44] for an excellent textbook describing the physics of SAS.

In lieu of a crystal, X-ray scattering from nucleic acid samples in solution can provide essential structural information on the time-averaged ensemble structure. Information about the size, shape, compactness, and molecular weight of the scattering molecules are readily obtained from the scattering data. Beginning in the late 1980s, as the methods of analysis and image reconstruction technologies became more accessible and sophisticated (see the ATSAS software package [45]), so too did the functional insights and applications. Thanks to advances in computational capabilities and instrumentation, particularly with the increased flux available now at synchrotron sources, SAS has developed into a powerful structural tool that complements and enhances other structural information to provide a more complete understanding of the structure-function relationship. For example, Wang and co-workers used SAS to describe an unusual topological structure that the HIV-1 (human immunodeficiency virus) uses to recognize its own messenger ribonucleic acid (mRNA) [46]. SANS and contrast variation techniques are ideally suited to examining the conformational changes within the protein and its nucleic acid binding partner upon complexation with one another. Recently, Sonntag et al. [47] demonstrated the power of contrast variation and SANS in resolving ambiguities and improving the interpretation of complementary SAXS and NMR data on a ternary protein-RNA complex involved in alternate splicing.

Of importance in extending these SAS technologies to study NANPs specifically, SAS provides not only information on the sizes and shapes of particles but also information on the internal structures of disordered and partially disordered systems. Rambo and Tainer [48,49] have improved and tested the SAS computational tools and technologies specifically for applications to the inherently flexible nucleic acid and related structures. Their SAXS results have discovered and demonstrated that conformational variation is a general functional feature of macromolecules. Importantly, SAS can tolerate a variety of measurement conditions, thus allowing rapid comparison of the effects of environmental changes on the detected structural properties. Moreover, extraction of meaningful 3D details from 1D scattering data via molecular modeling techniques has become increasingly sophisticated [50,51], allowing for the development of experimentally constrained structural models that can be further interpreted or constrained by other types of structural knowledge on the system being studied (for recent reviews see [35,52]). Indeed, a major concern in interpreting resultant SAS-based models is that there may be several structures that produce similar scattering patterns. One must always keep in mind that these models represent the time-averaged ensemble, which could include a mixture of dynamic conformations and/or intermolecular interactions. For this reason, complementary data from other structural techniques is essential to proper interpretation.

Small-Angle Scattering Methodology: Light scattering, in general, is useful for studying the state of association or conformation of biological macromolecules in solution [53]. Both static (elastic) and dynamic (quasi-elastic) light scattering techniques are generally easy to perform and can be done on solutions with relatively low concentrations of analyte. The static light scattering (SLS) experiment monitors the total light scattering intensity averaged over time and can provide information on the "apparent" molecular weight (M_{app}) and the radius of gyration (R_g) of the macromolecule in solution. Dynamic light scattering (DLS) experiments monitor fluctuations in the intensity of light scattered by small volume elements in solution, which are directly related to the Brownian motion of the solutes, thereby providing information on the hydrodynamic radius (R_H), which also can be related to an apparent molecular weight. In either case, light scattering techniques can be used as an initial probe of the NANP conformations to monitor aggregation or conformational changes in varying solution environments. Determining particle size and shape, however, requires a light source with much smaller wavelengths, such as X-rays or neutrons.

X-ray and neutron SAS represents a major tool for obtaining global information on the size and shape of folding intermediates of RNA molecules in solution, since it provides quantitative characterization of mixtures by measuring the radius of gyration and maximum linear dimension of the molecules to ~1–10 nm resolution. Typical experimental set-up and analysis is shown in Figure 1. A sample containing randomly oriented molecules in solution is placed in an X-ray or neutron beam with wavelengths between 1–6 Å. The coherent scattering, $I(Q)$, from a homogeneous solution of monodisperse particles can be expressed mathematically as:

$$I(Q) = \langle\, |\, \int [\rho(r) - \rho s] \bullet \exp(-iQ\bullet r)\, dr\, |\, 2\, \rangle \tag{1}$$

The integration is taken over the volume of the particle and $\langle\,\rangle$ denotes the average over all particle orientations. Q is the momentum transfer or scattering vector and can be expressed as $4\pi(sin\theta)/\lambda$, where θ is half the scattering angle and λ is the wavelength of the scattered radiation. $\rho(r)$ and ρs are the scattering length densities for the particle and solvent, respectively. Structural information is derived from a measurement of the intensity of the scattered X-ray ($I(Q)$) as a function of scattering angle (Q). Analysis of these data is accomplished initially with a Guinier approximation by fitting the data in the low Q region (where $Q \cdot Rg < 1.3$). This approximation can be done for globular or for rod-like particles and yields a direct estimation of the molecule's R_g or cross-sectional radius (R_c), respectively. For well-folded samples, a Kratky plot can be used to estimate the hydrated volume, or Porod volume. Comparative changes in a Kratky plot can reveal flexibility, unfolding, or a conformational change.

More detailed structural information may be obtained from analysis of the pair-distance distribution function, P(r). An inverse Fourier transformation of the scattering data yields the probable distribution of atom-pair distances (r) weighted by the product of their scattering powers, and is typically represented as a 1-dimensional P(r) versus r profile. For well-behaved samples, the P(r) will approach zero at the maximum linear dimension, d_{max}, of the scattering particle. The zeroth and second moments of the P(r) give forward scatter, I_0, and radius of gyration, R_g, respectively. The forward scatter, I_0, is directly proportional to the molecular weight squared of the scattering molecule and thus is very sensitive to changes in the size of the scattering particle due to, for example, complex formation, specific oligomerization, or aggregation. P(r) is sensitive to the symmetry of the scattering particle and to the relationships between domains or repeating structures. This effect is demonstrated in Figure 1, which shows the P(r) functions for various one- and two- domain structures of uniform scattering density. It is worth noting how the asymmetry of the P(r) function increases with the asymmetry of the shape of the object. Determining the 3-dimensional shape that gives rise to a measured SAS (intensity versus Q) profile is recognized as an 'underdetermined' problem (as a result of rotational averaging of the scattered intensity arising from the random orientation of molecules in solution). Nonetheless, molecular modeling of these data can be highly informative, particularly if the models are interpreted by utilizing other known structural constraints [54]. One interesting approach to assessing the ambiguity in SAS profiles has been reported [55]. An accepted practice is to generate multiple solutions using Monte Carlo-based minimization methods and simple constraints, such as connectivity and compactness, and then to evaluate the variability and range of potential solutions. The software for completing this type of analysis is available in the popular ATSAS analysis package [45].

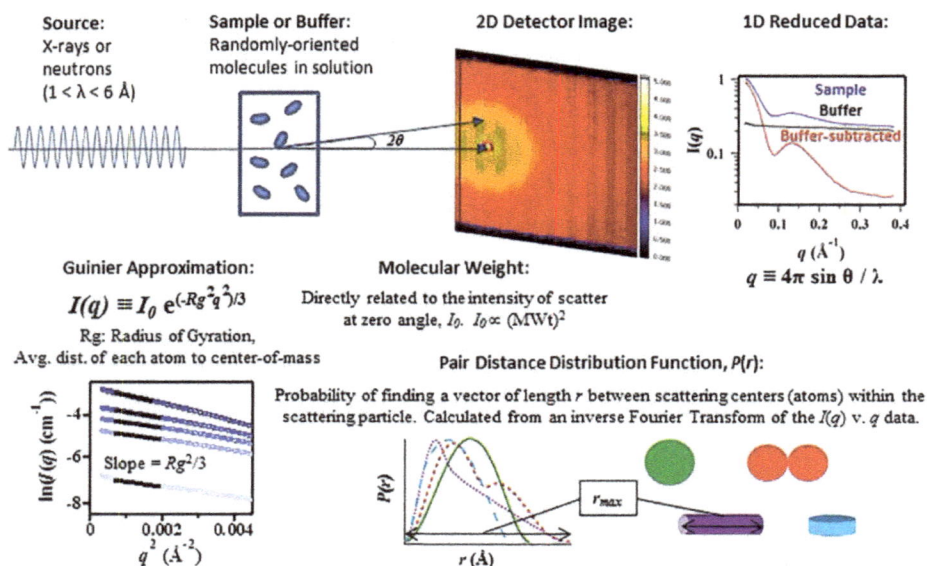

Figure 1. Typical small-angle X-ray and neutron scattering (SAS) experimental set-up and data analysis. I(q) is the intensity of the scattered light as a function of momentum scattering vector, q, as defined above. I(0) is the intensity of the scatter at zero angle and is directly proportional to the square of the molecular weight of the biomolecule $(MWt)^2$. R_g is the biomolecule's Radius of Gyration, and is defined as the average distance of each scattering center, atom, from the center-of-mass. P(r) is the pair distance distribution function, calculated as an inverse Fourier Transform of the scattering data and representative of the probability of finding a vector of length r between the atoms within the biomolecule.

Neutron (SANS) Methodology: Examination of the individual component structures within the context of larger macromolecular assemblies (NA:protein or NA:lipid:protein structures, for instance) can be achieved by collecting neutron scattering data on the complexes while varying the solvent contrast (for a recent review see [56]). Scattering length densities (SLDs) are calculated by summing the scattering amplitudes of each atom within a volume and dividing by that volume. From Equation (1) it can be readily seen that the intensity of the scattering from a particle in solution depends upon the difference in scattering density between the particle and the solvent, i.e., its "contrast". The SLD of a particle is a function of its elemental composition and the associated atomic scattering lengths (specifically the coherent scattering lengths, b_{coh}), which are a measure of the strength of the interaction of an X-ray or neutron with an atom. The fact that hydrogen and its isotope deuterium have dramatically different scattering lengths ($b_{coh} = -3.74 \times 10^{-15}$ m and 6.67×10^{-15} m, respectively) empowers a neutron scattering contrast variation technique for structural biology. The fraction of D_2O substitution for H_2O in aqueous buffers provides a continuous spectrum of values for the solvent's SLD. The true utility of being able to change the SLD of the solvent relative to that of the scattering particle becomes evident when working with structures composed of materials which have different SLDs, such as proteins, lipids, and NAs. These various biomolecules have an inherently dissimilar elemental composition and thus different average scattering lengths, so each will be 'visible' (or 'invisible') at unique solvent contrasts. Furthermore, the production of deuterium-enriched biological materials makes possible the reconstitution of multi-component structures with selectively deuterated subunits. Example SLDs of various biological macromolecules, including examples involving deuterium-labeled material, are shown in Figure 2 as a function of the H_2O/D_2O mixture in the background solution.

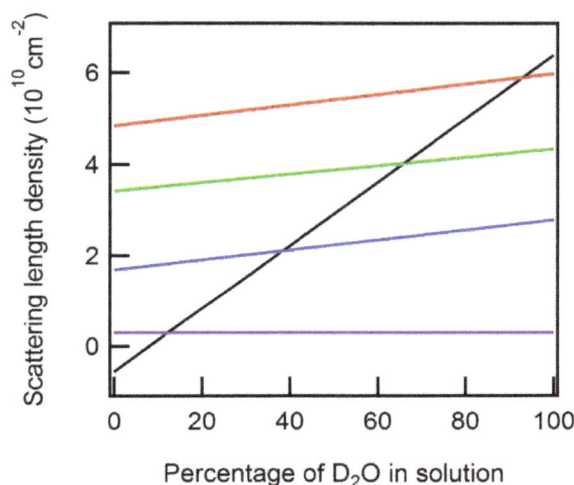

Figure 2. Scattering length densities of various biological macromolecules plotted as a function of percent D_2O in the solvent: a hydrogenated protein (blue), a deuterated protein with 65% of the non-exchangeable protons replaced by deuterium (red), messenger ribonucleic acid (mRNA) (green), and lipid (purple). The black line corresponds to the scattering length density of the background solution. The "match" point of these macromolecules (circled) is found at the percent D_2O where the scattering light density (SLD) of the solvent equals that of the molecule.

Of particular interest in this plot are the intersections where the scattering length density of the solvent (black line) matches that of the various biomolecules. At these points (referred to as contrast match points), the contrast between the molecule and background (solvent), and therefore the measured intensity of that molecule, is zero. The measured scattering intensity I(Q) from a multi-subunit assembly containing a subunit(s) with a solvent contrast-matched SLD would only reflect the remaining subunit(s), which have a nonzero contrast. The result is structural information on individual subunit components within the macromolecular complex. This particular kind of experiment is known as a contrast-matching experiment. An extension of the contrast-matching experiment is a contrast variation experiment. In a contrast variation experiment, the total scattering of the complex is measured at several solvent contrasts (fractions of D_2O) and then mathematically extrapolated to yield the scattering profile of the individual components.

SANS Applications: A classic set of examples for the application of SANS with contrast variation involves the study of various ribosomes. The earliest studies probed the internal structure of the 30S ribosome [57–59]. Contrast-variation methods were used to determine the relative distances between subunits in this multi-subunit complex. Ultimately, this research led to a structural model for the disposition of these subunits in space [60]. These early studies were followed by subsequent studies of the larger 50S and 70S ribosomes. A map of the distribution of protein and RNA within the 50S ribosome from *Escherichia coli* was generated using SANS with contrast variation data and shape restoration by spherical harmonics [61–63].

More recently, SANS has demonstrated the structural influence that ionic strength and temperature have on the corona structures found in DNA-capped gold nanoparticles [64] (Figure 3). These data will assist in customizing tailor-made corona structures for designer materials and devices. X-ray data has provided information on the inorganic cores of these nanoparticles but the complementary neutron data has expanded the structural scope, revealing the 15-mer DNA capped corona structures and the formation of ionic strength- and temperature-dependent aggregate species.

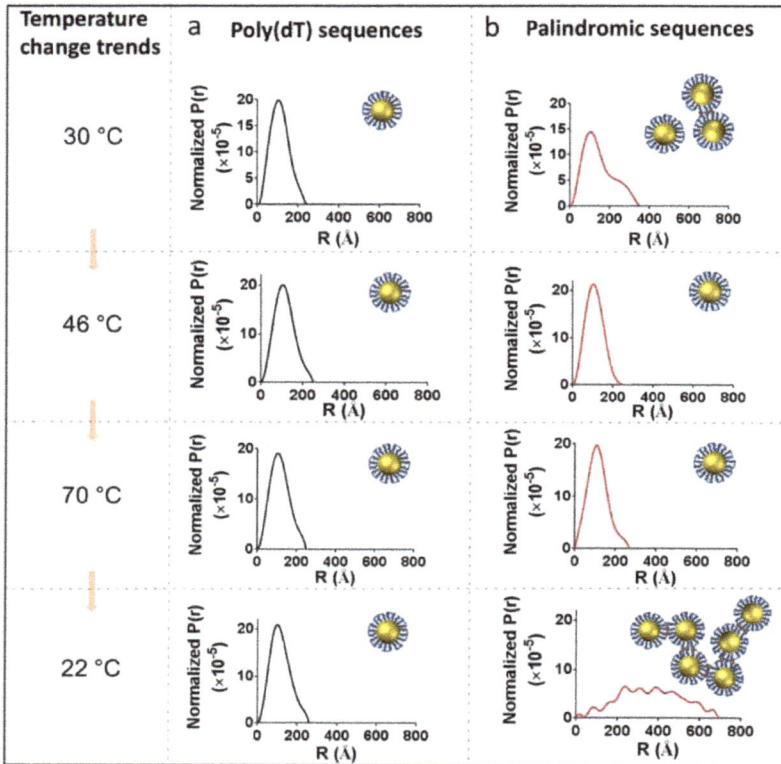

Figure 3. The pair distribution function P(r) is shown for two DNA-capped nanoparticles, T15 (**a**) and T7−8 (**b**) conjugates computed at various temperatures (30 °C, 46 °C, 70 °C, and 22 °C) in 0.5 M salt buffer. Insets are the scheme of temperature effect on poly(dT) sequenced deoxyribosenucleic acid (DNA) and palindromic sequenced DNA. Reprinted (adapted) with permission from (Yang, W.; et al. Probing Soft Corona Structures of DNA-Capped Nanoparticles by Small Angle Neutron Scattering. *J. Phys. Chem. C* 2015, *119*, 18773–18778). Copyright (2015) American Chemical Society.

SAXS Applications: The structures of small fragments of functional RNAs have been successfully solved using SAXS and confirmed by other techniques [65,66]. This approach allows for an investigation of the influence of the size, composition, and shape of functionalized NANPs on their ability to be delivered to diseased cells and to further their functional efficiency. Structural models built from the solid-state, i.e., X-ray crystal diffraction or cryo-EM, can be used to generate an expected scattering curve, which can then be directly compared with the measured solution scattering data to detect differences in the solution state of the particle. One ultimate goal might be to utilize SAXS under varying solvent conditions (e.g., ionic strength, pH, or binding partners, etc.) in order to gain a better understanding of structure-to-function relationships in various synthetic and natural RNAs. These results would assist in refining computational-experimental protocols for functional RNA nanoparticle pipeline production. Additionally, time-resolved methods using bright synchrotron sources could provide kinetic insights into their assembly. These methods have been successfully used to provide kinetic data on ribozyme folding [67], transfer ribonucleic acid (tRNA) assembly [68], and riboswitches [69,70], so precedents exist for using them to examine the assembly dynamics of NANPs.

Conformational changes, flexibility, and self-assembly [71,72] processes of DNA nanostructures are being investigated using SAS techniques with increased frequency. The structural features determined by these solution-based techniques offer structural insights that are distinct from those provided by techniques that require the nanostructure to be fixed onto a substrate. For example, an X-shaped DNA-based molecular switch has been examined through SAXS, solution fluorescence resonance energy transfer (FRET), single-molecule FRET, and transmission electron microscopy (TEM) to determine the population of the two distinct conformational states (Figure 4). The switch's conformation, which closes to form a linear rod-like structure in high ionic strength environments, was shown through SAXS and solution FRET to have a statistically significantly lower population of molecules in the linear conformation than was determined by single molecule FRET or TEM. It is suggested that the fixation to surface, dyeing, and/or the manual assignment of conformations of TEM images may bias these experimental methods towards a closed conformation, while the solution-based techniques gave more accurate assessments of the particle conformations [72]. The increased availability of SAS instruments located off high flux, synchrotron sources allowed these measurements to be made with reasonable signal to noise profiles on samples at concentrations of only 25–100 nM. They have been able to detect conformational changes triggered by changes in the solution environment. These studies were followed up with time-resolved SAXS [74] to monitor this large-scale conformational transition and it was found that it switches from its open to closed conformation upon addition of $MgCl_2$ within milliseconds, which is close to the theoretical diffusive speed limit. The construction of functional NANPs will likely require dynamic structures that can undergo controllable conformational changes and SAXS is well placed as a tool for resultant structural kinetic studies. DNA devices based on shape complementary stacking interactions have been demonstrated to undergo reversible conformational changes triggered by changes in ionic environments or temperature. In another, unrelated experiment, molecular dynamics and SANS were used in combination to predict and test the gelling properties of tetravalent DNA nanostars as a function of temperature [71]. The time-resolved growth of the DNA nanostar gel was monitored by following changes in a signature peak intensity, and, thus, these studies allowed for kinetic and thermodynamic measurements of the nanostar structural formation.

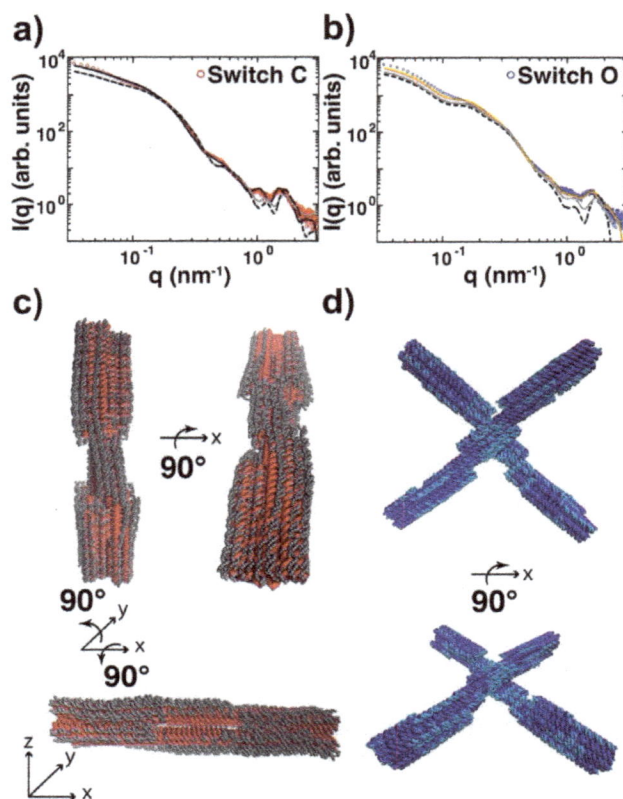

Figure 4. (**a**) and (**b**) depict the mode-based refinement of a DNA switch structure against small-angle X-ray scattering (SAXS) data. (**a**) Experimental SAXS data for the DNA switch in the closed, linear conformation and (**b**) the open, X-shaped conformation are shown as red and blue circles, respectively, against the scattering profile predicted by preliminary models in CanDo (dashed black lines) and CRYSOL software (gray lines) [73]. The preliminary structures of the (**c**) open and (**d**) closed switch conformations are shown as red and blue cylinders with the refined structures shown as gray and light-blue orbs, respectively. Reprinted (adapted) with permission from Bruetzel, L.K.; et al. Conformational Changes and Flexibility of DNA Devices Observed by Small-Angle X-ray Scattering. *Nano Lett.* 2016, *16*, 4871–4879. Copyright (2016) American Chemical Society.

SAS Vision Application for Nucleic Acid Architectures: One of the examples where SAS can be readily employed may be seen in the recent achievements of RNA nanotechnology, where two orthogonal NANP designing strategies (exemplified by RNA nanocubes [4,8,12,13,23,25,75] and RNA nanorings [7,13,24,41,75–77]) were introduced, with potential for broad use in nanotechnology and biomedical applications. In one strategy (nanocubes) the RNA strands are specifically designed to only form intermolecular bondings with their cognate partners while avoiding the formations of any intramolecular secondary structures. Another strategy (nanorings), takes advantage of RNA long-range tertiary interacting motifs that require the formation of specific secondary structures of individual monomers, and the intermolecular interactions are activated in the presence of magnesium ions. Both NANPS were tested against several different cancer- and HIV-infected cell lines and showed a significant therapeutic effect. Furthermore, the desired activity of these functional NANPs was demonstrated remarkably in vivo [7]. Importantly, the immunorecognition of NANPs by human peripheral blood mononuclear cells strongly depends on the type of NANPs and the extent of their functionalization with siRNAs [13,78,79]. In vitro characterization with cryo-EM microscopy has revealed that the structure of the nanoring functionalized with six siRNAs has a pinwheel-like crown shape (Figure 5a). That topology, if accurately reflecting the in vivo state, may affect the efficiency of the intracellular release of siRNAs through 'dicing', due to steric issues, and influence their interactions with the pathogen recognition receptors of the immune system. The issues of imprecisely predicted and verified topology may become even more evident in the case of 3D polyhedral self-assembled functional nanostructures such as nanocubes [4]. Figure 5b provides the calculated SAXS profiles for several of these predicted structures, demonstrating that SAXS data is sufficient to differentiate between the various architectures. It is possible that further optimization of the existing designs is needed (such as an extension of dicable siRNA-containing arms, modification of the 5'-end of the scaffold, changes in base composition, or introduction of additional RNA structural motifs, etc.) to ensure the enhancement of siRNA release and processivity.

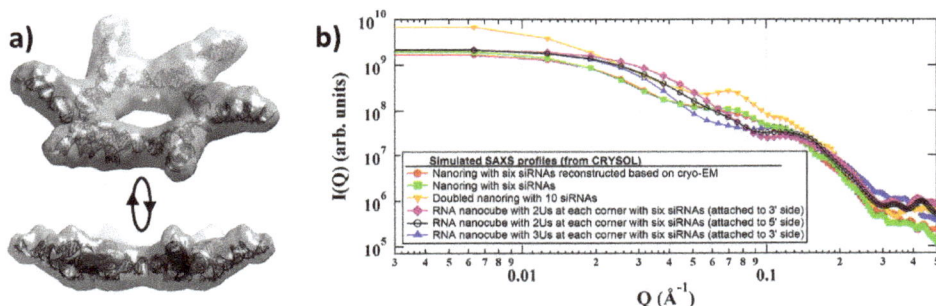

Figure 5. (a) Cryogenic-electron microscopy (cryo-EM) reconstruction of RNA nanoring functionalized with six short interfering RNAs (siRNAs) and (b) its calculated SAXS profile (red circles) compared to the calculated SAXS profile for other predicted NA nanoparticle structures. Profiles were calculated from models based on cryo-EM pdb files using the program CRYSOL [73].

Moreover, 3D structures of other individual, relatively bulky groups, such as aptamers or antibodies introduced for targeting, are not known, and their function may be attenuated due to steric clashes within NANPs. Therefore, alternative approaches that can provide complementary data about 3D orientations and shapes of the NANPs and their individual components in solution are needed. Utilizing natural contrast between the scattering components in these systems, neutron scattering will allow for determination of the structural parameters of the NANPs bound to any functional groups. Another vision for the use of SAS in structural characterization of NANPs is to extend the SAS profile collected to include a larger angle scattering region (WAXS). These data may be useful for investigating the Ag-Ag distances within DNA-based assemblies of fluorescent silver nanoclusters [80].

Additionally, delivery of NANP-based therapeutics in vivo is one of the most challenging tasks due to RNA's negative charge, chemical instability, and stimulation of immune system responses. Investigating different potential carriers, such as lipids or cell-penetrating peptides, for in vivo delivery of RNA therapeutics is therefore one area of RNA nanotechnology that would benefit from SAS-based approaches. Experiments with NANPs employing various carriers such as magnetic nanoparticles [81], lipids [82], mesoporous silica-based nanoparticles [79], polysilsesquioxane [83], and bolaamphiphiles or 'bolas' [84,85] have already been successfully initiated. The use of SANS can significantly improve our current understanding of the interactions between the NANPs and carriers, which can further improve NANP delivery in vivo. For example, bolas consist of positively charged acetylcholine head groups on each side of a hydrophobic chain. In aqueous solution, these bolas form micelles and are efficiently associated with siRNAs for their further delivery in vivo. It was also recently demonstrated that bolas can form vesicles, rather than micelles, and can be used for delivery of encapsulated analgesic peptides and small molecules within a mouse brain [86,87]. These vesicles may become strong candidates for the delivery of functional RNA nanoparticles in vivo, especially across the blood-brain barrier to glioblastomas. Preliminary, unpublished results indicate formation of stable siRNA/bola vesicle complexes and cryo-EM images show changes in the shape of the particle upon siRNA addition. Constraints for the formation of functional RNA nanoparticles/bola vesicle complexes must be directed by the architecture of the components including shape, size, and total charge. Therefore, comparison of different shapes for functional RNA nanoparticles and different RNA-to-bola ratios will be necessary to maximize the RNA-bola interaction capacity. The self-assembly of similar, stable monomolecular nucleic acid lipid particles has been studied by SAXS and complemented by SANS and TEM [88]. These SAXS data confirmed the overall size and spherical shape of the particles, whereas the inherent contrast between nucleic acid and lipid moieties' neutron scattering allowed for a more detailed structural description of the core shell-like structure of these particles.

3. Conclusions

Small-angle X-ray and neutron scattering are structural characterization techniques that are well placed to determine the conformation of nucleic acid nanoparticles under varying solution conditions, thus allowing for optimization of their design. SAS results should complement and extend the structural information obtained through direct imaging techniques and other high-resolution structures. SAS experiments provide information on the overall shapes and particle dimensions of macromolecules and are ideal for following conformational changes and dynamics of the molecular ensemble as it behaves in solution. In addition, the inherent differences in the neutron scattering of nucleic acids, lipids, and proteins, as well as the different neutron scattering properties of the isotopes of hydrogen, combined with the ability to uniformly label biological macromolecules with deuterium, allow for the characterization of the conformations and relative dispositions of the individual components within an assembly of biomolecules.

Funding: This work was supported by National Institute of General Medical Sciences of the National Institutes of Health grant R15GM128100 (JKK co-PI)).

Acknowledgments: We would like to thank Maya Hunter for her critical review and helpful editing suggestions during the assembly of this article.

Conflicts of Interest: The authors declare no conflict of interest.

References

1. Guo, P.X. The Emerging Field of RNA Nanotechnology. *Nat. Nanotechnol.* **2010**, *5*, 833–842. [CrossRef] [PubMed]
2. Shukla, G.C.; Haque, F.; Tor, Y.; Wilhelmsson, L.M.; Toulmé, J.J.; Isambert, H.; Guo, P.; Rossi, J.J.; Tenenbaum, S.A.; Shapiro, B.A. A Boost for the Emerging Field of RNA Nanotechnology. *ACS Nano* **2011**, *5*, 3405–3418. [CrossRef]
3. Li, H.; Lee, T.; Dziubla, T.; Pi, F.; Guo, S.; Xu, J.; Li, C.; Haque, F.; Liang, X.J.; Guo, P. RNA as a Stable Polymer to Build Controllable and Defined Nanostructures for Material and Biomedical Applications. *Nano Today* **2015**, *10*, 631–655. [CrossRef] [PubMed]
4. Kirill, A.A.; Kasprzak, W.; Bindewald, E.; Praneet, S.; Puppala Alex, R.; Diehl Kenneth, T.; Hall Tae Jin Kim Michael, T.; Zimmermann Robert, L.; Jernigan, L.J.; Shapiro, B.A. Computational and Experimental Characterization of RNA Cubic Nanoscaffolds. *Methods* **2014**, *67*, 256–265.
5. Kirill, A.A.; Kasprzak, W.K.; Bindewald, E.; Kireeva, M.; Viard, M.; Kashlev, M.; Shapiro, B.A. In Silico Design and Enzymatic Synthesis of Functional RNA Nanoparticles. *Account. Chem. Res.* **2014**, *47*, 1731–1741.
6. Afonin, K.A.; Viard, M.; Kagiampakis, I.; Case, C.L.; Dobrovolskaia, M.A.; Hofmann, J.; Vrzak, A.; Kireeva, M.; Kasprzak, W.K.; KewalRamani, V.N.; et al. Triggering of RNA Interference with RNA-RNA, RNA-DNA, and DNA-RNA Nanoparticles. *ACS Nano* **2015**, *9*, 251–259. [CrossRef]

7. Afonin, K.A.; Viard, M.; Koyfman, A.Y.; Martins, A.N.; Kasprzak, W.K.; Panigaj, M.; Desai, R.; Santhanam, A.; Grabow, W.W.; Jaeger, L.; et al. Multifunctional RNA Nanoparticles. *Nano Lett.* **2014**, *14*, 5662–5671. [CrossRef] [PubMed]

8. Dao, B.N.; Viard, M.; Martins, A.N.; Kasprzak, W.K.; Shapiro, B.A.; Afonin, K.A. Triggering Rnai with Multifunctional RNA Nanoparticles and Their Delivery. *DNA RNA Nanotechnol.* **2015**, *1*, 27–38. [CrossRef]

9. Ohno, H.; Kobayashi, T.; Kabata, R.; Endo, K.; Iwasa, T.; Yoshimura, S.H.; Takeyasu, K.; Inoue, T.; Saito, H. Synthetic RNA-Protein Complex Shaped Like an Equilateral Triangle. *Nat. Nanotechnol.* **2011**, *6*, 116–120. [CrossRef]

10. Shibata, T.; Fujita, Y.; Ohno, H.; Suzuki, Y.; Hayashi, K.; Komatsu, K.R.; Kawasaki, S.; Hidaka, K.; Yonehara, S.; Sugiyama, H.; et al. Protein-Driven RNA Nanostructured Devices That Function in Vitro and Control Mammalian Cell Fate. *Nat. Commun.* **2017**, *8*, 540. [CrossRef]

11. Lee, H.; Lytton-Jean, A.K.; Chen, Y.; Love, K.T.; Park, A.I.; Karagiannis, E.D.; Sehgal, A.; Querbes, W.; Zurenko, C.S.; Jayaraman, M.; et al. Molecularly Self-Assembled Nucleic Acid Nanoparticles for Targeted in Vivo Sirna Delivery. *Nat. Nanotechnol.* **2012**, *7*, 389–393. [CrossRef]

12. Halman, J.R.; Satterwhite, E.; Roark, B.; Chandler, M.; Viard, M.; Ivanina, A.; Bindewald, E.; Kasprzak, W.K.; Panigaj, M.; Bui, M.N.; et al. Functionally-Interdependent Shape-Switching Nanoparticles with Controllable Properties. *Nucleic Acids Res.* **2017**, *45*, 2210–2220. [CrossRef]

13. Hong, E.P.; JHalman, R.; Shah, A.B.; Khisamutdinov, E.F.; Dobrovolskaia, M.A.; Afonin, K.A. Structure and Composition Define Immunorecognition of Nucleic Acid Nanoparticles. *Nano Lett.* **2018**, *18*, 4309–4321. [CrossRef]

14. Douglas, S.M.; Bachelet, I.; Church, G.M. A Logic-Gated Nanorobot for Targeted Transport of Molecular Payloads. *Science* **2012**, *335*, 831–834. [CrossRef]

15. Bindewald, E.; Afonin, K.A.; Viard, M.; Zakrevsky, P.; Kim, T.; Shapiro, B.A. Multistrand Structure Prediction of Nucleic Acid Assemblies and Design of RNA Switches. *Nano Lett.* **2016**, *16*, 1726–1735. [CrossRef]

16. Roark, B.K.; Tan, L.A.; Ivanina, A.; Chandler, M.; Castaneda, J.; Kim, H.S.; Jawahar, S.; Viard, M.; Talic, S.; Wustholz, K.L.; et al. Fluorescence Blinking as an Output Signal for Biosensing. *ACS Sens.* **2016**, *1*, 1295–1300. [CrossRef] [PubMed]

17. Andersen, E.S.; Dong, M.; Nielsen, M.M.; Jahn, K.; Subramani, R.; Mamdouh, W.; Golas, M.M.; Sander, B.; Stark, H.; Oliveira, C.L.; et al. Self-Assembly of a Nanoscale DNA Box with a Controllable Lid. *Nature* **2009**, *459*, 73–76. [CrossRef]

18. Binzel, D.W.; Shu, Y.; Li, H.; Sun, M.; Zhang, Q.; Shu, D.; Guo, B.; Guo, P. Specific Delivery of Mirna for High Efficient Inhibition of Prostate Cancer by RNA Nanotechnology. *Mol. Ther.* **2016**, *24*, 1267–1277. [CrossRef]

19. Shu, D.; Li, H.; Shu, Y.; Xiong, G.; Carson, W.E.; Haque, F.; Xu, R.; Guo, P. Systemic Delivery of Anti-Mirna for Suppression of Triple Negative Breast Cancer Utilizing RNA Nanotechnology. *ACS Nano* **2015**, *9*, 9731–9740. [CrossRef]

20. Feng, L.; Li, S.K.; Liu, H.; Liu, C.Y.; LaSance, K.; Haque, F.; Shu, D.; Guo, P. Ocular Delivery of Prna Nanoparticles: Distribution and Clearance after Subconjunctival Injection. *Pharm. Res.* **2014**, *31*, 1046–1058. [CrossRef]

21. Shu, Y.; Shu, D.; Haque, F.; Guo, P. Fabrication of Prna Nanoparticles to Deliver Therapeutic Rnas and Bioactive Compounds into Tumor Cells. *Nat. Protoc.* **2013**, *8*, 1635–1659. [CrossRef]

22. Afonin, K.A.; Viard, M.; Martins, A.N.; Lockett, S.J.; Maciag, A.E.; Freed, E.O.; Heldman, E.; Jaeger, L.; Blumenthal, R.; Shapiro, B.A. Activation of Different Split Functionalities on Re-Association of RNA-DNA Hybrids. *Nat. Nanotechnol.* **2013**, *8*, 296–304. [CrossRef] [PubMed]

23. Kirill, A.A.; Grabow, W.W.; Walker, F.M.; Bindewald, E.; Dobrovolskaia, M.A.; Shapiro, B.A.; Jaeger, L. Design and Self-Assembly of Sirna-Functionalized RNA Nanoparticles for Use in Automated Nanomedicine. *Nat. Protoc.* **2011**, *6*, 2022–2034.

24. Sajja, S.; Chandler, M.; Fedorov, D.; Kasprzak, W.K.; Lushnikov, A.; Viard, M.; Shah, A.; Dang, D.; Dahl, J.; Worku, B.; et al. Dynamic Behavior of RNA Nanoparticles Analyzed by Afm on a Mica/Air Interface. *Langmuir* **2018**, *34*, 15099–15108. [CrossRef]

25. Afonin, K.A.; Bindewald, E.; Yaghoubian, A.J.; Voss, N.; Jacovetty, E.; Shapiro, B.A.; Jaeger, L. In Vitro Assembly of Cubic RNA-Based Scaffolds Designed in Silico. *Nat. Nanotechnol.* **2010**, *5*, 676–682. [CrossRef] [PubMed]

26. Davis, J.H.; Tonelli, M.; Scott, L.G.; Jaeger, L.; Williamson, J.R.; Butcher, S.E. RNA Helical Packing in Solution: Nmr Structure of a 30 Kda Gaaa Tetraloop-Receptor Complex. *J. Mol. Biol.* **2005**, *351*, 371–382. [CrossRef]

27. Dibrov, S.M.; McLean, J.; Parsons, J.; Hermann, T. Self-Assembling RNA Square. *Proc. Natl. Acad. Sci. USA* **2011**, *108*, 6405–6408. [CrossRef]

28. Manaia, E.B.; Abuçafy, M.P.; Chiari-Andréo, B.G.; Silva, B.L.; Junior JA, O.; Chiavacci, L.A. Physicochemical Characterization of Drug Nanocarriers. *Int. J. Nanomed.* **2017**, *12*, 4991–5011. [CrossRef] [PubMed]

29. Xavier, P.L.; Chandrasekaran, A.R. DNA-Based Construction at the Nanoscale: Emerging Trends and Applications. *Nanotechnology* **2018**, *29*, 062001. [CrossRef]

30. Ukraintsev, V.; Banke, B. Review of Reference Metrology for Nanotechnology: Significance, Challenges, and Solutions. *J. Micro Nanolithogr. MEMS MOEMS* **2012**, *11*, 011010. [CrossRef]

31. Bald, I.; Keller, A. Molecular Processes Studied at a Single-Molecule Level Using DNA Origami Nanostructures and Atomic Force Microscopy. *Molecular* **2014**, *19*, 13803–13823. [CrossRef]

32. Song, Y.; Chen, S. Janus Nanoparticles: Preparation, Characterization, and Applications. *Chem. Asian J.* **2014**, *9*, 418–430. [CrossRef]

33. Svergun, D.I.; Koch, M.H. Small-Angle Scattering Studies of Biological Macromolecules in Solution. *Rep. Prog. Phys.* **2013**, *66*, 1735–1782. [CrossRef]

34. Jacques, D.A.; Trewhella, J. Small-Angle Scattering for Structural Biology-Expanding the Frontier While Avoiding the Pitfalls. *Protein Sci.* **2010**, *19*, 642–657. [CrossRef]

35. Vestergaard, B.; Sayers, Z. Investigating Increasingly Complex Macromolecular Systems with Small-Angle X-Ray Scattering. *IUCrJ* **2014**, *1*, 523–529. [CrossRef]

36. Heller, W.T. Small-Angle Neutron Scattering and Contrast Variation: A Powerful Combination for Studying Biological Structures. *Acta Crystallogr. Sect. D Biol. Crystallogr.* **2010**, *66*, 1213–1217. [CrossRef]

37. Heller, W.T.; Littrell, K.C. Small-Angle Neutron Scattering for Molecular Biology: Basics and Instrumentation. *Micro Nano Technol. Bioanal. Methods Protoc.* **2009**, *544*, 293–305.

38. Hong, E.; Halman, J.R.; Shah, A.; Cedrone, E.; Truong, N.; Afonin, K.A.; Dobrovolskaia, M.A. Toll-Like Receptor-Mediated Recognition of Nucleic Acid Nanoparticles (Nanps) in Human Primary Blood Cells. *Molecules* **2019**, *24*, 1094. [CrossRef]

39. Chandler, M.; Afonin, K.A. Smart-Responsive Nucleic Acid Nanoparticles (Nanps) with the Potential to Modulate Immune Behavior. *Nanomaterials* **2019**, *9*, 611. [CrossRef]

40. Rychahou, P.; Haque, F.; Shu, Y.; Zaytseva, Y.; Weiss, H.L.; Lee, E.Y.; Mustain, W.; Valentino, J.; Guo, P.; Evers, B.M. Delivery of RNA Nanoparticles into Colorectal Cancer Metastases Following Systemic Administration. *ACS Nano* **2015**, *9*, 1108–1116. [CrossRef] [PubMed]

41. Grabow, W.W.; Zakrevsky, P.; Afonin, K.A.; Chworos, A.; Shapiro, B.A.; Jaeger, L. Self-Assembling RNA Nanorings Based on RNAi/Ii Inverse Kissing Complexes. *Nano Lett.* **2011**, *11*, 878–887. [CrossRef]

42. Wu, P.W.; Yu, Y.; McGhee, C.E.; Tan, L.H.; Lu, Y. Applications of Synchrotron-Based Spectroscopic Techniques in Studying Nucleic Acids and Nucleic Acid-Functionalized Nanomaterials. *Adv. Mater.* **2014**, *26*, 7849–7872. [CrossRef] [PubMed]

43. Dervichian, D.G.; Fournet, G.; Guinier, A. X-Ray Scattering Study of the Modifications Which Certain Proteins Undergo. *Biochim. Biophys. Acta* **1952**, *8*, 145–149. [CrossRef]

44. Glatter, O.; Kratky, O. *Small Angle X-Ray Scattering*; Academic Press: London, UK; New York, NY, USA, 1982.

45. Franke, D.; Petoukhov, M.V.; Konarev, P.V.; Panjkovich, A.; Tuukkanen, A.; Mertens, H.D.T.; Kikhney, A.G.; Hajizadeh, N.R.; Franklin, J.M.; Jeffries, C.M.; et al. Atsas 2.8: A Comprehensive Data Analysis Suite for Small-Angle Scattering from Macromolecular Solutions. *J. Appl. Crystallogr.* **2017**, *50*, 1212–1225. [CrossRef]

46. Fang, X.; Wang, J.; O'Carroll, I.P.; Mitchell, M.; Zuo, X.; Wang, Y.; Yu, P.; Liu, Y.; Rausch, J.W.; Dyba, M.A.; et al. An Unusual Topological Structure of the Hiv-1 Rev Response Element. *Cell* **2013**, *155*, 594–605. [CrossRef]

47. Hennig, J.; Wang, I.; Sonntag, M.; Gabel, F.; Sattler, M. Combining Nmr and Small Angle X-Ray and Neutron Scattering in the Structural Analysis of a Ternary Protein-RNA Complex. *J. Biomol. NMR* **2013**, *56*, 17–30. [CrossRef] [PubMed]

48. Rambo, R.P.; Tainer, J.A. Bridging the Solution Divide: Comprehensive Structural Analyses of Dynamic RNA, DNA, and Protein Assemblies by Small-Angle X-Ray Scattering. *Curr. Opin. Struct. Biol.* **2010**, *20*, 128–137. [CrossRef]

49. Rambo, R.P.; Tainer, J.A. Improving Small-Angle X-Ray Scattering Data for Structural Analyses of the RNA World. *RNA* **2010**, *16*, 638–646. [CrossRef] [PubMed]

50. Etoukhov, M.V.; Konarev, P.V.; Kikhney, A.G.; Svergun, D.I. Atsas 2.1—Towards Automated and Web-Supported Small-Angle Scattering Data Analysis. *J. Appl. Crystallogr.* **2007**, *40*, S223–S228. [CrossRef]

51. Svergun, D.I. Restoring low resolution structure of biological macromolecules from solution scattering using simulated annealing. *Biophys. J.* **1999**, *76*, 2879–2886. [CrossRef]

52. Yang, S. Methods for Saxs-Based Structure Determination of Biomolecular Complexes. *Adv. Mater.* **2014**, *26*, 7902–7910. [CrossRef] [PubMed]

53. Maguire, C.M.; Rosslein, M.; Wick, P.; Prina-Mello, A. Characterisation of Particles in Solution—A Perspective on Light Scattering and Comparative Technologies. *Sci. Technol. Adv. Mater.* **2018**, *19*, 732–745. [CrossRef]

54. Trewhella, J. Small-Angle Scattering and 3d Structure Interpretation. *Curr. Opin. Struct. Biol.* **2016**, *40*, 1–7. [CrossRef] [PubMed]

55. Petoukhov, M.V.; Svergun, D.I. Ambiguity Assessment of Small-Angle Scattering Curves from Monodisperse Systems. *Acta Crystallogr. Sect. D* **2015**, *71*, 1051–1058. [CrossRef]

56. Mahieu, E.; Gabel, F. Biological Small-Angle Neutron Scattering: Recent Results and Development. *Acta Crystallogr. D Struct. Biol.* **2018**, *74*, 715–726. [CrossRef] [PubMed]

57. Ramakrishnan, V.; Capel, M.; Kjeldgaard, M.; Engelman, D.M.; Moore, P.B. Positions of Protein-S14, Protein-S18 and Protein-S20 in the 30-S Ribosomal-Subunit of Escherichia-Coli. *J. Mol. Biol.* **1984**, *174*, 265–284. [CrossRef]

58. Ramakrishnan, V.R.; Yabuki, S.; Sillers, I.Y.; Schindler, D.G.; Engelman, D.M.; Moore, P.B. Positions of Proteins S6, S11 and S15 in the 30-S Ribosomal-Subunit of Escherichia-Coli. *J. Mol. Biol.* **1981**, *153*, 739–760. [CrossRef]

59. Ramakrishnan, V.; Engelman, D.M.; Moore, P.B. 3-Dimensional Localization of 12 Proteins of the 30s Ribosome of Escherichia-Coli. *Fed. Proc.* **1980**, *39*, 2122.

60. Capel, M.S.; Engelman, D.M.; Freeborn, B.R.; Kjeldgaard, M.; Langer, J.A.; Ramakrishnan, V.; Schindler, D.G.; Schneider, D.K.; Schoenborn, B.P.; Sillers, I.Y.; et al. A Complete Mapping of the Proteins in the Small Ribosomal-Subunit of Escherichia-Coli. *Science* **1987**, *238*, 1403–1406. [CrossRef]

61. Svergun, D.I.; Koch, M.H.J.; Pedersen, J.S.; Serdyuk, I.N. Structural Model of the 50-S Subunit of Escherichia-Coli Ribosomes from Solution Scattering.2. Neutron-Scattering Study. *J. Mol. Biol.* **1994**, *240*, 78–86. [CrossRef]

62. Svergun, D.I.; Pedersen, J.S. Propagating Errors in Small-Angle Scattering Data Treatment. *J. Appl. Crystallogr.* **1994**, *27*, 241–248. [CrossRef]

63. Svergun, D.I.; Pedersen, J.S.; Serdyuk, I.N.; Koch, M.H.J. Solution Scattering from 50s Ribosomal-Subunit Resolves Inconsistency between Electron-Microscopic Models. *Proc. Natl. Acad. Sci. USA* **1994**, *91*, 11826–11830. [CrossRef]

64. Yang, W.; Lu, J.; Gilbert, E.P.; Knott, R.; He, L.; Cheng, W. Probing Soft Corona Structures of DNA-Capped Nanoparticles by Small Angle Neutron Scattering. *J. Phys. Chem. C* **2015**, *119*, 18773–18778. [CrossRef]

65. Zuo, X.B.; Wang, J.B.; Yu, P.; Eyler, D.; Xu, H.; Starich, M.R.; Tiede, D.M.; Simon, A.E.; Kasprzak, W.; Schwieters, C.D.; et al. Solution Structure of the Cap-Independent Translational Enhancer and Ribosome-Binding Element in the 3'Utr of Turnip Crinkle Virus. *Proc. Natl. Acad. Sci. USA* **2010**, *107*, 1385–1390. [CrossRef]

66. Wang, J.B.; Zuo, X.B.; Yu, P.; Xu, H.; Starich, M.R.; Tiede, D.M.; Shapiro, B.A.; Schwieters, C.D.; Wang, Y.X. A Method for Helical RNA Global Structure Determination in Solution Using Small-Angle X-Ray Scattering and Nmr Measurements. *J. Mol. Biol.* **2009**, *393*, 717–734. [CrossRef]

67. Sosnick, T.; Pan, T.; Fang, X.W.; Shelton, V.; Thiyagarajan, P.; Littrel, K. Metal Ions and the Thermodynamics and Kinetics of Tertiary RNA Folding. *Indian J. Chem. Sect. A Inorg. Bio Inorg. Phys. Theor. Anal. Chem.* **2002**, *41*, 54–64.

68. Fang, X.W.; Littrell, K.; Yang, X.; Henderson, S.J.; Siefert, S.; Thiyagarajan, P.; Pan, T.; Sosnick, T.R. Mg2+-Dependent Compaction and Folding of Yeast Trna(Phe) and the Catalytic Domain of the B-Subtilis Rnase P RNA Determined by Small-Angle X-Ray Scattering. *Biochemistry* **2000**, *39*, 11107–11113. [CrossRef]

69. Serganov, A.; Nudler, E. A Decade of Riboswitches. *Cell* **2013**, *152*, 17–24. [CrossRef]

70. Stoddard, C.D.; Montange, R.K.; Hennelly, S.P.; Rambo, R.P.; Sanbonmatsu, K.Y.; Batey, R.T. Free State Conformational Sampling of the Sam-I Riboswitch Aptamer Domain. *Structure* **2010**, *18*, 787–797. [CrossRef]

71. Fernandez-Castanon, J.; Bomboi, F.; Rovigatti, L.; Zanatta, M.; Paciaroni, A.; Comez, L.; Porcar, L.; Jafta, C.J.; Fadda, G.C.; Bellini, T.; et al. Small-Angle Neutron Scattering and Molecular Dynamics Structural Study of Gelling DNA Nanostars. *J. Chem. Phys.* **2016**, *145*, 084910. [CrossRef] [PubMed]

72. Bruetzel, L.K.; Gerling, T.; Sedlak, S.M.; Walker, P.U.; Zheng, W.; Dietz, H.; Lipfert, J. Conformational Changes and Flexibility of DNA Devices Observed by Small-Angle X-Ray Scattering. *Nano Lett.* **2016**, *16*, 4871–4879. [CrossRef]

73. Svergun, D.; Barberato, C.; Koch, M.H.J. Crysol—A Program to Evaluate X-Ray Solution Scattering of Biological Macromolecules from Atomic Coordinates. *J. Appl. Crystallogr.* **1995**, *28*, 768–773. [CrossRef]

74. Bruetzel, L.K.; Walker, P.U.; Gerling, T.; Dietz, H.; Lipfert, J. Time-Resolved Small-Angle X-Ray Scattering Reveals Millisecond Transitions of a DNA Origami Switch. *Nano Lett.* **2018**, *18*, 2672–2676. [CrossRef]

75. Afonin, K.A.; Kireeva, M.; Grabow, W.W.; Kashlev, M.; Jaeger, L.; Shapiro, B.A. Co-Transcriptional Assembly of Chemically Modified RNA Nanoparticles Functionalized with Sirnas. *Nano Lett.* **2012**, *12*, 5192–5195. [CrossRef]

76. Afonin, K.A.; Viard, M.; Tedbury, P.; Bindewald, E.; Parlea, L.; Howington, M.; Valdman, M.; Johns-Boehme, A.; Brainerd, C.; Freed, E.O.; et al. The Use of Minimal RNA Toeholds to Trigger the Activation of Multiple Functionalities. *Nano Lett.* **2016**, *16*, 1746–1753. [CrossRef]

77. Yingling, Y.G.; Shapiro, B.A. Computational Design of an RNA Hexagonal Nanoring and an RNA Nanotube. *Nano Lett.* **2007**, *7*, 2328–2334. [CrossRef] [PubMed]

78. Ke, W.; Hong, E.; Saito, R.F.; Rangel, M.C.; Wang, J.; Viard, M.; Richardson, M.; Khisamutdinov, E.F.; Panigaj, M.; Dokholyan, N.V.; et al. RNA-DNA Fibers and Polygons with Controlled Immunorecognition Activate Rnai, Fret and Transcriptional Regulation of Nf-Kappab in Human Cells. *Nucleic Acids Res.* **2018**, *47*, 1350–1361. [CrossRef] [PubMed]

79. Rackley, L.; Stewart, J.M.; Salotti, J.; Krokhotin, A.; Shah, A.; Halman, J.; Juneja, R.; Smollett, J.; Roark, B.; Viard, M.; et al. RNA Fibers as Optimized Nanoscaffolds for Sirna Coordination and Reduced Immunological Recognition. *Adv. Funct. Mater.* **2018**, *28*, 1805959. [CrossRef]

80. Yourston, L.E.; Lushnikov, A.Y.; Shevchenko, O.A.; Afonin, K.A.; Krasnoslobodtsev, A.V. First Step Towards Larger DNA-Based Assemblies of Fluorescent Silver Nanoclusters: Template Design and Detailed Characterization of Optical Properties. *Nanomaterials* **2019**, *9*, 613. [CrossRef]

81. Cruz-Acuna, M.; Halman, J.R.; Afonin, K.A.; Dobson, J.; Rinaldi, C. Magnetic Nanoparticles Loaded with Functional RNA Nanoparticles. *Nanoscale* **2018**, *10*, 17761–17770. [CrossRef]

82. Gupta, K.; Mattingly, S.J.; Knipp, R.J.; Afonin, K.A.; Viard, M.; Bergman, J.T.; Stepler, M.; Nantz, M.H.; Puri, A.; Shapiro, B.A. Oxime Ether Lipids Containing Hydroxylated Head Groups Are More Superior Sirna Delivery Agents Than Their Nonhydroxylated Counterparts. *Nanomedicine* **2015**, *10*, 2805–2818. [CrossRef]

83. Juneja, R.; Lyles, Z.; Vadarevu, H.; Afonin, K.A.; Vivero-Escoto, J.L. Multimodal Polysilsesquioxane Nanoparticles for Combinatorial Therapy and Gene Delivery in Triple-Negative Breast Cancer. *ACS Appl. Mater. Interfaces* **2019**, *11*, 12308–12320. [CrossRef] [PubMed]

84. Kim, T.; Afonin, K.A.; Viard, M.; Koyfman, A.Y.; Sparks, S.; Heldman, E.; Grinberg, S.; Linder, C.; Blumenthal, R.P.; Shapiro, B.A. In Silico, in Vitro, and in Vivo Studies Indicate the Potential Use of Bolaamphiphiles for Therapeutic Sirnas Delivery. *Mol. Ther. Nucleic Acids* **2013**, *2*, e80. [CrossRef] [PubMed]

85. Gupta, K.; Afonin, K.A.; Viard, M.; Herrero, V.; Kasprzak, W.; Kagiampakis, I.; Kim, T.; Koyfman, A.Y.; Puri, A.; Stepler, M.; et al. Bolaamphiphiles as Carriers for Sirna Delivery: From Chemical Syntheses to Practical Applications. *J. Control. Release* **2015**, *213*, 142–151. [CrossRef] [PubMed]

86. Dakwar, G.R.; Hammad, I.A.; Popov, M.; Linder, C.; Grinberg, S.; Heldman, E.; Stepensky, D. Delivery of Proteins to the Brain by Bolaamphiphilic Nano-Sized Vesicles. *J. Control. Release* **2012**, *160*, 315–321. [CrossRef] [PubMed]

87. Popov, M.; Grinberg, S.; Linder, C.; Waner, T.; Levi-Hevroni, B.; Deckelbaum, R.J.; Heldman, E. Site-Directed Decapsulation of Bolaamphiphilic Vesicles with Enzymatic Cleavable Surface Groups. *J. Control. Release* **2012**, *160*, 306–314. [CrossRef]

88. Rudorf, S.; Radler, J.O. Self-Assembly of Stable Monomolecular Nucleic Acid Lipid Particles with a Size of 30 Nm. *J. Am. Chem. Soc.* **2012**, *134*, 11652–11658. [CrossRef]

nanomaterials

MDPI

Article

Structural and Functional Stability of DNA Nanopores in Biological Media

Jonathan R. Burns [1],* and Stefan Howorka [1,2,*

[1] Department of Chemistry, Institute of Structural Molecular Biology, University College London, London WC1H 0AJ, UK
[2] Institute of Biophysics, Johannes Kepler University, A-4020 Linz, Austria
* Correspondence: jonathan.burns@ucl.ac.uk (J.R.B.); s.howorka@ucl.ac.uk (S.H.)

Received: 12 March 2019; Accepted: 23 March 2019; Published: 29 March 2019

Abstract: DNA nanopores offer a unique nano-scale foothold at the membrane interface that can help advance the life sciences as biophysical research tools or gate-keepers for drug delivery. Biological applications require sufficient physiological stability and membrane activity for viable biological action. In this report, we determine essential parameters for efficient nanopore folding and membrane binding in biocompatible cell media. The parameters are identified for an archetypal DNA nanopore composed of six interwoven strands carrying cholesterol lipid anchors. Using gel electrophoresis and fluorescence spectroscopy, the nanostructures are found to assemble efficiently in cell media, such as LB and DMEM, and remain structurally stable at physiological temperatures. Furthermore, the pores' oligomerization state is monitored using fluorescence spectroscopy and confocal microscopy. The pores remain predominately water-soluble over 24 h in all buffer systems, and were able to bind to lipid vesicles after 24 h to confirm membrane activity. However, the addition of fetal bovine serum to DMEM causes a significant reduction in nanopore activity. Serum proteins complex rapidly to the pore, most likely via ionic interactions, to reduce the effective nanopore concentration in solution. Our findings outline crucial conditions for maintaining lipidated DNA nanodevices, structurally and functionally intact in cell media, and pave the way for biological studies in the future.

Keywords: DNA nanotechnology; nanopores; biological media; serum; stability; aggregation

1. Introduction

DNA nanotechnology excels at the bottom-up fabrication of engineered nanostructures. DNA duplexes can be manipulated into user-defined shapes by exploiting the base-pairing rules for duplex formation [1–3]. Discrete nanostructures can be assembled in two and three dimensions with sub-nanometer control using dedicated design software [4,5]. Chemical diversity and functionality can be incorporated into structures site-specifically using, for example, solid phase DNA synthesis [6,7], or non-specifically via intercalation [8,9], or electrostatic interactions [10,11]. This rapidly evolving field has transformed materials science with wide-ranging applications, including the generation of DNA origami devices for optical sensing [12], controlled single molecule synthesis using a lab-on-a-chip DNA board [13], computation devices [14,15], and finite sub-nm movement of DNA-based robots using DNA ligands [16].

DNA nanotechnology applied to the life sciences is gaining traction. DNA nanostructures can help control processes within cells, or at the membrane interface to advance biological understanding [17–20]. This progress includes the generation of novel diagnostic tools [21–23], the enhancement of existing drugs [24], and devices with novel therapeutic action [25]. Recently, intracellular DNA-based delivery vehicles have been used to transport biomolecules. Engineered DNA cages that encapsulate small molecule drugs [25], mRNA [26], peptides, and proteins [27,28] have been developed to deliver biomolecular cargo. DNA nanostructures can be internalized in specific mammalian cells, although the nature of the design appears to play an important role [29,30]. Coating the nanostructures in certain chemical groups can improve cellular uptake [10,31,32].

To fulfill desirable biomolecular functions, the DNA nanostructures have to be stable. DNA origami-based nanostructures have been studied previously in vitro and in vivo [33]. Generally, the origami constructs withstand diverse biology conditions under short time durations. Yan and colleagues have recently shown intact and functional DNA origami in the renal system of a mouse model [34]. However, other reports have identified significant degradation and unfolding of DNA origami structures in biological media [35,36]. This instability has been attributed to the low level of Mg^{2+} ions—essential to stabilize DNA origami nanostructures—and digestion from enzymes including DNAses. The susceptibility to degradation appears to be design-specific, with tubular designs proving more resilient [37]. Other strategies can be employed to help stabilize the nanostructures, including chemical ligation of DNA nicks [38], the introduction of non-native base pairs, such as LNA, PNA, and XNA [39]. Alternatively, cationic peptides [10], polymers [32], or intercalators [9] can be used to improve structural stability.

DNA-based nanopores are the most recent class of membrane channels which can potentially offer a unique degree of control at the membrane interface [40–44]. Naturally occurring nanopores are usually composed of proteins or peptides to help regulate ion transport across cell membranes [45]. However, it can be challenging to de novo design amino acid-based nanopores due to unexpected protein misfolding [46]. In contrast, utilizing DNA as a construction material can help overcome this issue. To date, DNA nanotechnology has produced nanopores with highly customizable properties, including channel diameter, length, functionalized groups within the lumen, and ligand-controlled pore opening [40–44]. For future biological applications, including pore-mediated drug delivery, nanopore stability and solubility within biological media must be maintained.

To investigate DNA nanopore stability, this study employs the DNA nanopore NP-3C (Figure 1a) [42]. The pore is assembled from six single strands (Tables S1 and S2, Figures S1 and S2), which form six interwoven DNA duplexes to generate a six helical barrel. Three cholesterol lipid anchors are site-specifically incorporated to the exterior of the bundle to facilitate membrane binding and nanopore behavior. The assembled pore punctures the membrane to generate a toroidal pore to enable ion transport across the lipid bilayer (Figure 1b) [44,47,48]. However, the hydrophobic lipid anchors can also mediate other undesired behavior, including intermolecular oligomerization (Figure 1c) [49,50]. To help distinguish the lipid anchor effect, a cholesterol-free version, NP-0C, was assembled to serve as a negative control (Tables S1 and S2, Figure S1).

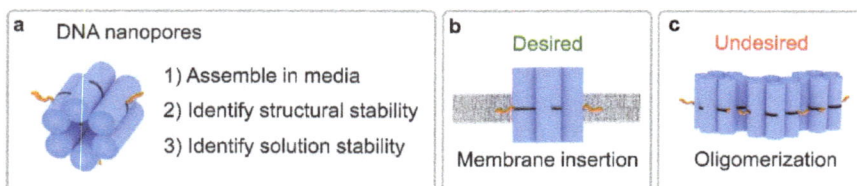

Figure 1. Identifying the formation, structural and solution-phase stability of amphiphilic DNA nanopores in biological media. (**a**) Depiction of six helical bundle nanopore (**blue cylinders**) containing cholesterol lipid anchors (**orange**) and the parameters monitored within; (**b**) the desired monomer membrane binding action in vivo; (**c**) and the undesired hydrophobic lipid anchor-mediated oligomerization which can prevent membrane binding and reduce the active nanopore concentration.

Cell media is composed of complex ions and nutrients which help to maintain cell homeostasis and phenotype. For biological applications, the amphiphilic DNA nanostructures must remain structurally stable within the used medium. Therefore, this report assayed biologically compatible media to identify the pore's structural stability and membrane activity, including phosphate-buffered saline (PBS), bacterial growth medium lysogeny broth (LB), mammalian cell media Dulbecco's modified Eagle medium (DMEM), and DMEM supplemented with 10% v/v fetal bovine serum (FBS) (Table 1) [51,52]. Serum is required for specific cell types to maintain cell function, and is composed of a wide array of entities including proteins, hormones, and electrolytes. The total protein concentration in FBS is ~0.3–0.5 mg/mL [53]. Albumin, globulin, and fibrinogen make up the majority of proteins found in serum, at approximately 55, 38, and 7%, respectively. In addition, over a thousand other regulatory proteins exist at much smaller levels. Metal ions must also be considered. Positively charged metal ions coordinate with DNA ionically to stabilize duplexes. Therefore, a range of metal cations was assayed to identify the counterion stabilization on the nanostructures [33]. We tested monovalent sodium and potassium ions typically used for single channel current recordings used to study nanopores [54], and divalent magnesium ions, conventionally used for the stabilization of DNA origami constructs. The nanopore formation was determined using gel electrophoresis. To identify the thermal stability at physiological temperatures in biological media, the melting temperatures of the constructs were identified using fluorescence spectroscopy [55]. Further, our study identified the aggregation extent of the nanostructures using fluorescence spectroscopy and confocal microscopy over time. Finally, to confirm membrane activity of the nanopore, binding to model membranes was determined using fluorescence microscopy. With the knowledge gained using our approach, new pore formulations and folding protocols can be established which should help provide insights for future applications across the life sciences.

Table 1. Buffer solutions and biological media including their ionic strength.

Abbreviation	Na	K	Mg	PBS	LB	D	FBS
Salt/media	NaCl	KCl	MgCl$_2$ TAE	Phosphate buffered-saline	Lysogeny broth	Dulbecco's modified Eagle medium	D + 10% fetal bovine serum
Ionic strength	0.32 M	0.32 M	0.11 M	0.17 M	0.17 M	0.17 M	0.19 M

2. Materials and Methods

All reagents were purchased from Sigma Aldrich (UK) unless stated otherwise. The DNA nanopore was published previously (information on the sequences, including 2D maps and dimensions is provided in the Supporting Information) [42]. The DNA nanopores were assembled by mixing an equimolar mixture of the component DNA strands (0.5 μM, unless stated otherwise) (Integrated DNA Technologies, Coralville, IA, USA) containing the stated buffer or media. The nanopores were folded by heating the solution from 95 °C for 2 min, and cooling to 20 °C at a rate of 5 °C per min. The folded DNA nanopore constructs were stored at room temperature, and vortexed for 2 s before use. Where stated, *n*-octyl-oligo-oxyethylene (OPOE) (Enzo Life Sciences, Exeter, UK) was added to the folding mixture prior to nanopore assembly (1.5% v/v).

Buffer and reagents. Na: NaCl 300 mmol/L, tris 15 mmol/L, pH 8.0. K: KCl 300 mmol/L, tris 15 mmol/L, pH 8.0. Mg: MgCl$_2$ 14 mmol/L, tris 40 mmol/L, acetic acid 20 mmol/L, ethylenediaminetetraacetic 1 mmol/L, pH 8.3. PBS: NaCl 137 mmol/L, KCl 2.7 mmol/L, Na$_2$HPO$_4$ 8 mmol/L, KH$_2$PO$_4$ 2 mmol/L, pH 7.4. LB: tryptone 10 mg/mL; yeast extract 5 mg/mL; NaCl 10 mg/mL values taken from [56]. D components include CaCl$_2$ 2.4 mmol/L, MgSO$_4$ 0.8 mmol/L, KCl 5.4 mmol/L, NaHCO$_3$ 44.0 mmol/L, NaCl 109.5 mmol/L, NaH$_2$PO$_4$ 0.9 mmol/L. Neat fetal bovine serum components include bilirubin 2.4 mg/L; Cholesterol 340 mg/L; creatinine 27.3 mg/L; urea 260 mg/L; Na$^+$ 142 mmol/L; Cl$^-$ 155.5 mmol/L; K$^+$ 8 mmol/L; Ca^{2+} 3 mmol/L; Mg^{2+} 1.1 mmol/L; PO$_4^{3-}$ 2.3 mmol/L; Fe 1.6 mg/L; glucose 550 mg/L; protein 36 g/L; albumin 17 g/L; α-globulin 17 g/L; β-globulin 2 g/L; γ-globulin 1 g/L, values taken from [57].

DNA nanopore folding was characterized using 12% sodium dodecyl sulfate polyacrylamide gel electrophoresis (SDS-PAGE) (Bio-Rad, Watford, UK) with standardized buffers typically applied to proteins. The gel was thermally equilibrated at 8 °C prior to loading. The gel was run at 140 V for 70 min. The bands were visualized by first removing SDS with deionized water, then staining using ethidium bromide solution. A 100 base-pair DNA marker (New England Biolabs, Hitchin, UK) was used as a reference.

The Förster resonance energy transfer (FRET) characteristics of the fluorescein (FAM) and Cy3 labeled nanopore constructs were identified using a Varian Eclipse fluorescence spectrophotometer (Agilent, Stockport, UK). 20 μL of the various DNA nanostructures (folded in PBS at 1 μM) (see Tables S1 and S2 for strand information) was added to PBS (180 μL) in a quartz cuvette with a path length of 1 cm. The samples were analyzed by excitation at 495 nm, and the emission monitored between 505–700 nm. A 5 nm slit width and 600 PMT voltage was applied, along with a scanning rate of 600 nm per min; the scan was performed 3 times and averaged.

The melting transitions of the DNA nanostructures were identified using a MyIQ real-time PCR (Bio-Rad, Watford, UK). The nanostructures were assembled containing FAM and Cy3 FRET pairs (folded at 1 μM in PBS). The DNA constructs were diluted into the stated buffer systems to give a final DNA concentration of 0.1 μM (total volume of 25 μL) in a 96-well thin wall fluorescence plate (Bio-Rad, Watford, UK). Optical quality sealing tape (Bio-Rad, Watford, UK) was placed on top to prevent evaporation. The sample was heated from 30–85 °C at a rate of 0.5 °C per min. The melting temperature was determined from taking the derivative of the donor fluorescence profile. Errors were identified from 3 independent experiments.

Fluorescence spectroscopic analysis was performed on Cy3-modified DNA nanostructures using a Varian Eclipse fluorescence spectrophotometer (Agilent, Stockport, UK) with a fluorescence cuvette. The samples were analyzed by excitation at 540 nm, and by monitoring the emission from 550–600 nm, using a 10 nm slit width, 800 PMT voltage, scanning at 600 nm per min and taking the average of 3 repeat scans. The DNA nanostructures (2 μL, folded at 0.5 μM) in the stated buffers were scanned once the folding temperature reached 40 °C by diluting in the buffer systems (200 μL final volume). At the designated time points, the samples were centrifuged for 10 min at 16k revolutions per min at room temperature (Eppendorf, Stevenage, UK), and the supernatant was carefully extracted and the fluorescence monitored using the same dilution and settings as described.

Confocal laser scanning microscopy (CLSM) images were collected using a 60× oil objective FV-1000 Olympus microscope. Images were analyzed using ImageJ software. To image the DNA nanopore constructs, the folded pore containing a Cy3 dye (10 μL, 0.5 μM in PBS) was deposited on a fluorodish (World Precision Instruments, Sarasota, FL, USA), and left to settle for 20 min prior to imaging. For the vesicle-binding assays, 1-palmitoyl-2-oleoyl-sn-glycero-3-phosphocholine (POPC) giant unilamellar vesicles (GUVs) were prepared by modifying a published protocol [48,58]. POPC (150 μL, 10 mM) in chloroform was added to a 1 mL glass vial, the solvent was removed under vacuum, and underwent rotation using a rotary evaporator. The thin film generated was resuspended in mineral oil (150 μL) by vortexing and sonicating for 10 min. Green fluorescent protein (5 μL, 10 μM in PBS) was mixed with sucrose solution (20 μL, 400 mM), followed by addition of mineral oil (150 μL). The suspension was vortexed and sonicated for 10 min at room temperature, then carefully added to the top of a glucose solution (1 mL, 400 mM) in a plastic vial (1 mL). The vesicles were generated by centrifuging at 16K RPM at 4 °C for 10 min. The mineral oil top layer and the majority of the sucrose layer (~800 μL) were carefully removed. The remaining solution containing the pelleted vesicles was gently mixed with a pipettor, then transferred to a clean plastic vial. The POPC GUVs (5 μL) were added to a KCl solution (5 μL, 0.5 M), and then Cy3-labeled NP-3C (2 μL folded at 0.25 μM) in the stated buffers was mixed, and the solution deposited on the confocal slide and used within 24 h. For the serum time series assay, NP-3C (10 μL, 0.25 μM folded in PBS) was added to FBS (10 μL) for the stated time durations. The NP-3C-FBS solution (4 μL) was added to the GUV solution as described above. All images were collected after 20 min using identical settings.

3. Results

3.1. Determining Nanopore Formation in Media

The folding efficiency of the DNA nanopores in biological media was analyzed using SDS-PAGE (Figure 2). First, the formation of the folded barrel was confirmed by assembling NP-0C, as well as versions missing some of the component strands (Figure 2a). Combining all six strands yielded the slowest migrating band slightly above the 1517 base pair marker band, as indicated by the top arrow. Removal of a single strand from the folding mixture resulted in an increase in band mobility. The 5-component construct migrated towards the 500 base pair marker band, as indicated by the bottom arrow. The large shift in band migration between a fully and partially assembled barrel is consistent with the formation of a higher order tertiary nanostructure. However, the strand combinations 1–4 and 1–3 also gave rise to a band migrating aligned to the 500 bp marker band. This result indicates that the addition of strands 4 and 5 to the pooled mixture did not successfully incorporate within the assembled bundle. Comparing component strands 1–2 and 1 gave the expected step-wise change indicating successful assembly.

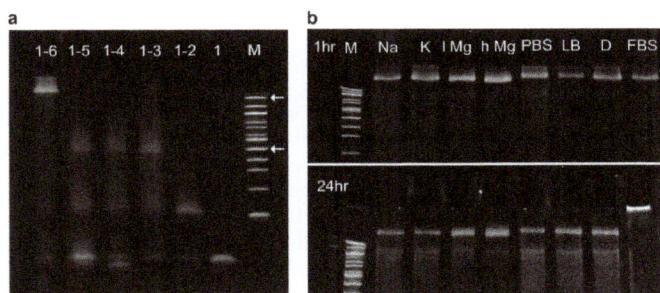

Figure 2. The DNA nanopores fold efficiently and remain structurally stable in a range of salts and buffer conditions as shown by gel electrophoresis. Sodium dodecyl sulfate polyacrylamide gel electrophoresis (SDS-PAGE) reveals (**a**) step-wise assembly of NP-0C, left to right, fully assembled barrel strands (**1–6**), followed by component strands (**1–5**), (**1–4**), (**1–3**), (**1–2**), (**1**), and 100 base pair marker (**M**), the arrows indicate the 1517 (**top**) and 500 base pair (**bottom**) marker bands; (**b**) 1 h (**top**) and 24 h (**bottom**) after folding of NP-0C assembled in a variety of conditions; 100 base pair marker (**M**), 0.3 M sodium chloride (**Na**), 0.3 M potassium chloride (**K**), 14 mM magnesium chloride (**l Mg**), 140 mM magnesium chloride (**h Mg**), phosphate buffered solution (**PBS**), Lysogeny Broth (**LB**), Dulbecco's modified Eagle medium (**D**), and Dulbecco's modified Eagle medium supplemented with 10% fetal bovine serum (**FBS**).

The assembled NP-0C construct folded efficiently in all biological buffers as assayed by SDS-PAGE (Figure 2b). This result confirmed stable DNA nanostructure formation in diverse media over short time durations, even in the presence of low salt conditions. After 24 h the gel was repeated, all bands showed very similar behavior—except for the media containing FBS—indicating that the pores are generally stable under these varied conditions. Adding protein-containing FBS led to the formation of protein-DNA complexes that did not migrate into the gel. The surfactant in the gel buffer, sodium dodecyl sulfate (SDS), was not able to disrupt the protein binding to the DNA nanostructure.

3.2. Identifying Nanopore Melting Temperatures in Biological Media

The thermal stability of the pores was established using DNA nanostructures labeled with a fluorophore pair for Förster resonance energy transfer (FRET). FAM (fluorescein) (donor) and Cy3 (acceptor) FRET pairs [59] were incorporated into the nanostructures on strands 2 and 6, respectively (see the Supporting Information for details for the DNA strands) (Figure 3a). Successful FRET was confirmed by scanning the donor emission in the absence and presence of the acceptor in assembled NP-3C (Figure 3b). Next, the donor emission was monitored upon heating (Figure 3c). The donor profile gave rise to a sigmoidal curve and its derivative yielded the melting transition. All constructs in the media types displayed melting transitions significantly above physiological temperatures (Table 2). Monovalent sodium and potassium gave very similar melting transitions for NP-3C, at 51.3 °C and 52.2 °C, respectively. However, divalent magnesium gave rise to a 1.6 °C enhancement, even though the counterion concentration was significantly lower. Biologically compatible PBS reduced the structural stability by 4 °C, possibly due to the lower concentration of monovalent sodium (137 mM). However, the overall high thermal stability of all the nanostructures in all the media conditions confirmed their suitability for biological applications from a structural perspective.

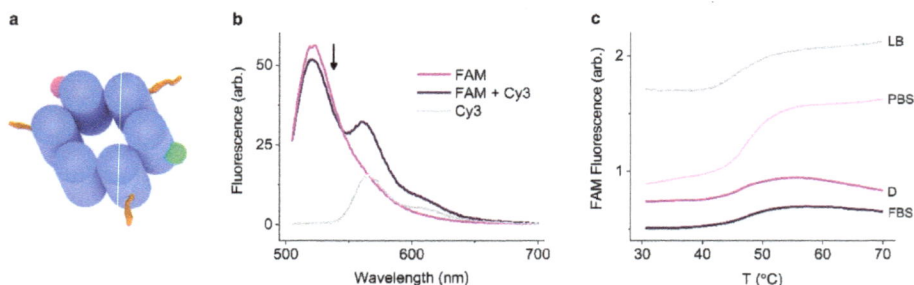

Figure 3. Determining the thermal stability of DNA nanopores in biological media using the Förster resonance energy transfer (FRET) pair labeled DNA nanopores. (**a**) Representation of fluorescein (FAM) (**purple**) and Cy3 (**green**) fluorophores incorporated into the DNA nanopore constructs; (**b**) fluorescence emission spectra of FAM (donor) and Cy3 (acceptor) labeled DNA nanopores, excitation at 495 nm, the donor emission is decreased in the presence of the acceptor in the assemble pore construct; (**c**) fluorescence donor emission thermal melting profiles of NP-3C in the stated buffers. The different fluorescence intensities reflect how the applied buffer system influence fluorophore emission.

Table 2. Melting temperatures of DNA nanopore constructs in stated salt and media systems. Errors were identified from three independent experiments.

Construct	Na	K	Mg	PBS	LB	D	FBS
NP-0C	49.7 ± 0.3	50.9 ± 1.2	52.7 ± 0.3	46.4 ± 0.9	46.6 ± 0.3	45.8 ± 0.3	45.7 ± 0.3
NP-3C	51.3 ± 0.6	52.2 ± 0.8	53.8 ± 1.5	46.7 ± 0.3	45.0 ± 1.0	46.8 ± 0.3	47.2 ± 0.3

3.3. Identifying Time-Dependent Nanopore Water-Solubility

We tested the water solubility of constructs in the different media conditions. Centrifugation was used to pellet and separate any large NP-3C clusters from smaller water-soluble fractions. The fluorescence in the supernatant was quantified using a fluorometer (Figure 4a), and fluorescence in the pellet using confocal laser scanning microscopy (CLSM) (Figure 4b). The cholesterol-free construct remained predominantly water-soluble over the course of 48 h in all assayed media. The cholesterol labeled nanopore, NP-3C, showed some aggregation and pelleting after 24 h. However, the majority of the oligomerized and monomeric form remained water soluble (>75%) even after 48 h. The protein-DNA complexes generated by FBS resulted in a noticeable increase in the pelleting fraction; however, the majority of complexed NP-3C in FBS remained water-soluble. By comparison, NP-0C showed no aggregation either in the supernatant, or CLSM images after 48 h across all conditions. This result confirms the cholesterol lipid anchors were responsible for the detectable pelleting observed.

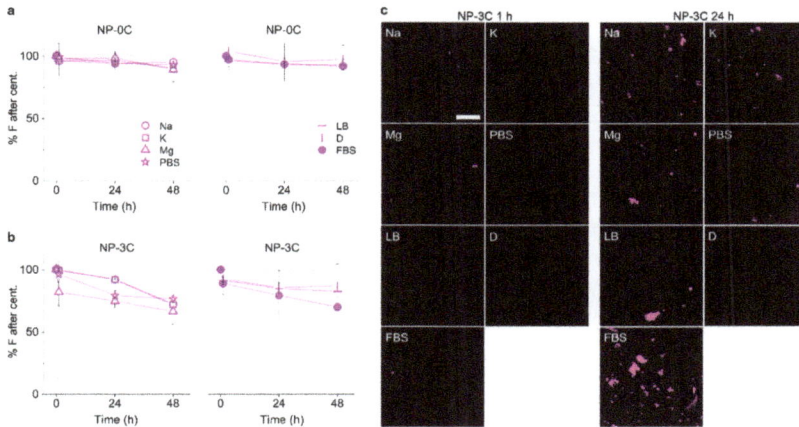

Figure 4. Identifying aggregation of DNA nanopores over time. Relative fluorescence intensities of Cy3-labeled (**a**) NP-0C and (**b**) NP-3C remaining in solution after centrifugation at 1, 24, and 48 h, in the stated buffers; (**c**) CLSM images of Cy3-labeled NP-3C in the stated buffers at 1 and 24 h. Scale bar 50 μm.

3.4. Identifying Nanopore Membrane Binding Activity in Media

We tested the ability of DNA nanopores to bind to vesicles in media. The DNA nanopores binding towards giant unilamellar vesicles (GUVs) was identified using CLSM (Figure 5). Green fluorescence protein (GFP) was encapsulated inside the GUVs to aid visualization. Cy3-labeled NP-3C in PBS, LB, or DMEM was diluted into the GUV solution to minimize the effect different salts and buffers have on dye fluorescence. Extensive membrane binding to GUVs was observed for these combinations, as shown by intense membrane halos around the vesicles' perimeter. In contrast, NP-3C in serum-containing media blocked the nanopores' binding event significantly. In agreement with SDS-PAGE analysis described above, FBS proteins complexed to the pore, via the generation of a higher molecular weight complex which prevents cholesterol-mediated binding. However, this problem can be circumvented by minimizing the pores exposure to FBS prior to vesicle addition. Adding the pores to FBS for 20 min, or directly from PBS, followed by addition to GUVs resulted in significant membrane binding, comparable to the other media conditions. This is an important finding for future lipidated DNA nanodevices employed in serum-containing media.

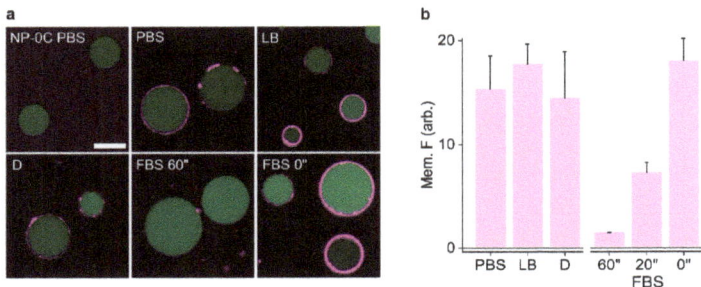

Figure 5. Confocal microscopy shows NP-3C binding to GUVs in biological buffers while FBS complexes to NP-3C to block membrane tethering. (**a**) Cy3-labeled NP-3C (**magenta**) mixed in the stated buffers for 1 h, and NP-3C added to FBS for different time periods, then added to GUVs containing GFP (**green**), image collected after 20mins of pore-vesicle incubation on a glass slide; (**b**) relative membrane fluorescence intensities from (**a**). All images collected under identical conditions. Error bars represent the averages of three experiments. Scale bar 10 μm.

3.5. Adding Detergent Prevents Nanopore Aggregation

We employed the non-ionic surfactant *n*-octyl-oligo-oxyethylene (OPOE) to generate a DNA nanopore-detergent complex to improve solubility and prevent aggregation. Naturally occurring membrane proteins are amphiphilic and can aggregate due to poor aqueous solubility. To help overcome this issue surfactants are routinely utilized to extract, solubilize, and stabilize membrane proteins [60]. OPOE was previously shown to aid DNA nanopore insertion into membranes during single channel current recordings [42]. We tested whether the surfactant can reverse pre-aggregation of nanopores using the centrifugation assay described above. The surfactant was added in high concentrations above the critical micelle concentration (CMC). OPOE was added after incubating the pore 48 h in FBS. CLSM revealed NP-3C aggregates after the addition of OPOE (Figure 6a) suggesting that the surfactant was not able to disrupt the pre-formed NP-3C aggregate. However, when the detergent was added to the pooled DNA prior to assembly, no aggregates were observed either by confocal microscopy, or a decrease in the supernatant fluorescence after 48 h (Figure 6a,b). These results confirm that the mild surfactant was able to solubilize the pores and prevent higher-order assembly for long durations. These results indicate that future folding protocols should include detergent within the folding mixture, prior to, or shortly after nanopore folding to help prevent lipidated DNA nanostructure aggregation.

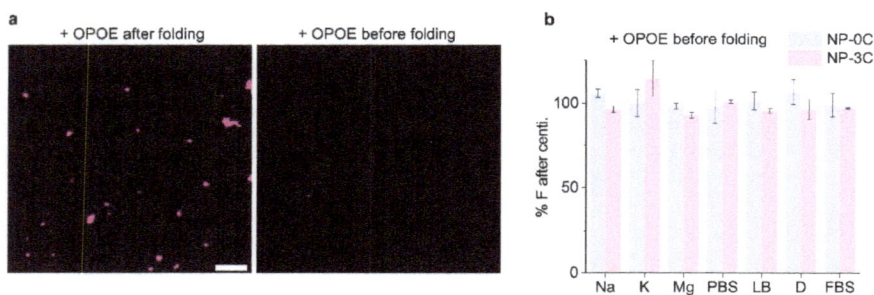

Figure 6. Adding surfactant prevents nanopore aggregation and precipitation. (**a**) CLSM images of Cy3-labeled NP-3C (**magenta**) in FBS 48 h after folding, with OPOE added at 48 h (**left**), or prior to folding (**right**), all images collected under identical conditions, scale bar 50 µm; (**b**) fluorescence analysis of the supernatant fraction after centrifugation of NP-0C (**blue**) and NP-3C (**magenta**) folded in the presence of OPOE. Error bars represent the averages of three independent experiments.

4. Discussion

DNA-based nanopores are a recent and exciting class of synthetic membrane channel. This construction approach provides a unique level of biophysical control across lipid bilayers. For DNA nanopores to provide functionality in the life sciences, for example, as biosensors, drug delivery vehicles, or cytotoxic-inducing agents, biocompatibility needs to be addressed. Essential parameters before in vitro and in vivo testing include confirming structural stability and solution-phase solubility in biologically compatible media. We have employed an archetypal DNA nanopore and tested its folding capabilities in diverse biological media routinely used to culture mammalian or bacterial cells. The conditions were deliberately chosen to be stringent as folding was performed in the cell media. It would also have been possible to fold pores in protein-free buffers and then add them to the protein-containing media. This differential treatment would have, however, made a fair comparison across all buffers and media more difficult. Our results suggest that pores folded efficiently and remained structurally stable in all media assayed. Importantly, the pore constructs displayed melting temperatures above physiological temperatures, even in the absence of divalent magnesium ions. With the exception of FBS-containing media, the DNA nanopores remained predominately water soluble in all conditions tested over 48 h. These results reinforce the suitability of DNA nanotechnology as a good building material for use in biological studies.

Serum proteins cause aggregation most likely due to complexation towards the DNA's negatively charged phosphate backbone. Albumin serum proteins have been shown to complex to antisense oligonucleotides, which in some instances enhances the half-life of intravenously injected DNA [61,62]. However, hydrophobic groups can further increase the protein complexation extent [63]. In the case of amphiphilic DNA nanopores used in this study, the resultant complexes caused significant aggregation, drastically reducing the amount of membrane tethering action. This result is an important finding in the field of lipidated DNA nanostructures, and if the use of FBS is unavoidable, the DNA nanopores should be applied for short time periods of less than 1 h, or the cells transferred temporarily into other media, such as phosphate buffered saline. Alternatively, vesicle delivery agents may be employed to shield the nanostructures from aggregation. Other strategies include using DNA masking groups, such as coating the structures in polyethylene glycol, or carboxylic acids, both are known to improve the circulation time of biomolecules in the bloodstream. Cationic groups such as lysine and arginine-rich peptides [32], polyamines [64], and metal cations can be used to block the serum proteins from binding. In addition, charge-neutral peptide nucleic acids nanopore equivalents [39] can be developed to prevent serum complexation. We expect lipidated DNA nanodevices and DNA nanopores to provide a useful foothold cell biology, and should find use as drug delivery gate-keepers, to function as drug molecules, such as immunosuppressants or immunoactivators, or act as novel tools in diagnostics and sensing.

Supplementary Materials: The following are available online at http://www.mdpi.com/2079-4991/9/4/490/s1, Figure S1: 2D maps and Figure S2: graphical representation of constructs, Table S1: DNA sequences and Table S2: DNA strand combinations.

Author Contributions: Conceptualization, J.R.B., S.H.; methodology and data acquisition J.R.B., manuscript written by J.R.B. with input from S.H.

Funding: This work was funded by the EPSRC (EP/N009282/1), the BBSRC (BB/M025373/1 and BB/N017331/1), the Leverhulme Trust (RPG-2017-015), and the Austrian Science Fund (FWF) (P 30368-N28).

Conflicts of Interest: The authors declare no conflict of interest.

References

1. Rothemund, P.W. Folding DNA to create nanoscale shapes and patterns. *Nature* **2006**, *440*, 297–302. [CrossRef]
2. Dietz, H.; Douglas, S.M.; Shih, W.M. Folding DNA into twisted and curved nanoscale shapes. *Science* **2009**, *325*, 725–730. [CrossRef]
3. Seeman, N.C.; Sleiman, H.F. DNA nanotechnology. *Nat. Rev. Mater.* **2017**, *3*, 17068. [CrossRef]
4. Douglas, S.M.; Marblestone, A.H.; Teerapittayanon, S.; Vazquez, A.; Church, G.M.; Shih, W.M. Rapid prototyping of 3D DNA-origami shapes with caDNAno. *Nucleic Acids Res.* **2009**, *37*, 5001–5006. [CrossRef]
5. Benson, E.; Mohammed, A.; Gardell, J.; Masich, S.; Czeizler, E.; Orponen, P.; Hogberg, B. DNA rendering of polyhedral meshes at the nanoscale. *Nature* **2015**, *523*, 441–444. [CrossRef] [PubMed]
6. Verma, S.; Eckstein, F. Modified oligonucleotides: Synthesis and strategy for users. *Ann. Rev. Biochem.* **1998**, *67*, 99–134. [CrossRef] [PubMed]
7. Edwardson, T.G.W.; Carneiro, K.M.M.; Serpell, C.J.; Sleiman, H.F. An efficient and modular route to sequence-defined polymers appended to DNA. *Angew. Chem. Int. Ed.* **2014**, *53*, 4567–4571. [CrossRef]
8. Zhang, Q.; Jiang, Q.; Li, N.; Dai, L.; Liu, Q.; Song, L.; Wang, J.; Li, Y.; Tian, J.; Ding, B.; et al. DNA origami as an in vivo drug delivery vehicle for cancer therapy. *ACS Nano* **2014**, *8*, 6633–6643. [CrossRef] [PubMed]
9. Halley, P.D.; Lucas, C.R.; McWilliams, E.M.; Webber, M.J.; Patton, R.A.; Kural, C.; Lucas, D.M.; Byrd, J.C.; Castro, C.E. Daunorubicin-loaded DNA origami nanostructures circumvent drug-resistance mechanisms in a leukemia model. *Small* **2016**, *12*, 308–320. [CrossRef]
10. Mikkila, J.; Eskelinen, A.P.; Niemela, E.H.; Linko, V.; Frilander, M.J.; Torma, P.; Kostiainen, M.A. Virus-encapsulated DNA origami nanostructures for cellular delivery. *Nano Lett.* **2014**, *14*, 2196–2200. [CrossRef]

11. Kiviaho, J.K.; Linko, V.; Ora, A.; Tiainen, T.; Järvihaavisto, E.; Mikkilä, J.; Tenhu, H.; Nonappa; Kostiainen, M.A. Cationic polymers for DNA origami coating—Examining their binding efficiency and tuning the enzymatic reaction rates. *Nanoscale* **2016**, *8*, 11674–11680. [CrossRef] [PubMed]

12. Strauss, M.T.; Schueder, F.; Haas, D.; Nickels, P.C.; Jungmann, R. Quantifying absolute addressability in DNA origami with molecular resolution. *Nat. Commun.* **2018**, *9*, 1600. [CrossRef] [PubMed]

13. Voigt, N.V.; Torring, T.; Rotaru, A.; Jacobsen, M.F.; Ravnsbaek, J.B.; Subramani, R.; Mamdouh, W.; Kjems, J.; Mokhir, A.; Besenbacher, F.; et al. Single-molecule chemical reactions on DNA origami. *Nat. Nanotechnol.* **2010**, *5*, 200–203. [CrossRef]

14. Woo, S.; Rothemund, P.W. Programmable molecular recognition based on the geometry of DNA nanostructures. *Nat. Chem.* **2011**, *3*, 620–627. [CrossRef]

15. Li, J.; Green, A.A.; Yan, H.; Fan, C. Engineering nucleic acid structures for programmable molecular circuitry and intracellular biocomputation. *Nat. Chem.* **2017**, *9*, 1056–1067. [CrossRef]

16. Gerling, T.; Wagenbauer, K.F.; Neuner, A.M.; Dietz, H. Dynamic DNA devices and assemblies formed by shape-complementary, non-base pairing 3D components. *Science* **2015**, *347*, 1446–1452. [CrossRef]

17. Chen, Y.J.; Groves, B.; Muscat, R.A.; Seelig, G. DNA nanotechnology from the test tube to the cell. *Nat. Nanotechnol.* **2015**, *10*, 748–760. [CrossRef] [PubMed]

18. Hu, Q.; Li, H.; Wang, L.; Gu, H.; Fan, C. DNA Nanotechnology-Enabled Drug Delivery Systems. *Chem. Rev.* **2018**. [CrossRef] [PubMed]

19. Hong, E.; Halman, J.R.; Shah, A.B.; Khisamutdinov, E.F.; Dobrovolskaia, M.A.; Afonin, K.A. Structure and composition define immunorecognition of nucleic acid nanoparticles. *Nano Lett.* **2018**, *18*, 4309–4321. [CrossRef]

20. Ke, W.; Hong, E.; Saito, R.F.; Rangel, M.C.; Wang, J.; Viard, M.; Richardson, M.; Khisamutdinov, E.F.; Panigaj, M.; Dokholyan, N.V.; et al. RNA-DNA fibers and polygons with controlled immunorecognition activate RNAi, FRET and transcriptional regulation of NF-kappaB in human cells. *Nucleic Acids Res.* **2018**. [CrossRef]

21. Saha, S.; Prakash, V.; Halder, S.; Chakraborty, K.; Krishnan, Y. A pH-independent DNA nanodevice for quantifying chloride transport in organelles of living cells. *Nat. Nanotechnol.* **2015**, *10*, 645–651. [CrossRef] [PubMed]

22. Yang, D.; Hartman, M.R.; Derrien, T.L.; Hamada, S.; An, D.; Yancey, K.G.; Cheng, R.; Ma, M.; Luo, D. DNA materials: Bridging nanotechnology and biotechnology. *Acc. Chem. Res.* **2014**, *47*, 1902–1911. [CrossRef] [PubMed]

23. Afonin, K.A.; Grabow, W.W.; Walker, F.M.; Bindewald, E.; Dobrovolskaia, M.A.; Shapiro, B.A.; Jaeger, L. Design and self-assembly of siRNA-functionalized RNA nanoparticles for use in automated nanomedicine. *Nat. Protoc.* **2011**, *6*, 2022–2034. [CrossRef]

24. Jiang, Q.; Song, C.; Nangreave, J.; Liu, X.; Lin, L.; Qiu, D.; Wang, Z.-G.; Zou, G.; Liang, X.; Yan, H.; et al. DNA origami as a carrier for circumvention of drug resistance. *J. Am. Chem. Soc.* **2012**, *134*, 13396–13403. [CrossRef] [PubMed]

25. Veetil, A.T.; Chakraborty, K.; Xiao, K.; Minter, M.R.; Sisodia, S.S.; Krishnan, Y. Cell-targetable DNA nanocapsules for spatiotemporal release of caged bioactive small molecules. *Nat. Nanotechnol.* **2017**, *12*, 1183–1189. [CrossRef] [PubMed]

26. Li, D.; Mo, F.; Wu, J.; Huang, Y.; Zhou, H.; Ding, S.; Chen, W. A multifunctional DNA nano-scorpion for highly efficient targeted delivery of mRNA therapeutics. *Sci. Rep.* **2018**, *8*, 10196. [CrossRef]

27. Zhao, Z.; Fu, J.; Dhakal, S.; Johnson-Buck, A.; Liu, M.; Zhang, T.; Woodbury, N.W.; Liu, Y.; Walter, N.G.; Yan, H. Nanocaged enzymes with enhanced catalytic activity and increased stability against protease digestion. *Nat. Commun.* **2016**, *7*, 10619. [CrossRef] [PubMed]

28. Burns, J.R.; Lamarre, B.; Pyne, A.L.B.; Noble, J.E.; Ryadnov, M.G. DNA origami inside-out viruses. *ACS Synth. Biol.* **2018**, *7*, 767–773. [CrossRef]

29. Bastings, M.M.C.; Anastassacos, F.M.; Ponnuswamy, N.; Leifer, F.G.; Cuneo, G.; Lin, C.; Ingber, D.E.; Ryu, J.H.; Shih, W.M. Modulation of the cellular uptake of DNA origami through control over mass and shape. *Nano Lett.* **2018**, *18*, 3557–3564. [CrossRef]

30. Wang, P.; Rahman, M.A.; Zhao, Z.; Weiss, K.; Zhang, C.; Chen, Z.; Hurwitz, S.J.; Chen, Z.G.; Shin, D.M.; Ke, Y. Visualization of the cellular uptake and trafficking of DNA origami nanostructures in cancer cells. *J. Am. Chem. Soc.* **2018**, *140*, 2478–2484. [CrossRef]

31. Ko, S.; Liu, H.; Chen, Y.; Mao, C. DNA nanotubes as combinatorial vehicles for cellular delivery. *Biomacromolecules* **2008**, *9*, 3039–3043. [CrossRef]

32. Ponnuswamy, N.; Bastings, M.M.C.; Nathwani, B.; Ryu, J.H.; Chou, L.Y.T.; Vinther, M.; Li, W.A.; Anastassacos, F.M.; Mooney, D.J.; Shih, W.M. Oligolysine-based coating protects DNA nanostructures from low-salt denaturation and nuclease degradation. *Nat. Commun.* **2017**, *8*, 15654. [CrossRef] [PubMed]

33. Ramakrishnan, S.; Ijas, H.; Linko, V.; Keller, A. Structural stability of DNA origami nanostructures under application-specific conditions. *Comput. Struct. Biotechnol. J.* **2018**, *16*, 342–349. [CrossRef]

34. Jiang, D.; Ge, Z.; Im, H.J.; England, C.G.; Ni, D.; Hou, J.; Zhang, L.; Kutyreff, C.J.; Yan, Y.; Liu, Y.; et al. DNA origami nanostructures can exhibit preferential renal uptake and alleviate acute kidney injury. *Nat. Biomed. Eng.* **2018**, *2*, 865–877. [CrossRef] [PubMed]

35. Hahn, J.; Wickham, S.F.; Shih, W.M.; Perrault, S.D. Addressing the instability of DNA nanostructures in tissue culture. *ACS Nano* **2014**, *8*, 8765–8775. [CrossRef] [PubMed]

36. Kocabey, S.; Meinl, H.; MacPherson, S.I.; Cassinelli, V.; Manetto, A.; Rothenfusser, S.; Liedl, T.; Lichtenegger, S.F. Cellular uptake of tile-assembled DNA nanotubes. *Nanomaterials* **2015**, *5*, 47–60. [CrossRef]

37. Kielar, C.; Xin, Y.; Shen, B.; Kostiainen, M.A.; Grundmeier, G.; Linko, V.; Keller, A. On the stability of DNA origami nanostructures in low-magnesium buffers. *Angew. Chem. Int. Ed.* **2018**, *57*, 9470–9474. [CrossRef] [PubMed]

38. O'Neill, P.; Rothemund, P.W.K.; Kumar, A.; Fygenson, D.K. Sturdier DNA nanotubes via ligation. *Nano Lett.* **2006**, *6*, 1379–1383. [CrossRef]

39. Michelotti, N.; Johnson-Buck, A.; Manzo, A.J.; Walter, N.G. Beyond DNA origami: The unfolding prospects of nucleic acid nanotechnology. *Wiley Interdiscip. Rev. Nanomed.* **2012**, *4*, 139–152. [CrossRef] [PubMed]

40. Langecker, M.; Arnaut, V.; Martin, T.G.; List, J.; Renner, S.; Mayer, M.; Dietz, H.; Simmel, F.C. Synthetic lipid membrane channels formed by designed DNA nanostructures. *Science* **2012**, *338*, 932–936. [CrossRef]

41. Burns, J.R.; Stulz, E.; Howorka, S. Self-assembled DNA nanopores that span lipid bilayers. *Nano Lett.* **2013**, *13*, 2351–2356. [CrossRef] [PubMed]

42. Burns, J.R.; Seifert, A.; Fertig, N.; Howorka, S. A biomimetic DNA-based channel for the ligand-controlled transport of charged molecular cargo across a biological membrane. *Nat. Nanotechnol.* **2016**, *11*, 152–156. [CrossRef] [PubMed]

43. Gopfrich, K.; Li, C.Y.; Ricci, M.; Bhamidimarri, S.P.; Yoo, J.; Gyenes, B.; Ohmann, A.; Winterhalter, M.; Aksimentiev, A.; Keyser, U.F. Large-conductance transmembrane porin made from DNA origami. *ACS Nano* **2016**, *10*, 8207–8214. [CrossRef]

44. Birkholz, O.; Burns, J.R.; Richter, C.P.; Psathaki, O.E.; Howorka, S.; Piehler, J. Multi-functional DNA nanostructures that puncture and remodel lipid membranes into hybrid materials. *Nat. Commun.* **2018**, *9*, 1521. [CrossRef] [PubMed]

45. Howorka, S. Building membrane nanopores. *Nat. Nanotechnol.* **2017**, *12*, 619–630. [CrossRef]

46. Pugh, G.C.; Burns, J.R.; Howorka, S. Comparing proteins and nucleic acids for next-generation biomolecular engineering. *Nat. Rev. Chem.* **2018**, *2*, 113–130. [CrossRef]

47. Ohmann, A.; Li, C.-Y.; Maffeo, C.; Al Nahas, K.; Baumann, K.N.; Göpfrich, K.; Yoo, J.; Keyser, U.F.; Aksimentiev, A. A synthetic enzyme built from DNA flips 107 lipids per second in biological membranes. *Nat. Commun.* **2018**, *9*, 2426. [CrossRef]

48. Burns, J.R.; Howorka, S. Defined bilayer interactions of DNA nanopores revealed with a nuclease-based nanoprobe strategy. *ACS Nano* **2018**, *12*, 3263–3271. [CrossRef] [PubMed]

49. List, J.; Weber, M.; Simmel, F.C. Hydrophobic actuation of a DNA origami bilayer structure. *Angew. Chem. Int. Ed.* **2014**, *53*, 4236–4239. [CrossRef]

50. Edwardson, T.G.W.; Carneiro, K.M.M.; McLaughlin, C.K.; Serpell, C.J.; Sleiman, H.F. Site-specific positioning of dendritic alkyl chains on DNA cages enables their geometry-dependent self-assembly. *Nat. Chem.* **2013**, *5*, 868. [CrossRef]

51. Arora, M. Cell Culture Media: A Review. *Mater. Methods* **2013**, *3*. [CrossRef]

52. Moore, T.L.; Rodriguez-Lorenzo, L.; Hirsch, V.; Balog, S.; Urban, D.; Jud, C.; Rothen-Rutishauser, B.; Lattuada, M.; Petri-Fink, A. Nanoparticle colloidal stability in cell culture media and impact on cellular interactions. *Chem. Soc. Rev.* **2015**, *44*, 6287–6305. [CrossRef] [PubMed]

53. Smith, G.S.; Walter, G.L.; Walker, R.M. Clinical pathology in non-clinical toxicology testing. In *Haschek and Rousseaux's Handbook of Toxicologic Pathology*; Academic Press: Cambridge, MA, USA, 2013; pp. 565–594. [CrossRef]

54. Seifert, A.; Gopfrich, K.; Burns, J.R.; Fertig, N.; Keyser, U.F.; Howorka, S. Bilayer-spanning DNA nanopores with voltage-switching between open and closed state. *ACS Nano* **2015**, *9*, 1117–1126. [CrossRef] [PubMed]

55. Wei, X.; Nangreave, J.; Jiang, S.; Yan, H.; Liu, Y. Mapping the thermal behavior of DNA origami nanostructures. *J. Am. Chem. Soc.* **2013**, *135*, 6165–6176. [CrossRef]

56. Sambrook, J.; Russell, D.W. *Molecular Cloning: A Laboratory Manual*, 3rd ed.; Cold Spring Harbor Laboratory Press: Cold Spring Harbor, NY, USA, 2001.

57. Branzoi, I.V.; Iordoc, M.; Branzoi, F.; Vasilescu-Mirea, R.; Sbarcea, G. Influence of diamond-like carbon coating on the corrosion resistance of the NITINOL shape memory alloy. *Surf. Interface Anal.* **2010**, *42*, 502–509. [CrossRef]

58. Krishnan, S.; Ziegler, D.; Arnaut, V.; Martin, T.G.; Kapsner, K.; Henneberg, K.; Bausch, A.R.; Dietz, H.; Simmel, F.C. Molecular transport through large-diameter DNA nanopores. *Nat. Commun.* **2016**, *7*, 12787. [CrossRef] [PubMed]

59. Norman, D.G.; Grainger, R.J.; Uhrín, D.; Lilley, D.M.J. Location of Cyanine-3 on Double-Stranded DNA: Importance for Fluorescence Resonance Energy Transfer Studies. *Biochemistry* **2000**, *39*, 6317–6324. [CrossRef]

60. Wolfe, A.J.; Si, W.; Zhang, Z.; Blanden, A.R.; Hsueh, Y.C.; Gugel, J.F.; Pham, B.; Chen, M.; Loh, S.N.; Rozovsky, S.; et al. Quantification of membrane protein-detergent complex interactions. *J. Phys. Chem. B* **2017**, *121*, 10228–10241. [CrossRef]

61. Geary, R.S.; Norris, D.; Yu, R.; Bennett, C.F. Pharmacokinetics, biodistribution and cell uptake of antisense oligonucleotides. *Adv. Drug Deliv. Rev.* **2015**, *87*, 46–51. [CrossRef] [PubMed]

62. Kuhlmann, M.; Hamming, J.B.R.; Voldum, A.; Tsakiridou, G.; Larsen, M.T.; Schmokel, J.S.; Sohn, E.; Bienk, K.; Schaffert, D.; Sorensen, E.S.; et al. An albumin-oligonucleotide assembly for potential combinatorial drug delivery and half-life extension applications. *Mol. Ther. Nucleic Acids* **2017**, *9*, 284–293. [CrossRef]

63. Osborn, M.F.; Coles, A.H.; Biscans, A.; Haraszti, R.A.; Roux, L.; Davis, S.; Ly, S.; Echeverria, D.; Hassler, M.R.; Godinho, B.; et al. Hydrophobicity drives the systemic distribution of lipid-conjugated siRNAs via lipid transport pathways. *Nucleic Acids Res.* **2018**. [CrossRef] [PubMed]

64. Ahmadi, Y.; De Llano, E.; Barisic, I. (Poly)cation-induced protection of conventional and wireframe DNA origami nanostructures. *Nanoscale* **2018**, *10*, 7494–7504. [CrossRef] [PubMed]

nanomaterials

MDPI

Article

Fluorogenic RNA Aptamers: A Nano-platform for Fabrication of Simple and Combinatorial Logic Gates

Victoria Goldsworthy, Geneva LaForceSeth Abels and Emil F. Khisamutdinov *

Department of Chemistry, Ball State University, Muncie, IN 47304, USA; vgoldsworthy@alex.k12.in.us (V.G.); grlaforce@bsu.edu (G.L.); sethabels1982@gmail.com (S.A.)
* Correspondence: kemil@bsu.edu; Tel: 765-285-8084

Received: 25 October 2018; Accepted: 20 November 2018; Published: 28 November 2018

Abstract: RNA aptamers that bind non-fluorescent dyes and activate their fluorescence are highly sensitive, nonperturbing, and convenient probes in the field of synthetic biology. These RNA molecules, referred to as light-up aptamers, operate as molecular nanoswitches that alter folding and fluorescence function in response to ligand binding, which is important in biosensing and molecular computing. Herein, we demonstrate a conceptually new generation of smart RNA nano-devices based on malachite green (MG)-binding RNA aptamer, which fluorescence output controlled by addition of short DNA oligonucleotides inputs. Four types of RNA switches possessing AND, OR, NAND, and NOR Boolean logic functions were created in modular form, allowing MG dye binding affinity to be changed by altering 3D conformation of the RNA aptamer. It is essential to develop higher-level logic circuits for the production of multi-task nanodevices for data processing, typically requiring combinatorial logic gates. Therefore, we further designed and synthetized higher-level half adder logic circuit by "in parallel" integration of two logic gates XOR and AND within a single RNA nanoparticle. The design utilizes fluorescence emissions from two different RNA aptamers: MG-binding RNA aptamer (AND gate) and Broccoli RNA aptamer that binds DFHBI dye (XOR gate). All computationally designed RNA devices were synthesized and experimentally tested in vitro. The ability to design smart nanodevices based on RNA binding aptamers offers a new route to engineer "label-free" ligand-sensing regulatory circuits, nucleic acid detection systems, and gene control elements.

Keywords: logic gates; nucleic acid computing; RNA aptamers; RNA nanotechnology

1. Introduction

The progression in the field of RNA nanotechnology makes RNA molecules the most promising candidate to fabricate bio-computers due to their variable folding properties as well as their catalytic functions [1,2]. Numerous non-canonical nucleotide interactions, found only in RNA [3,4], enable this biopolymer to self-assemble into various shapes and dimensions as exemplified by naturally occurring ribosomal RNA [5] and ribozymes [6,7] as well as by artificially constructed RNA polygons [8–12], prisms, and cubes [13–15]. This diverse structural capability of RNA led to the development of aptamer technology almost 30 years ago [16,17]. Aptamers are single-stranded RNA or DNA oligonucleotides, with typical length of no more than 100 nts that were artificially selected from combinatorial libraries for high binding affinities to specific molecular targets. Since their development, aptamers have revolutionized the field of biosensing by enabling scientists to rationally generate different aptamers targeting a diverse range of ligands [18–20]. RNA-based fluorogenic modules are of particular interest [21–23] since they have applications in monitoring gene expression [24,25] and new drug screening pipelines using microarrays developed to sense target molecules of variable size [26]. This florescence module includes a light-up RNA aptamer and fluorogen. The light-up

RNA aptamers are selected to specifically bind to small organic molecules exhibiting minimal to no fluorescent emission when free in solution (fluorogen or fluorogenic dyes) and trigger its florescence. The well-studied examples include malachite green (MG)-binding RNA aptamer [27], and 3,5-difluoro-4-hydroxybenzylidene imidazolinone (DFHBI)-binding RNA aptamers [28,29] among others [30,31].

In RNA nanotechnology, the development and implementation of RNA-based nanodevices that respond to biomolecular inputs by generating output signals in accordance with logic gate behavior has attracted considerable attention [32–36]. Computing using both RNA and DNA molecules integrates biochemistry and molecular biology disciplines to achieve a certain goal through designing algorithmic processes embedded within polynucleotide structures. The concept of using nucleic acids (NAs) for computation was proven in 1994 when Leonard Adleman demonstrated the ability of synthetic DNA oligonucleotides to solve a seven-point Hamiltonian path problem [37] and, since then, many studies have reviewed the possibility of developing a new generation of molecular logic gates and molecular computers based on nucleic acids [38,39]. In contrast to silicon-based computers, NA computers implement concentrations of specific molecules, such as metal ions, small organic dyes, single stranded DNA or RNA oligonucleotides, peptides or proteins, as inputs to derive certain signals, e.g., switching between RNA conformations, activation or deactivation of ribozyme activity, down- or up-regulation of certain genes, etc. [40–45]. This relies on the algorithmic processes carefully designed and encompassed within a nucleic acid complex (referred to as logic gates (LGs)) that are capable of performing simple AND, OR, NAND, and NOR logic operations, as well as more sophisticated logic circuits.

DNA has been routinely used for the development of biochemical circuits and all basic logic operations, including INHIBIT, IMPLICATION, and XOR have been mimicked with DNA as a template [46–50]. There are also various classes of functional RNA molecules, such as ribozymes, riboswitches, miRNA, siRNA, and orthogonal ribosomes, that enable the fabrication of computational systems [51–54] and simple RNA fluorogenic biosensors [55,56]. However, it is often essential to develop higher-level logic circuits for the production of multi-task nanodevices for data processing, which usually require combinatorial logic gates [57,58]. For example, a half adder can perform an addition operation on two binary digits by integration of an XOR gate and an AND gate in parallel to generate a SUM (S) output and a CARRY (C) output, respectively. To the best of our knowledge, the development of combinatorial RNA logic gates based on light-up RNA aptamer fluorogenic systems has yet to be realized, and would represent a label-free oligonucleotide bio-sensing platform with potential applications in biocomputing and biosensing.

Herein, we report the design and assembly of a conceptually new generation of molecular logic gates that possess simple AND, OR, NAND, and NOR logic operations implementing the light-up MG-binding RNA aptamer. Single stranded DNA (ssDNA) oligonucleotides were used as inputs to trigger conformational changes in the RNA aptamer. The corresponding output values of OFF (0) and ON (1) are obtained by low and high fluorescence emissions, respectively (Figure 1A).

Furthermore, we developed a basic half adder computing platform based on the RNA light-up aptamer strategy (Figure 1B). The design utilizes fluorescence emissions from two distinct fluorogenic RNA modules, MG-binding RNA aptamer [27,59] and DFHBI-binding Broccoli RNA aptamer [29], fused with the previously reported tetragon RNA nanoparticle [8,9].

The function of the half adder is triggered by the same two ssDNA inputs. The ssDNA inputs alter the conformation of the RNA aptamers in such a manner as to either permit or deny fluorogen dye binding to the aptamer, resulting in fluorescent (ON) or non-fluorescent (OFF) states within one nanostructure. This RNA logic gate system demonstrates the great potential of light-up RNA aptamers as an arithmetic tool for molecular programming and will open a way to further development concerning well-regulated molecular electronic devices and biosensors.

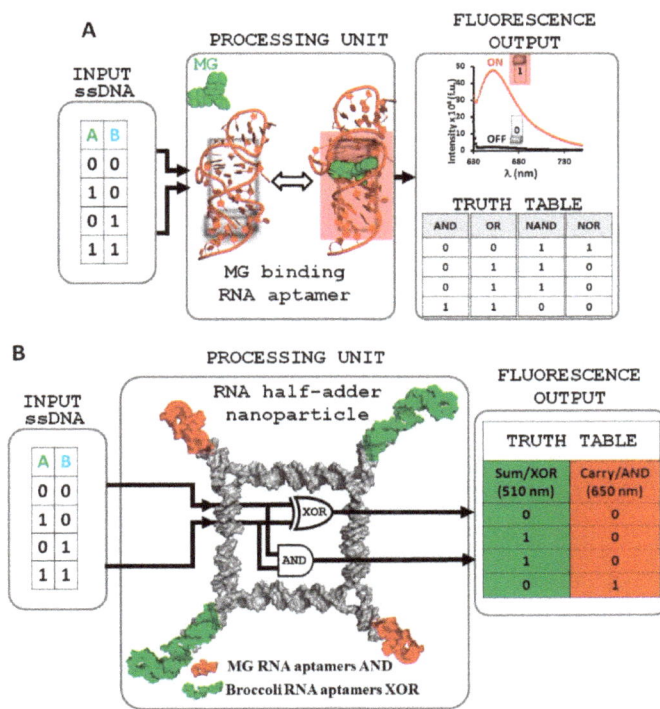

Figure 1. Logic gate design strategy based on light-up RNA aptamers. (**A**) malachite green (MG)-binding RNA aptamer used to design four simple AND, OR, NAND, and NOR logic gates. (**B**) Illustration of RNA half adder system based on MG RNA and Broccoli RNA aptamers conjugated with RNA tetragonal nanoparticle. Fluorophores MG and DFHBI in their unbound state exhibit low fluorescence (0), the emission of these chromophores increases upon binding to their corresponding RNA aptamers (1).

2. Results and Discussion

2.1. Design and Fabrication of a Logic Gate Possessing AND, OR, NAND and NOR Boolean Functions

RNA molecules featuring aptamers that bind a fluorogenic dye and activate its fluorescence have the potential to be highly sensitive and convenient probes in the field of synthetic biology. The initial system is comprised of an RNA hairpin molecule containing the MG RNA binding aptamer sequence and the MG fluorogen dye [60]. The fluorescence of the MG dye is negligible when free in solution (OFF state) and is increased significantly upon binding to its RNA aptamer (ON state) (Figure 1). The principle of the LG design is based on the structural manipulation of the RNA aptamer, due to the fact that binding affinity of the fluorogenic dye to its light-up RNA aptamer depends on the correctly folded RNA structure. Thus, the fluorescence emission can be precisely turned ON (correctly folded RNA structure) and OFF (disrupted conformation). The four logic gates AND, OR, NAND, and NOR were designed using this highly effective and modular approach (Figure 2).

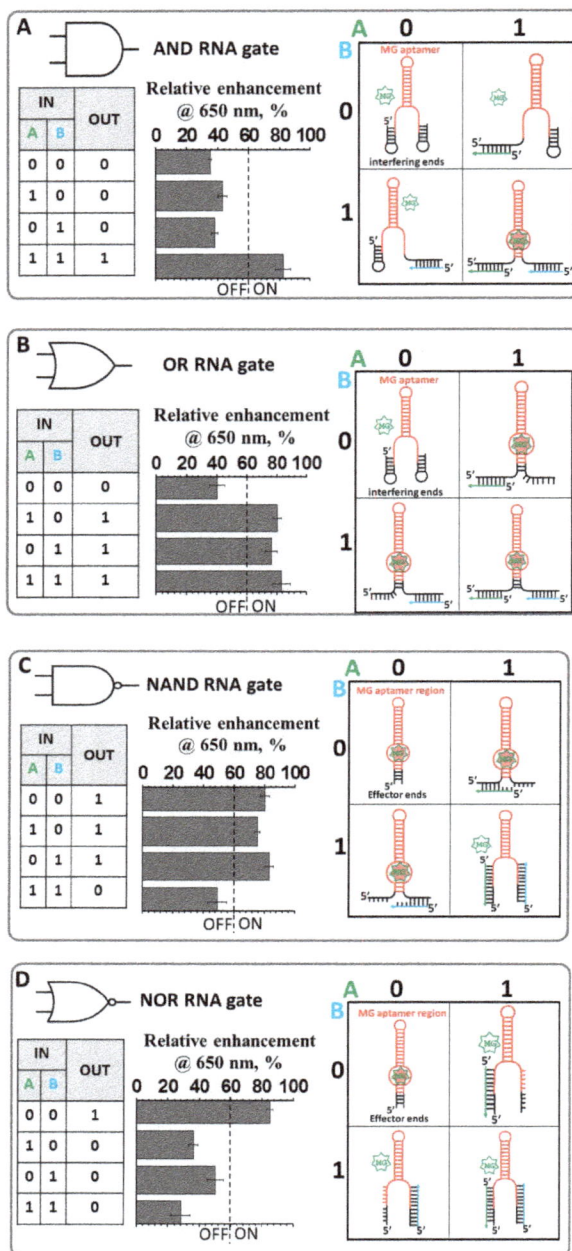

Figure 2. Logic gate design principles for an AND gate (**A**), OR gate (**B**), NAND gate (**C**), NOR gate (**D**). The predicted 2D conformations of the RNA aptamers in the presence and absence of individual or both inputs are shown to the right. The rules specified by each gate are shown in truth tables to the left. Normalized fluorescence enhancement of the gates is displayed in the middle. The fluorescence enhancement data are reported with ± standard error of the mean (SEM) bars; measurements have been reproduced from least three repetitive trials.

Individual gates employed light-up RNA aptamers consisting of two extended sequences localized at both the 5'- and 3'-ends highlighted in Figure 2 in black. This will further be referred to as the interfering end for the AND and OR gates or the non-interfering ends for the NAND and NOR gates. The MG RNA aptamer core sequence is highlighted in red. The end sequences are tailored to bind the ssDNA oligonucleotides which serve as inputs. All the RNA gates were designed in silico relying on secondary structure prediction algorithms encompassed within NUPAC [61] or mfold programs [62] to confirm the secondary structures of each RNA sequence prior to synthesis. If the calculated lowest free-energy secondary structure corresponded to the desired RNA conformation, and no other secondary structure was closer than 20% in energy to the lowest energy structure, the sequence was used without alternations. Otherwise, minor changes were made to Watson–Crick base-paired positions to destabilize competing conformations. The fluorescence was measured at 22 °C in 1× TMS buffer as described in the materials and methods. Each RNA logic gate was designed to have complementary regions at the 5'- and 3'- ends to ssDNA oligonucleotide inputs with lengths ranging from 15 to 27 nucleotides (Supplementary Table S1). Each gate processes a different pair of oligonucleotide inputs—for instance, input A of the AND gate is not the same as input A of the OR gate—and the terms A and B were merely applied across the table for simplicity. However, the sequences and therefore 2D structures for AND and OR gates were designed to be identical as both initial structures, at the no inputs setting, should have 0 output or exist in the OFF state according to the truth table (Figure 2A,B).

In a similar manner, initial structures for the NAND and NOR gates were chosen to share identical nucleotides as the ON state is a requirement for both structures at the no inputs condition. For each logic gate, the fluorescence intensities were normalized throughout the experiments. The threshold value was determined to be 60% where an intensity greater than this value yields an output = 1, while an intensity below this value yields an output = 0.

The AND and OR gates with default setting (0-0; no inputs present) was designed using interfering ends. The purpose of this was to form a complementary base pairing with the RNA aptamer MG-binding region as illustrated in Figure 2A,B. The interfering sequences at both ends were chosen to form weak hairpin-like 2D structures (2D structures available in Supplementary Figure S1). These hairpins are formed by involving core nucleotides that are responsible for forming a binding pocket for the MG fluorogen (Supplementary Figure S1). Thus, the presence of the hairpins prevents the light-up aptamer from binding to its MG chromogen and ultimately diminishing the fluorescent signal. The AND operation is achieved by carefully designing two short ssDNA inputs (A and B) to bind to the 5' and the 3' interfering ends, respectively. The structural rearrangement to the ON conformation occurs only when there are two inputs present at the same time. The presence of the inputs releases the core MG nucleotides (nts) thus allowing them to fold into the proper ON conformation. The selection of these particular ssDNA inputs was achieved by varying their length using RNA structure prediction programs (Supplementary Figure S2). The goal was to select the appropriate length of the inputs that would hybridize (or not hybridize at all) with its target hairpin without perturbing the conformation of the other hairpin when only one input is added. The selection of the DNA inputs was based on the computed melting temperature (T_m), the temperature at which 50% of double stranded nucleic acid is converted to single-standard form. The desired Tm values for inputs hybridization were chosen to be slightly greater than 22 °C, the temperature at which the experiment was performed. Such input selection relies on the equilibrium between duplex (formed by ssDNA and hairpin nts) and hairpin structures where equilibrium shifts slightly in favor of the hairpin structure (Figure S2, AND GATE). However, inclusion of both inputs favors the formation of two duplexes triggering structural rearrangement of the overall complex, which leads to the liberation of the core nts accountable for MG-binding site formation. As shown in Figure 2A, resulting fluorescence intensity measured for four annealed samples (0-0; 0-1; 1-0, and 1-1) clearly indicates the effectiveness of the AND gate function.

The OR construct, which yields output = 0 for scenario 0-0 and output = 1 for scenarios 1-0, 0-1, and 1-1, shares the identical RNA sequence as the AND gate. However, the ssDNA inputs were designed to have several extra nucleotides to achieve an equilibrium state where inclusion of at least one input will trigger conformational change in favor of the correctly folded MG-binding RNA aptamer. The design criterion is based on a strand displacement reaction. Upon binding to its corresponding hairpin, input A or B will create a stable hybrid RNA/DNA duplex. The DNA inputs were selected to contain longer sequences with much higher Tm values as compared to the AND gate, so that disrupted hairpin nts initiate a conformational change of the whole construct favoring formation of the ON aptamer state (Figure S2, OR GATE). The measured fluorescence emission values in solution demonstrates OR behavior of the RNA construct (Figure 2B).

In contrast to the AND and OR gates, the NAND and NOR gates were designed so the default structures possess correctly folded light-up MG RNA aptamers and the extensions at the 5'- and 3'-ends do not interfere with the core structure. To produce the corresponding NAND operation, the non-interfering ends must be able to bind input A or input B without sacrificing the conformation of the aptamer. However, when both A and B are presented, the conformation of the aptamer needs to be sufficiently distorted to register an output = 0. Figure 2C summarizes the fluorescence enhancement measurement for the NAND gate. Interestingly, while input A alone (1-0) decreases fluorescence as compared to the 0-0 state, input B alone (0-1) increases it slightly. Inputs A and B in tandem (1-1) triggered a noticeable decrease in fluorescent intensity (Figure 2C).

The NOR logic gate was constructed utilizing the NAND RNA molecule. Figure 2D shows an obvious increase in fluorescence intensity between output = 0 and output = 1, owing to the nature of the RNA NAND logic gate. The designed ssDNA inputs that complimentarily pair with the RNA gate can significantly disrupt the conformation of the RNA molecule, rendering MG binding impossible. Hence, the output was "1" when only neither DNA input was added (Figure 2D).

Collectively, the modular approach to the fabrication of RNA Boolean logic gates based on the light-up RNA aptamer was demonstrated. All designed gates produced the expected OFF or ON values corresponding to low or high fluorescence intensity at λ_{max} = 650 nm, respectively, in response to DNA oligonucleotide inputs. A threshold value of fluorescence enhancement of 60% was chosen to distinguish the OFF (any value bellow 60%) and ON (any value above 60%) states. Various concentrations of RNA molecular gates, inputs, and MG dye in solution were explored, with those yielding the greatest difference in fluorescence between output = 0 and output = 1 reported here. The extent to which these modular RNA logic gates can be used to probe three or more inputs simultaneously will depend on their reliability in tandem.

2.2. Implementing Logic Gates to Construct a Half-Adder Logic Circuit

The production of multi-functional nanodevices for data analysis or processing is extremely important and yet challenging due to requirements of multiple coordinated logic gates operations within a single unit. Based on the aforementioned results, we next integrated two different fluorogen-binding RNA aptamers: (i) MG RNA aptamer and (ii) the recently developed Broccoli RNA aptamer that binds DFHBI dye within one RNA complex. As a key building block, the half adder is used to construct more advanced computational circuits and is in high demand in information technology [63]. The representative secondary structure of this complex is demonstrated in Figure 3A. The differences in the emission properties of these two fluorogens (MG emits in the "red" region while DFHBI emits in the "green" region of the visible spectra range) were implemented to construct a half-adder logic circuit, which is a primary step in constructing a full adder, a basic arithmetic unit in computing.

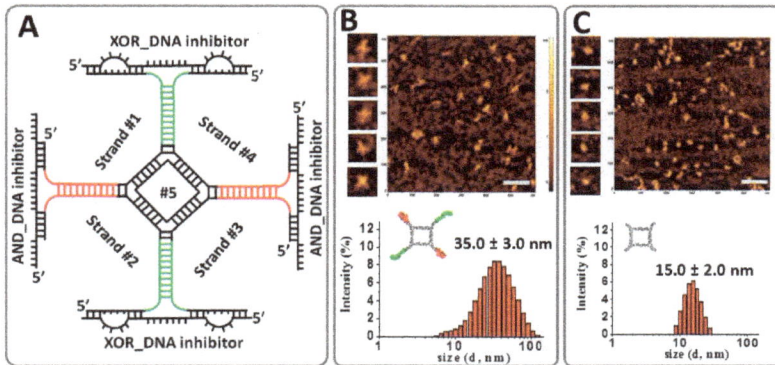

Figure 3. Implementation principle of the developed half adder. (**A**) Secondary structure of the RNA complex based on MG and Broccoli RNA aptamer conformations fused to tetragonal nano-scaffold. (**B**) Representative atomic force microscopy image (AFM) of the RNA half adder (bar scale = 100 nm) and Dynamic Light Scattering (DLS) data showing average size of the complex ± SEM. (**C**) AFM and DLS data for control RNA tetragonal nanoparticle, with the diameter of the nanoparticles is reported.

The half adder is composed of an AND gate with a MG-binding light-up RNA aptamer, and a XOR (eXclusive OR) gate based on the DFHBI-binding Broccoli light-up RNA aptamer [29]. The fluorescent intensity was measured in solution using fluorescent spectroscopy with excitation wavelengths of $\lambda_{ex/em}$ = 465/510 nm (corresponding to the XOR gate) and $\lambda_{ex/em}$ = 615/650 nm (corresponding to the AND gate). These gates were rationally designed to use two ssDNA inputs to output two fluorescence signals: SUM (λ_{em} = 510 nm) and CARRY (λ_{em} = 650 nm) generated by the AND and XOR gates, respectively (Figure 4). Both the MG RNA aptamer and Broccoli RNA aptamers were incorporated on alternating vertices to a previously developed RNA tetragon nanoparticle. The RNA half adder self-assembles from five RNA strands with a yield exceeding 80% (Supplementary Figures S3 and S4). The conformation of the assembled RNA tetragonal geometry was confirmed by atomic force microscopy and size was determined by dynamic light scattering (Figure 3B). Atomic Force Microscopy image (AFM) imagining revealed the extensions at each vertex in the designed RNA half adder as compared to the control RNA square nanoparticle. The size of the nanoparticle increases from 15 nm to approximately 35 nm with the addition of the RNA aptamers as shown by Dynamic Light Scattering (DLS) experiment. Also, the significant size variation between RNA tetragon and RNA half-adder nanoparticles was confirmed by native polyacrylamide gel electrophoresis (PAGE) (Figures S5 and S6).

To perform the XOR and AND logic operations using the same two ssDNA inputs, additional DNA inhibitor strands were introduced to bind complementary to the light-up RNA aptamers and interfere with their ON states or correctly folded structures. This tetragonal shaped half-adder RNA complex is designed according to the competitive hybridization and displacement principle of the DNA strands. The assembly experiments shown in Figure S5 confirm complexation of both inhibitors with the tetragonal nanoparticle. Importantly, the RNA tetragon containing 2 MG and 2 Broccoli RNA aptamers assembles with their corresponding DNA inhibitors at 1:2 ratio, i.e., one tetragon and 2 AND_DNA inhibitors and 2 XOR_DNA inhibitors.

The Broccoli RNA aptamer was designed to act as an XOR gate. To maintain the proper XOR gate function, an inhibitor DNA strand was designed (XOR_DNA inhibitor). The inhibitor bound to the aptamer, disrupting the binding of DFHBI and thus diminishing fluorescence in the presence of neither input. This XOR_DNA inhibitor contains two loop regions on either side of the aptamer. These internal loops contain eight unpaired nts and are designed to complement the ssDNA inputs. The addition of either ssDNA input destabilizes the RNA aptamer-inhibitor complex, separating the strands enough to reform the functional RNA aptamer shape allowing (1-0) or (0-1) truth values in fluorescence with an output of "1" (Figure 4A).

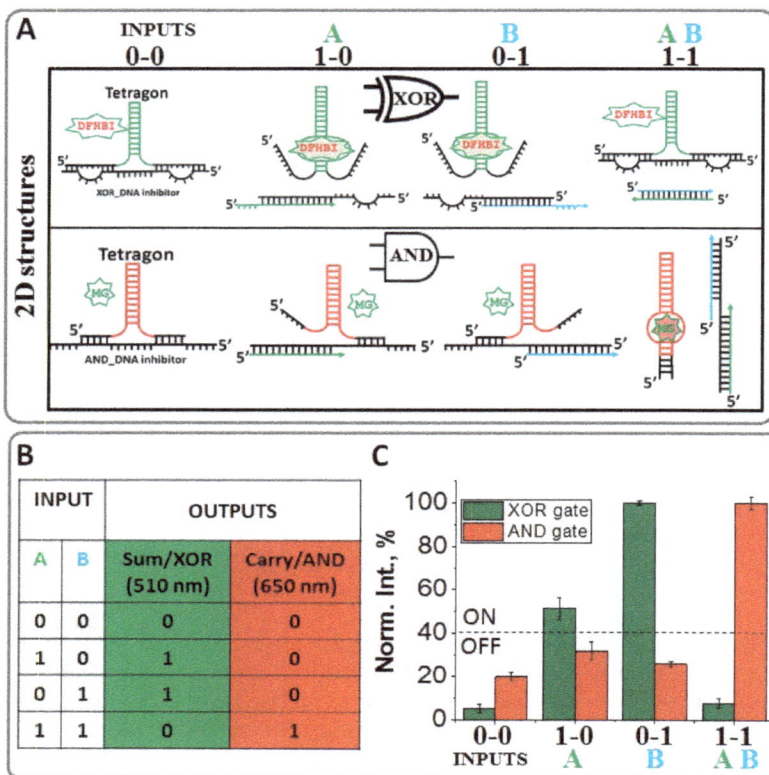

Figure 4. Design principles of the RNA AND and XOR gates using inhibitor DNA strand. (**A**) Predicted secondary structures of the nucleic acid displacement reactions within XOR and AND gates. (**B**) Truth table of a half adder. (**C**) The normalized fluorescence enhancement of the system at 510 nm and 650 nm as a function of the various inputs (Inputs A and B); the error bars indicate ± SEM from three independent measurements.

However, in the presence of both inputs, fluorescence is once again inhibited as the hybridization of inputs A and B is favored over hybridization with XOR_DNA inhibitor. To achieve this, inputs A and B were designed to bind more competitively to one another than to the XOR_DNA inhibitor through 17 "sticky" nts at the 5′ end of input A and 3′ end of input B (Table S1). Therefore, the XOR_DNA inhibitor paired with the RNA aptamer yielded a low fluorescent output signal. Figure 4C shows the normalized fluorescence intensity of the designed XOR system at 510 nm output readout in response to the ssDNA inputs. The presence of each input is defined as "1" (the absence is considered "0") and the output signal is defined as ON or OFF when the normalized fluorescence emission is higher or lower than 40%, respectively. The system exhibits ON in the presence of the individual inputs; otherwise, it remains OFF. The XOR logic operation performs the SUM digit function in the half adder as shown in the Truth Table (Figure 4B).

The AND logic operation of the half adder was designed utilizing the fluorescent properties of the MG-binding RNA aptamer system as the output signal. Similar to the XOR gate approach, a DNA inhibitor (AND_DNA) was used to disrupt the RNA aptamer conformation. The AND gate has an output of OFF or (0) in the absence of inputs (0-0) or in the presence of only one of the inputs (1-0 or 0-1). As the inputs are the same as those used for the Broccoli XOR gate, it is critical for the AND_DNA inhibitor to be complementary to the previously designed inputs. For this purpose, the AND_DNA inhibitor was designed to contain "sticky" nts at the 5′- and 3′-ends. These "sticky" nts are complementary to the ssDNA inputs. The addition of either input causes only a partial displacement of the AND_DNA inhibitor from the light-up RNA aptamer resulting in the low output value as demonstrated in Figure 4C. However, in the presence of both inputs, fluorescence increases significantly (ON state). This was accomplished by disassociating the AND_DNA inhibitor. The ssDNA inputs bind more competitively to the inhibitor to form a three-stranded DNA/DNA/RNA complex enabling the successful separation from the RNA light-up aptamer. Figure 4A (lower panel) summarizes the 2D structures computed for the AND logic system in the presence and absence of the ssDNA inputs. The normalized fluorescence intensities of the system at 650 nm as a function of the inputs are plotted in Figure 4C indicating that the system exhibits "1" only when both inputs coexist, indicative of an AND logic gate. The AND logic gate is responsible for the CARRY digit function in the half adder, as shown in the truth table in Figure 4B. To conclude, the AND and the XOR gates were implemented in parallel utilizing light-up RNA aptamers as a label-free fluorogenic platform. Both gates were triggered by the same set of inputs, satisfying the requirements for a half adder [64]. The further development of the full adder system based on the RNA high-up aptamers is currently under investigation. By definition, the full adder should perform an addition operation on three binary digits and similarly to the half adder, it generates a carry out to the next addition column. This development requires three inputs, which can be the same two ssDNA inputs and an additional carry-in DNA input to receive the carry signal from a previous stage.

3. Materials and Methods

3.1. Nucleic Acid Sequence Design, Synthesis, and Assembly

Polynucleotide sequence design was carried out using the multi-strand secondary structure prediction programs NUPACK and mfold [61,62]. To meet the requirements of the developed logic gates, the DNA and RNA sequences used in the experiments were first designed and then analyzed by the above 2D structure folding predicting software. According to the predicted 2D structures, experiments were performed to determine whether the designed ssDNA oligonucleotides were operational in the corresponding logic gate processing reactions. If the satisfying fluorescence readouts were not achieved, the DNA sequences were redesigned and the procedures were repeated until the desired DNA sequences obtained.

All DNA oligonucleotides were purchased from IDT DNA (Coralville, IA, USA) as desalted products and used without purification. RNA strands corresponding to individual logic gates and to the tetragonal half-adder complex were prepared by in vitro transcription using T7 RNA polymerase [8]. For this, synthetic DNA strands coding for the anti-sense sequence of the RNA strands were amplified by polymerase chain reaction (PCR) using primers containing the T7 RNA polymerase promoter. PCR products were purified using the QiaQuick PCR purification kit (Qiagen Sciences, Germantown, MD, USA). The transcribed RNA molecules were purified by denaturing 20% polyacrylamide gel electrophoresis containing 8M UREA.

The self-assembly of individual MG-based light-up RNA logic gate complexes AND, OR, NAND, and NOR was achieved by mixing equimolar oligonucleotide strands (1 μM) in TMS (50 mM TRIS pH = 8.0, 100 mM NaCl and 10 mM MgCl$_2$) buffer and heating the mixture to +80 °C and gradually cooling it down to +4 °C over a period of 1 h on a PCR thermocycler. Once the RNA aptamer self-assembly was achieved, a small amount (2 μM final concentrations) of the malachite green oxalate salt (Sigma Aldrich Co., St. Louis, MO, USA) was added to each RNA or RNA/DNA assembly. The mixture was left to incubate for an additional 30 min at 22 °C to reach proper binding equilibrium.

DNA inputs oligonucleotides were added to the assembled AND, OR, and NOR gates at the final concentrations of 2 μM (each) making the final stoichiometry of the complexes as follows:

1 GATE: 1 INPUT A: 1 INPUT B.

For the NAND system the optimal results achieved at stoichiometry:

1 NAND: 2 INPUT A: 2 INPUT B.

The self-assembly of the half-adder RNA construct was achieved by mixing corresponding RNA and DNA polynucleotides at 1:1 stoichiometric ratio. For example, the RNA half adder in the absence of inputs contained 1 μM of each RNA strands, 2 μM of AND_DNA inhibitor, and 2 μM XOR_DNA inhibitor. Malachite green and 3,5-difluoro-4-hydroxybenzylidene imidazolinone (DFHBI) (Sigma Aldrich Co., St. Louis, MO, USA) dyes were added to the corresponding complexes (0-0, 0-1, 1-0, 1-1) to make 5 μM final concentration. The resulting mixture was allowed to incubate for an additional 30 min at 22 °C. After reaching equilibrium, DNA inputs (5 μM each) were added in accordance to the truth table for the half-added RNA systems and fluorescence were recorded after additional 30 min incubation, which was necessary to achieve input driven strand displacement effect (Figure S8)

3.2. Fluorescence Measurements

Fluorescence was measured on a Fluoromax-3 (Hibora Jobin-Yvon, Horiba Scientific, Edison, NJ, USA) spectrofluorimeter using a Sub-Micro quartz fluorometer cell (Starna cells Inc., Atascadero, CA, USA). Fluorescence intensities were recorded separately for each dye. For the DFHBI-binding Broccoli RNA aptamer, the excitation wavelength centered at 465 nm and emission was collected in the range of 475–700 nm. For the MG-binding RNA aptamer, the excitation was centered at 615 nm and emission was recorded from the range of 630–750 nm.

The fluorescent enhancement was quantified by the ratio of the maximum emission of the fluorogenic dyes bound to its aptamers divided by the emission of the free dyes in solution. The fluorescence enhancement data were normalized after the experiments; a threshold value was chosen to be 60% for the MG based RNA logic gates, an intensity greater than this value yields an output = 1, while an intensity below this value yields an output = 0.

3.3. Dynamic Light Scattering

Hydrodynamic diameters of assembled half-adder RNA constructs and the control tetragon RNA nanoparticles were measured by a Zetasizer nano-ZS (Malvern Instrument Ltd., Malvern Panalytical Ltd., Malvern, UK) at 22 °C following previously described protocols [9].

3.4. Atomic Force Microscopy Imaging

The RNA tetragon and RNA half-adder complexes were imaged with MultiMode AFM NanoScope IV system (Veeco Instruments Inc., Plainview, NY, USA), following previous methods [65].

4. Conclusions

Molecular logic gates hold great potential for a wide range of biotechnological applications, including gene expression regulation, biosensors, therapeutic molecule design, metabolic reprogramming, studies of drug-nucleic acid interactions, and tools for elucidating cellular functions. The emergence of RNA nanotechnology offers great opportunities for applications of RNA-based logic gates. In this study, we have used a computational approach to design various oligonucleotide-responsive RNA logic gates (AND, OR, NAND and NOR) based on the MG-binding RNA aptamer. The structures of four logic gates were designed based on the general 2D architecture depicted in Figure 2 and all functioned as robust RNA switches that exhibit fluorescence emission once activated. The design process used here accounts for the thermodynamic stability of various base-paired structures in the absence or presence of input oligonucleotides. This functional design was possible due to the fact that nucleic acid secondary structure folding largely follows the simple rules of Watson-Crick base pairing, and the thermodynamic parameters for base-pair interactions are available. In addition, a half adder was successfully demonstrated by combining the hybridization and replacement of ssDNA strands. Specifically, introducing two light-up RNA aptamers MG and Broccoli into a half-adder system to modulate the output signal makes it flexible and enables the potential design of various other types of logic gates according to the requirements of the data processing. Although the developed individual logic gates and half adder are implemented in an experimental stage and exclusively in vitro, the demonstrated system presents great potential for the development of other RNA-light up based logic circuits as a universal arithmetic tool. To summarize, this work provides a novel light-up RNA aptamer-based platform for the design and assembly of higher-order circuits for arithmetic operations and opens the possibility to develop a new approach for constructing multicomponent devices on a single biomolecular nano-platform.

Supplementary Materials: The following are available online at http://www.mdpi.com/2079-4991/8/12/984/s1.

Author Contributions: Conceptualization, E.F.K. and S.A., V.G.; Methodology, E.F.K., V.G., G.L.; Software, E.F.K.; Validation, V.G., G.L., S.A.; Formal Analysis, E.F.K, V.G.; Investigation, E.F.K.; Resources, E.F.K.; Data Curation, V.G., G.L., S.A.; Writing—Original Draft Preparation, E.F.K.; Writing—Review & Editing, E.F.K.; Visualization, E.F.K., V.G., G.L., S.A.; Supervision, E.F.K.; Project Administration, E.F.K.; Funding Acquisition, E.F.K.

Funding: This research was funded by Chemistry Department Ball State University, start up grant to Emil F. Khisamutdinov.

Acknowledgments: The research was supported by Department of Chemistry BSU start-up funds and in part by IAS grant # G9000602A to Emil F. Khisamutdinov. Authors are thankful to Alexander Lushnikov and Alexey Krasnoslobodtsev for performing AFM imaging of the tetragon nanoparticles at the Nanoimaging core facility at the University of Nebraska Medical Center.

Conflicts of Interest: The authors declare no conflict of interest.

References

1. Guo, P. The emerging field of RNA nanotechnology. *Nat. Nanotechnol.* **2010**, *5*, 833–842. [CrossRef] [PubMed]
2. Shukla, G.C.; Haque, F.; Tor, Y.; Wilhelmsson, L.M.; Toulme, J.J.; Isambert, H.; Guo, P.; Rossi, J.J.; Tenenbaum, S.A.; Shapiro, B.A. A boost for the emerging field of RNA nanotechnology. *ACS Nano* **2011**, *5*, 3405–3418. [CrossRef] [PubMed]
3. Parlea, L.G.; Sweeney, B.A.; Hosseini-Asanjan, M.; Zirbel, C.L.; Leontis, N.B. The RNA 3D Motif Atlas: Computational methods for extraction, organization and evaluation of RNA motifs. *Methods* **2016**, *103*, 99–119. [CrossRef] [PubMed]
4. Sweeney, B.A.; Roy, P.; Leontis, N.B. An introduction to recurrent nucleotide interactions in RNA. *Wiley Interdiscip. Rev. RNA* **2015**, *6*, 17–45. [CrossRef] [PubMed]
5. Jenner, L.; Melnikov, S.; Garreau de Loubresse, N.; Ben-Shem, A.; Iskakova, M.; Urzhumtsev, A.; Meskauskas, A.; Dinman, J.; Yusupova, G.; Yusupov, M. Crystal structure of the 80S yeast ribosome. *Curr. Opin. Struct. Biol.* **2012**, *22*, 759–767. [CrossRef] [PubMed]
6. Rupert, P.B.; Ferre-D'Amare, A.R. Crystal structure of a hairpin ribozyme-inhibitor complex with implications for catalysis. *Nature* **2001**, *410*, 780–786. [CrossRef] [PubMed]
7. Scott, W.G.; Finch, J.T.; Klug, A. The Crystal-Structure of an All-Rna Hammerhead Ribozyme—A Proposed Mechanism for Rna Catalytic Cleavage. *Cell* **1995**, *81*, 991–1002. [CrossRef]
8. Bui, M.N.; Brittany Johnson, M.; Viard, M.; Satterwhite, E.; Martins, A.N.; Li, Z.; Marriott, I.; Afonin, K.A.; Khisamutdinov, E.F. Versatile RNA tetra-U helix linking motif as a toolkit for nucleic acid nanotechnology. *Nanomed. Nanotechnol. Biol. Med.* **2017**, *13*, 1137–1146. [CrossRef] [PubMed]
9. Johnson, M.B.; Halman, J.R.; Satterwhite, E.; Zakharov, A.V.; Bui, M.N.; Benkato, K.; Goldsworthy, V.; Kim, T.; Hong, E.; Dobrovolskaia, M.A.; et al. Programmable Nucleic Acid Based Polygons with Controlled Neuroimmunomodulatory Properties for Predictive QSAR Modeling. *Small* **2017**, *13*. [CrossRef] [PubMed]
10. Khisamutdinov, E.F.; Bui, M.N.; Jasinski, D.; Zhao, Z.; Cui, Z.; Guo, P. Simple Method for Constructing RNA Triangle, Square, Pentagon by Tuning Interior RNA 3WJ Angle from 60 degrees to 90 degrees or 108 degrees. *MIMB* **2015**, *1316*, 181–193. [CrossRef]

11. Severcan, I.; Geary, C.; Verzemnieks, E.; Chworos, A.; Jaeger, L. Square-shaped RNA particles from different RNA folds. *Nano Lett.* **2009**, *9*, 1270–1277. [CrossRef] [PubMed]
12. Hong, E.; Halman, J.R.; Shah, A.B.; Khisamutdinov, E.F.; Dobrovolskaia, M.A.; Afonin, K.A. Structure and Composition Define Immunorecognition of Nucleic Acid Nanoparticles. *Nano Lett.* **2018**, *18*, 4309–4321. [CrossRef] [PubMed]
13. Khisamutdinov, E.F.; Jasinski, D.L.; Li, H.; Zhang, K.; Chiu, W.; Guo, P. Fabrication of RNA 3D Nanoprisms for Loading and Protection of Small RNAs and Model Drugs. *PANS* **2016**, *28*, 10079–10087. [CrossRef] [PubMed]
14. Afonin, K.A.; Bindewald, E.; Yaghoubian, A.J.; Voss, N.; Jacovetty, E.; Shapiro, B.A.; Jaeger, L. In vitro assembly of cubic RNA-based scaffolds designed in silico. *Nat. Nanotechnol.* **2010**, *5*, 676–682. [CrossRef] [PubMed]
15. Halman, J.R.; Satterwhite, E.; Roark, B.; Chandler, M.; Viard, M.; Ivanina, A.; Bindewald, E.; Kasprzak, W.K.; Panigaj, M.; Bui, M.N.; et al. Functionally-interdependent shape-switching nanoparticles with controllable properties. *Nucleic Acids Res.* **2017**, *45*, 2210–2220. [CrossRef] [PubMed]
16. Tuerk, C.; Gold, L. Systematic evolution of ligands by exponential enrichment: RNA ligands to bacteriophage T4 DNA polymerase. *Science* **1990**, *249*, 505–510. [CrossRef] [PubMed]
17. Ellington, A.D.; Szostak, J.W. In vitro selection of RNA molecules that bind specific ligands. *Nature* **1990**, *346*, 818–822. [CrossRef] [PubMed]
18. Bruno, J.G. A review of therapeutic aptamer conjugates with emphasis on new approaches. *Pharmaceuticals* **2013**, *6*, 340–357. [CrossRef] [PubMed]
19. Dehghani, S.; Nosrati, R.; Yousefi, M.; Nezami, A.; Soltani, F.; Taghdisi, S.M.; Abnous, K.; Alibolandi, M.; Ramezani, M. Aptamer-based biosensors and nanosensors for the detection of vascular endothelial growth factor (VEGF): A review. *Biosens. Bioelectron.* **2018**, *110*, 23–37. [CrossRef] [PubMed]
20. Farzin, L.; Shamsipur, M.; Sheibani, S. A review: Aptamer-based analytical strategies using the nanomaterials for environmental and human monitoring of toxic heavy metals. *Talanta* **2017**, *174*, 619–627. [CrossRef] [PubMed]
21. Dolgosheina, E.V.; Unrau, P.J. Fluorophore-binding RNA aptamers and their applications. *Wiley Interdiscip. Rev. RNA* **2016**, *7*, 843–851. [CrossRef] [PubMed]
22. Eydeler, K.; Magbanua, E.; Werner, A.; Ziegelmuller, P.; Hahn, U. Fluorophore binding aptamers as a tool for RNA visualization. *Biophys. J.* **2009**, *96*, 3703–3707. [CrossRef] [PubMed]
23. Stojanovic, M.N.; Kolpashchikov, D.M. Modular aptameric sensors. *JACS* **2004**, *126*, 9266–9270. [CrossRef] [PubMed]
24. Guet, D.; Burns, L.T.; Maji, S.; Boulanger, J.; Hersen, P.; Wente, S.R.; Salamero, J.; Dargemont, C. Combining Spinach-tagged RNA and gene localization to image gene expression in live yeast. *Nat. Commun.* **2015**, *6*, 8882. [CrossRef] [PubMed]
25. Nilaratanakul, V.; Hauer, D.A.; Griffin, D.E. Development and characterization of Sindbis virus with encoded fluorescent RNA aptamer Spinach2 for imaging of replication and immune-mediated changes in intracellular viral RNA. *J. Gen. Virol.* **2017**, *98*, 992–1003. [CrossRef] [PubMed]
26. DasGupta, S.; Shelke, S.A.; Li, N.S.; Piccirilli, J.A. Spinach RNA aptamer detects lead(II) with high selectivity. *Chem. Commun.* **2015**, *51*, 9034–9037. [CrossRef] [PubMed]
27. Babendure, J.R.; Adams, S.R.; Tsien, R.Y. Aptamers switch on fluorescence of triphenylmethane dyes. *JACS* **2003**, *125*, 14716–14717. [CrossRef] [PubMed]
28. Paige, J.S.; Wu, K.Y.; Jaffrey, S.R. RNA mimics of green fluorescent protein. *Science* **2011**, *333*, 642–646. [CrossRef] [PubMed]
29. Filonov, G.S.; Moon, J.D.; Svensen, N.; Jaffrey, S.R. Broccoli: Rapid selection of an RNA mimic of green fluorescent protein by fluorescence-based selection and directed evolution. *JACS* **2014**, *136*, 16299–16308. [CrossRef] [PubMed]
30. Sando, S.; Narita, A.; Hayami, M.; Aoyama, Y. Transcription monitoring using fused RNA with a dye-binding light-up aptamer as a tag: A blue fluorescent RNA. *Chem. Commun.* **2008**, 3858–3860. [CrossRef] [PubMed]
31. Dolgosheina, E.V.; Jeng, S.C.; Panchapakesan, S.S.; Cojocaru, R.; Chen, P.S.; Wilson, P.D.; Hawkins, N.; Wiggins, P.A.; Unrau, P.J. RNA mango aptamer-fluorophore: A bright, high-affinity complex for RNA labeling and tracking. *ACS Chem. Biol.* **2014**, *9*, 2412–2420. [CrossRef] [PubMed]

32. Qiu, M.K.; Khisamutdinov, E.; Zhao, Z.Y.; Pan, C.; Choi, J.W.; Leontis, N.B.; Guo, P.X. RNA nanotechnology for computer design and in vivo computation. *Philos Trans. R. Soc. A* **2013**, *371*. [CrossRef] [PubMed]

33. Shapiro, E.; Gil, B. Cell biology. RNA computing in a living cell. *Science* **2008**, *322*, 387–388. [CrossRef] [PubMed]

34. Ogihara, M.; Ray, A. Molecular computation: DNA computing on a chip. *Nature* **2000**, *403*, 143–144. [CrossRef] [PubMed]

35. Benenson, Y.; Paz-Elizur, T.; Adar, R.; Keinan, E.; Livneh, Z.; Shapiro, E. Programmable and autonomous computing machine made of biomolecules. *Nature* **2001**, *414*, 430–434. [CrossRef] [PubMed]

36. Abels Seth, G.; Khisamutdinov Emil, F. Nucleic Acid Computing and its Potential to Transform Silicon-Based Technology. *DNA RNA Nanotechnol.* **2015**, *2*, 13–22. [CrossRef]

37. Adleman, L.M. Molecular Computation of Solutions to Combinatorial Problems. *Science* **1994**, *266*, 1021–1024. [CrossRef] [PubMed]

38. Carell, T. Molecular computing: DNA as a logic operator. *Nature* **2011**, *469*, 45–46. [CrossRef] [PubMed]

39. Normile, D. Molecular computing. DNA-based computer takes aim at genes. *Science* **2002**, *295*, 951. [CrossRef] [PubMed]

40. Seelig, G.; Soloveichik, D.; Zhang, D.Y.; Winfree, E. Enzyme-free nucleic acid logic circuits. *Science* **2006**, *314*, 1585–1588. [CrossRef] [PubMed]

41. Mao, C.; LaBean, T.H.; Relf, J.H.; Seeman, N.C. Logical computation using algorithmic self-assembly of DNA triple-crossover molecules. *Nature* **2000**, *407*, 493–496. [CrossRef] [PubMed]

42. Benenson, Y. RNA-based computation in live cells. *Curr. Opin. Biotechnol.* **2009**, *20*, 471–478. [CrossRef] [PubMed]

43. Benenson, Y. Biocomputers: From test tubes to live cells. *Mol. Biosyst.* **2009**, *5*, 675–685. [CrossRef] [PubMed]

44. Benenson, Y. Engineering RNAi circuits. *Meth. Enzymol.* **2011**, *497*, 187–205. [CrossRef] [PubMed]

45. Xie, Z.; Wroblewska, L.; Prochazka, L.; Weiss, R.; Benenson, Y. Multi-input RNAi-based logic circuit for identification of specific cancer cells. *Science* **2011**, *333*, 1307–1311. [CrossRef] [PubMed]

46. Zadegan, R.M.; Jepsen, M.D.; Hildebrandt, L.L.; Birkedal, V.; Kjems, J. Construction of a fuzzy and Boolean logic gates based on DNA. *Small* **2015**, *11*, 1811–1817. [CrossRef] [PubMed]

47. Campolongo, M.J.; Kahn, J.S.; Cheng, W.L.; Yang, D.Y.; Gupton-Campolongo, T.; Luo, D. Adaptive DNA-based materials for switching, sensing, and logic devices. *J. Mater. Chem.* **2011**, *21*, 6113–6121. [CrossRef]

48. Qian, L.; Winfree, E. Scaling Up Digital Circuit Computation with DNA Strand Displacement Cascades. *Science* **2011**, *332*, 1196–1201. [CrossRef] [PubMed]

49. Shlyahovsky, B.; Li, Y.; Lioubashevski, O.; Elbaz, J.; Willner, I. Logic Gates and Antisense DNA Devices Operating on a Translator Nucleic Acid Scaffold. *ACS Nano* **2009**, *3*, 1831–1843. [CrossRef] [PubMed]

50. Zhu, J.; Li, T.; Zhang, L.; Dong, S.; Wang, E. G-quadruplex DNAzyme based molecular catalytic beacon for label-free colorimetric logic gates. *Biomaterials* **2011**, *32*, 7318–7324. [CrossRef] [PubMed]

51. Kim, J.; Yin, P.; Green, A.A. Ribocomputing: Cellular Logic Computation Using RNA Devices. *Biochemistry* **2018**, *57*, 883–885. [CrossRef] [PubMed]

52. Papenfort, K.; Espinosa, E.; Casadesus, J.; Vogel, J. Small RNA-based feedforward loop with AND-gate logic regulates extrachromosomal DNA transfer in Salmonella. *PNAS* **2015**, *112*, E4772–E4781. [CrossRef] [PubMed]

53. Schaerli, Y.; Gili, M.; Isalan, M. A split intein T7 RNA polymerase for transcriptional AND-logic. *NAR* **2014**, *42*, 12322–12328. [CrossRef] [PubMed]

54. Penchovsky, R.; Breaker, R.R. Computational design and experimental validation of oligonucleotide-sensing allosteric ribozymes. *Nat. Biotechnol.* **2005**, *23*, 1424–1433. [CrossRef] [PubMed]

55. Su, Y.; Hickey, S.F.; Keyser, S.G.; Hammond, M.C. In Vitro and In Vivo Enzyme Activity Screening via RNA-Based Fluorescent Biosensors for S-Adenosyl-l-homocysteine (SAH). *JACS* **2016**, *138*, 7040–7047. [CrossRef] [PubMed]

56. Jaffrey, S.R. RNA-Based Fluorescent Biosensors for Detecting Metabolites in vitro and in Living Cells. *Adv. Pharmacol.* **2018**, *82*, 187–203. [CrossRef] [PubMed]

57. Kang, D.; White, R.J.; Xia, F.; Zuo, X.L.; Vallee-Belisle, A.; Plaxco, K.W. DNA biomolecular-electronic encoder and decoder devices constructed by multiplex biosensors. *NPG Asia Mater.* **2012**, *4*. [CrossRef]

58. Xu, S.L.; Li, H.L.; Miao, Y.Q.; Liu, Y.Q.; Wang, E.K. Implementation of half adder and half subtractor with a simple and universal DNA-based platform. *NPG Asia Mater.* **2013**, *5*. [CrossRef]

59. Nguyen, D.H.; DeFina, S.C.; Fink, W.H.; Dieckmann, T. Binding to an RNA aptamer changes the charge distribution and conformation of malachite green. *JACS* **2002**, *124*, 15081–15084. [CrossRef]

60. Baugh, C.; Grate, D.; Wilson, C. 2.8 A crystal structure of the malachite green aptamer. *J. Mol. Biol.* **2000**, *301*, 117–128. [CrossRef] [PubMed]

61. Zadeh, J.N.; Steenberg, C.D.; Bois, J.S.; Wolfe, B.R.; Pierce, M.B.; Khan, A.R.; Dirks, R.M.; Pierce, N.A. NUPACK: Analysis and Design of Nucleic Acid Systems. *J. Comput. Chem.* **2011**, *32*, 170–173. [CrossRef] [PubMed]

62. Zuker, M. Mfold web server for nucleic acid folding and hybridization prediction. *NAR* **2003**, *31*, 3406–3415. [CrossRef] [PubMed]

63. Elbaz, J.; Lioubashevski, O.; Wang, F.; Remacle, F.; Levine, R.D.; Willner, I. DNA computing circuits using libraries of DNAzyme subunits. *Nat. Nanotechnol.* **2010**, *5*, 417–422. [CrossRef] [PubMed]

64. Pischel, U. Chemical approaches to molecular logic elements for addition and subtraction. *Angew. Chem.* **2007**, *46*, 4026–4040. [CrossRef] [PubMed]

65. Shlyakhtenko, L.S.; Gall, A.A.; Lyubchenko, Y.L. Mica functionalization for imaging of DNA and protein-DNA complexes with atomic force microscopy. *MIMB* **2013**, *931*, 295–312. [CrossRef]

nanomaterials

MDPI

Article

A Suite of Therapeutically-Inspired Nucleic Acid Logic Systems for Conditional Generation of Single-Stranded and Double-Stranded Oligonucleotides

Paul Zakrevsky [1], Eckart Bindewald [2], Hadley Humbertson [1], Mathias Viard [2], Nomongo Dorjsuren [1] and Bruce A. Shapiro [1,*]

[1]   RNA Biology Laboratory, National Cancer Institute, Frederick, MD 21702, USA;
     paul.zakrevsky@nih.gov (P.Z.); hadley.humbertson@nih.gov (H.H.); nomiko.d@gmail.com (N.D.)
[2]   Basic Science Program, Frederick National Laboratory for Cancer Research, Frederick, MD 21702, USA;
     eckart@mail.nih.gov (E.B.); mathias.viard@nih.gov (M.V.)
*   Correspondence: shapirbr@mail.nih.gov; Tel.: +1-301-846-5536

Received: 15 December 2018; Accepted: 25 March 2019; Published: 15 April 2019

Abstract: Several varieties of small nucleic acid constructs are able to modulate gene expression via one of a number of different pathways and mechanisms. These constructs can be synthesized, assembled and delivered to cells where they are able to impart regulatory functions, presenting a potential avenue for the development of nucleic acid-based therapeutics. However, distinguishing aberrant cells in need of therapeutic treatment and limiting the activity of deliverable nucleic acid constructs to these specific cells remains a challenge. Here, we designed and characterized a collection of nucleic acids systems able to generate and/or release sequence-specific oligonucleotide constructs in a conditional manner based on the presence or absence of specific RNA trigger molecules. The conditional function of these systems utilizes the implementation of *AND* and *NOT* Boolean logic elements, which could ultimately be used to restrict the release of functionally relevant nucleic acid constructs to specific cellular environments defined by the high or low expression of particular RNA biomarkers. Each system is generalizable and designed with future therapeutic development in mind. Every construct assembles through nuclease-resistant RNA/DNA hybrid duplex formation, removing the need for additional 2'-modifications, while none contain any sequence restrictions on what can define the diagnostic trigger sequence or the functional oligonucleotide output.

Keywords: RNA; RNA logic; conditional activation; functional RNA; nucleic acid therapeutic

1. Introduction

Deliverable nucleic acid-based systems present powerful methods to modulate specific gene expression and have the potential to be developed for therapeutic purposes, however, restricting their activity to a subset of intended cells remains challenging [1–3]. Numerous methods that utilize relatively small synthetic nucleic acids to regulate endogenous gene expression have originated in recent years, providing several approaches for which targeted therapeutics can be developed. The use of single-stranded antisense oligonucleotides (AON) were among the first of these regulatory techniques, whereby an oligonucleotide complementary to an mRNA of interest could be used to regulate expression of that gene [4]. AONs were originally designed to inhibit gene expression through steric inhibition of the translation machinery but have evolved into several parallel regulatory approaches. These include targeted degradation of mRNA through RNaseH activity [5], as well as alteration of mRNA splicing patterns by limiting the accessibility of specific splice sites [6]. RNA interference (RNAi) methods then followed, again evolving from the general notion of double stranded RNA being able to silence a complementary target mRNA [7], to the design of synthetic short interfering RNA (siRNA) [8], Dicer substrate siRNA (DsiRNA) [9], and other related constructs [10,11]. More recently, analogous approaches have been developed that increase rather than downregulate expression of a target gene. Delivery of double-stranded siRNA-like duplexes termed short activating RNA (saRNA) has been observed to activate gene expression when targeted to promotor regions [12], while single-stranded antagomiRs (also referred to as antimiRs) can be designed to sequester mature endogenous microRNAs and inhibit their native regulatory function [13]. Despite the development of diverse regulatory approaches, the prospect of these functional nucleic acids altering gene expression in non-afflicted healthy cells can be a cause for concern and presents a roadblock towards clinical development.

Strategies to date for cell-specific functions of nucleic acid constructs can largely be divided into two categories: targeted delivery and conditional activation. Targeted delivery most often involves the conjugation of a functionally active nucleic acid to a small molecule, protein, aptamer or other targeting agent that interacts with a receptor specifically expressed on the target cells of interest [14–17]. This approach can be effective but requires the development or identification of a ligand for the particular target cells of interest, the conjugation of such a ligand to the functional nucleic acid, and necessitates that the ligand/receptor interaction induces internalization of the nucleic acid payload.

Conditional activation represents an opposite approach to targeted delivery, in which the functional nucleic acid could be systemically delivered in an inactive state and only performs its active function in a subset of specified cells. This can be achieved through the implementation of nucleic acid logic elements that promote generation of a functional oligonucleotide construct through recognition of a specific cellular environment. The occurrence of disease can often be the result of, as well as result in, the mis-regulation of gene expression, and these differentially expressed genes can be used as biomarkers to distinguish corrupted cells from healthy tissue [18–20]. Numerous systems have been devised that incorporate nucleic acid logic elements to perform a specified function or generate predetermined molecular outputs conditional on the presence of an input oligonucleotide sequence. These include systems such as molecular beacons [21,22] and allosteric ribozymes [23,24], as well as nucleic acid strand exchange events and strand displacement cascades [25–32]. Several incarnations of nucleic acid logic systems have been previously devised which are able to release an oligonucleotide product in a conditional manner. For such approaches to be amenable to practical therapeutic application and systemic delivery, they should ideally be robust in their design to accommodate great diversity in terms of input and output oligonucleotide sequences, protect their RNA components from ribonuclease degradation, and be cheap and efficient to synthesize and produce. Existing strand exchange systems often fulfill one or two of these requirements, but rarely meet these criteria in their entirety.

Here, we present the design of several new systems for the conditional release of single-stranded (ss) and double-stranded (ds) nucleic acid constructs that are specifically tailored to meet these criteria of ideal characteristics, which many existing systems fail to adequately satisfy. These designs are influenced by several existing nucleic acid technologies such as cognate RNA/DNA hybrids, molecular beacons, and trigger-responsive multi-stranded switch constructs [21,31–34], with the aim to take the most favorable characteristics from existing systems and create derivative systems with improved features [35]. Within each construct, the diagnostic region is structurally separated from the oligonucleotide payload, resulting in systems where input and output sequences are completely decoupled and impart no sequence constraints on one another. Additionally, any RNA strands are initially bound within RNA/DNA hybrid duplexes to provide resistance from ribonuclease degradation without the need of additional 2′-modifications [31], as these modifications can increase the costs and reduce efficiency of commercial oligonucleotide synthesis. Furthermore, we expand the degree of conditional control commonly observed in systems designed for conditional generation of sequence specific dsRNA by demonstrating that conditional dsRNA release can not only be induced but also repressed upon interaction with an RNA trigger, culminating in a cognate pair of RNA/DNA hybrid constructs for which dsRNA release is under the control of multiple input triggers. As a complete collection, these novel conditional systems provide an assortment of diversity in terms of addressing different scenarios for treatment (ss vs. ds oligo release) and diagnosis (biomarker mediated induction or repression). Ultimately, this assemblage of conditional nucleic acid systems can be modified to harbor components with biologically relevant function and developed to act as conditionally regulated therapeutics.

2. Materials and Methods

2.1. Computational Considerations and RNA/DNA Hybrid Construct Design

The computational folding of individual strands and assembly of DNA/RNA constructs was assessed using Hyperfold [32], a nucleic acid structure prediction algorithm capable of predicting multi-strand assemblies from combinations of RNA and DNA strands. All folding predictions were performed at strand concentrations of 1 μM at 37 °C. The visualization and depiction of resulting secondary structure predictions was performed using Ribosketch [36]. A detailed design description of RNA/DNA hybrid pairs can be found in supporting information.

2.2. Oligonucleotide Synthesis and Purification

The DNA and RNA oligonucleotides used to assemble the conditional RNA/DNA constructs, including those that were fluorescently labeled, were purchased from Integrated DNA Technologies (IDT, Coralville, IA, USA) and reconstituted in nuclease-free water (Quality Biological, Gaithersburg, MD, USA) for use. All AlexaFluor546, AlexaFluor488 and 6-carboxyfluorescein (6-FAM) fluorescently labeled oligonucleotides were purchased from IDT. For commercially purchased oligonucleotides, 10 nmol quantities were purified as needed by denaturing polyacrylamide gel electrophoresis (PAGE). Specifically, 10 nmol quantities were mixed with 100 μL urea loading buffer (6 M Urea, 20 mM ethylenediaminetetraacetic acid (EDTA), 10% glycerol, 0.05% bromophenol blue) and heated to 90 °C for 2 min prior to loading on an 8% or 10% 19:1 acrylamide/bis-acrylamide denaturing gel (1× TBE buffer (89 mM Tris, 89 mM boric acid, 2 mM EDTA), 6 M Urea) for purification. Following electrophoresis, bands were cut from the gel and eluted in an elution buffer (10 mM Tris pH 7.5, 200 mM NaCl, 0.5 mM EDTA) overnight at 4 °C while shaken at 850 rpm. Eluted oligonucleotides were ethanol precipitated and reconstituted in nuclease-free water.

RNA trigger oligonucleotides either purchased from IDT or prepared from in vitro runoff transcription using T7 RNA polymerase. DNA templates for transcription were amplified by PCR using primers purchased form IDT. PCR was performed using MyTaq 2× mix (Bioline, Memphis, TN, USA) and purified using DNA Clean & Concentrator (Zymo Research, Irvine, CA, USA). Transcription was performed in 10 mM Tris pH 7.0 containing 6 mM $MgCl_2$, 0.5 mM $MnCl_2$, 2.5 mM each NTP, 0.01 u/μL inorganic pyrophosphatase, 2 mM dithiothreitol, and 2 mM spermidine. Approximately 50 pmol of DNA template was added to the transcription mix along with an in-house produced T7 RNA polymerase and incubated at 37 °C for 4 h. Transcription was terminated by addition of DNase I (New England Biolabs, Ipswich, MA, USA) for 30 min. The transcription mix was combined with 1/2 volume of urea loading buffer and heated at 90 °C for 2 min before purification by denaturing PAGE and precipitation as described above.

2.3. RNA/DNA Construct Assembly

Conditional RNA/DNA constructs were assembled using equimolar concentrations of their component strands. Strands were combined in water, heated to 90 °C for 1.5 min, and then immediately placed on a 37 °C heat block for 5 min. After this, samples were briefly spun in a tabletop centrifuge to collect condensed solvent and assembly buffer was added to a final 1× concentration of 2 mM $Mg(OAc)_2$, 50 mM KCl, 1× TB (89 mM Tirs, 89 mM boric acid, pH 8.2). The assembly was then incubated an additional 25 min at 37 °C. Control dsRNA duplexes and RNA trigger molecules were assembled/folded using the same protocol.

2.4. Non-Denaturing PAGE Analysis of Conditional Oligonucleotide Release

Assembled constructs were examined for their ability to regulate conditional oligonucleotide release in the presence and absence of specific RNA trigger molecules. All constructs and triggers were initially prepared separately in 1× assembly buffer. From these bulk individual assemblies, various construct/trigger combinations were combined and incubated at 37 °C for either 30, 90 or 180 min. Individual controls were prepared from the same bulk assemblies and subjected to identical incubation conditions. Generally, the conditional constructs were present at a final concentration of 500 nM. In the case of the beacon switch and adjacent targeting hybrids, RNA triggers were present in a 1× concentration relative to the conditional constructs. For inducible and repressible hybrid systems, the RNA triggers were generally present at 2×–3× concentrations, as indicated in the text. Following this incubation, samples were transferred to ice, combined with 1/5 volume of loading buffer (1× assembly buffer, 50% glycerol) and were loaded on non-denaturing PAGE gels (8–12% 19:1 acrylamide/bis-acrylamide, 2 mM Mg(OAc)$_2$, 1× TB). Electrophoresis was generally performed at 6 W for 2–3 h at 10 °C. Acrylamide concentrations and duration of electrophoresis were optimized on a case-by-case basis to achieve the necessary separation of species. In some instances, gels were subjected to total nucleic acid staining with ethidium bromide. In other instances, an individual molecule within a construct was fluorescently labeled (~10% of total molecules used in an assembly). In both cases, gels were imaged using a Typhoon Trio variable mode imager (GE Healthcare, Chicago, IL, USA) using appropriate excitation and emission filters. The amount of fluorescently labeled dsRNA output released from conditional systems was quantitated using ImageQuant 5.1 (Molecular Dynamics (now GE Healthcare), Chicago, IL, USA). Unless otherwise noted, the fraction of dsRNA released for a given sample is reported as the ratio of fluorescence observed in the released dsRNA band to the total amount of fluorescence observed for the entire lane. Statistical significance between populations was determined by a two-tailed Student's *t*-test performed using values from three distinct replicate experiments.

2.5. Analysis of RNA/DNA Strand Exchange by Förster Resonance Energy Transfer (FRET)

RNA/DNA strand exchange between cognate partners of inducible and repressible hybrid systems were examined by FRET. Cognate hybrids were assembled separately, and pre-warmed to 37 °C. Hybrids were combined and added to the cuvette, at which point the RNA trigger molecule was spiked in, if appropriate. The cuvette was immediately placed in a FluoroMax-3 fluorimeter (Jobin Yvon Horiba, Kyoto, Japan) at 37 °C and measurement was started. For FRET experiments where a fluorescence spectrum was measured at a given time point, the sense hybrid was assembled with an RNA sense strand containing a 3′ 6-FAM donor fluorophore, while the antisense hybrid was assembled with an RNA antisense strand possessing a 5′ AlexaFluor546 acceptor fluorophore. Hybrids were prepared to a final concentration of 250 mM and the trigger molecule was in three-fold molar excess, when present. Excitation was performed at 475 nm and emission measured between 480–620 nm at 1 nm increments using 0.5 s integration times and 2 nm slit widths.

For FRET experiments where time courses were recorded, the 6-FAM donor fluorophore on the RNA sense strand was replaced with AlexaFluor488. Hybrid and trigger concentration mirrored the conditions of analogous non-denaturing PAGE experiments, with hybrids at a final concentration of 500 nM, and trigger concentrations in 2–3 fold molar excess, as indicated. Measurements were recorded every 60 s using excitation at 475 nm and emission was measured at 515 nm and 565 nm, using a signal integration time of 0.5 s and slit widths of 2 nm. Observed rate constants (k_{obs}) were obtained by fitting the decrease in measured AlexaFluor488 donor fluorescence as a function of time to the equation $y = y_0 + Ae^{-k^*t}$ for single exponential decay.

3. Results

3.1. Crafting a Collection of Diverse, Logic-Based Nucleic Acid Systems Geared Towards Therapeutic Development

Multiple nucleic acid systems were designed to specifically adhere to the ideal structural and functional criteria for an RNA-based conditional therapeutic that were outlined above. Each individual assembly is composed of only RNA and DNA oligonucleotides, with no additional 2'-modified nucleotides. Every initial-state assembly was designed to contain RNA/DNA hybrid duplexes to minimize potential ribonuclease cleavage of the RNA payloads, and the diagnostic components of each system are composed entirely of DNA to prevent its processing by ribonucleases, which would likely compromise its conditional function.

The conditional systems that follow display a continuous increase in design complexity, starting with a simple bimolecular switch that is able to detect a single input biomarker and release an ssRNA oligo when the RNA biomarker is present. From there, several pairs of cognate RNA/DNA hybrid constructs are characterized that perform conditional dsRNA release though differing diagnostic mechanisms. The first utilizes a diagnostic method whereby the two cognate constructs recognize neighboring sequence regions of a single input trigger to induce dsRNA release. The subsequent RNA/DNA hybrid pairs were all designed such that the diagnostic component responsible for the RNA biomarker that governed conditional function was completely contained within one of the two hybrids, while the cognate partner hybrid recognized a biomarker-dependent structural change of the first hybrid. Using this strategy, it is first demonstrated that dsRNA release can be induced by a single input, but then also that dsRNA release can be repressed following slight alterations to the design of the diagnostic component. Ultimately, a single cognate pair of RNA/DNA hybrid constructs are coupled that are able to detect multiple RNA triggers, with the release of dsRNA being dependent on the presence of one biomarker and the absence of a second. As a whole, the suite of conditional systems provides diversity in terms of the oligonucleotide output that can be generated, and the ability to either induce or repress oligonucleotide release based on the presence or absence of a specific RNA of interest.

3.2. A Beacon-Derived Conditional Switch Releases a Single-Stranded Oligonucleotide in the Presence of an RNA Trigger

Traditional molecular beacons act as a unimolecular diagnostic tool, giving a fluorescent output signal that changes as a result of the presence of a specific oligonucleotide trigger [21] (Figure 1A). Rather than use fluorescence as an output signal, we have re-engineered the beacon system as a bimolecular switch construct that is able to release a single-stranded oligonucleotide upon recognition of a specific trigger sequence. Whereas traditional molecular beacons contain complementary regions at the 5' and 3' ends resulting in a hairpin structure (Figure 1A), the beacon switch is designed such that the output oligonucleotide is complementary across its length to the 5' and 3' ends of the diagnostic strand, generating a structure that resembles the shape of a horseshoe (Figure 1B). As with traditional molecular beacons, the diagnostic strand contains a large loop that is complementary to the trigger and serves as an internal toehold. Hybridization between the internal toehold and the trigger RNA acts as a thermodynamic driver that is intended to disrupt the pairing between the output strand and the diagnostic strand, resulting in the release of the single-stranded output.

Since the internal toehold of the diagnostic strand does not need to overlap with the 5' and 3' regions that are bound to the output oligonucleotide, essentially any set of trigger and target sequences can be implemented. The single-stranded output of the beacon switch could be composed of RNA or DNA depending on the desired function of the output strand. This conditional system could find application in instances where an irregular or diseased cellular state can be identified by a high copy number of a specific endogenous RNA, and the use of an AON, antagomir or other short single-stranded RNA would have a significant impact on rectifying the irregular state or inducing cell death.

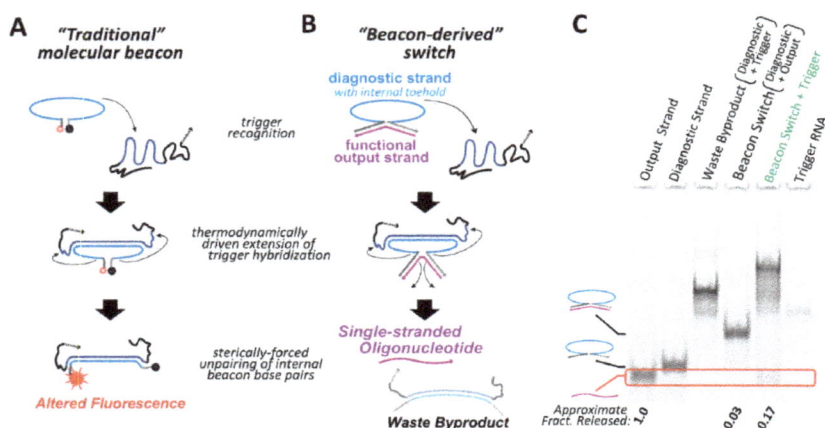

Figure 1. A conditional nucleic acid system based on the design of molecular beacons. (**A**) "Traditional" molecular beacons are fluorescence-based unimolecular diagnostic systems that adopt an initial hairpin structure. Hybridization of a trigger sequence complementary to the hairpin loop opens the hairpin and alters the fluorescence of the beacon by separating a fluorophore/quencher pair. (**B**) The "beacon-derived" switch is a biomolecular system composed of a diagnostic strand and an output strand. The output strand is hybridized to the 5′ and 3′ ends of the diagnostic strand creating a large bulge in the diagnostic strand. This bulge acts as an internal toehold. Hybridization of a trigger to this toehold region forms a persistent helix that outcompetes the internal pairing between the diagnostic and output strands, causing release of the output strand. (**C**) The conditional function of the beacon-derived switch was analyzed by 10% acrylamide non-denaturing PAGE and total staining with ethidium bromide. The beacon switch was assembled from the diagnostic and output strands. Addition of the trigger RNA to the pre-assembled beacon switch releases an output strand (red box) and shows generation of the expected waste byproduct. The fraction of output strand released was estimated by comparing the density of the output band to the output strand control lane of the same initial concentration. However, it should be noted that this approach is only semi-quantitative as it cannot be assumed that nucleic acid staining is completely uniform across the entirety of the gel. All samples were incubated for 30 min at 37 °C.

For proof-of-principle illustration, a beacon switch was designed to respond to a fragment of the Kirsten rat sarcoma proto-oncogene (KRAS) mRNA as a trigger and release an RNA antagomir output strand in a conditional fashion. Analysis of beacon switch assembly and conditional output release was performed by non-denaturing polyacrylamide gel electrophoresis (PAGE) (Figure 1C). Assembly between the diagnostic strand and output strand to form the beacon switch is extremely efficient based on non-denaturing PAGE and total nucleic acid staining, with only trace amounts of single-stranded output strand observed after assembly. Co-incubation of the assembled beacon switch with the KRAS trigger at 37 °C results in the release of the output strand and the appearance of a band corresponding to the expected waste product. A higher migrating band also appears in this lane, which is presumed to be a trinary molecular complex of the assembled beacon switch bound to the trigger. The amount of output strand observed to be released is likely a lower limit of the amount of single-stranded oligo that is actually newly accessible following interaction with the trigger RNA. This is because only one end of the single-stranded output needs to be released by the diagnostic strand to allow complete hybridization with the trigger oligonucleotide (Figure S1). However, even a partially released single-stranded output oligo should be accessible to hybridize to a target RNA and still be able to perform its intended regulatory function.

3.3. Strand Exchange between RNA/DNA Hybrid Duplexes Can Be Facilitated by Toeholds That Target Adjacent Sequence Regions of an RNA Trigger

Cognate RNA/DNA hybrid pairs were previously designed that harbor split functional RNAs, devised to release a recombined functional dsRNA through recognition of complementary single stranded toeholds (Figure 2A) [31,37,38]. The separated single strands composing the functional duplex can be referred to as the sense strand and the antisense strand, and each of these RNA strands were annealed to a complementary DNA oligo. These assembled RNA/DNA hybrids are denoted as the sense hybrid (*sH*) and the antisense hybrid (*aH*), respectively. The "traditional" approach to cognate hybrid design utilized complementary single stranded toeholds emanating from *sH* and *aH*, with hybridization of these toeholds to one another initiating RNA/DNA strand exchange. Here, we have redesigned the toeholds to be complementary to adjacent regions of an RNA trigger sequence, rather than complementary to one another. As the toeholds can no longer drive strand exchange by hybridization to one another, release of the dsRNA product is conditional on the presence of the RNA trigger molecule.

In this "adjacent targeting" incarnation of the RNA/DNA hybrid system, a fragment of the connective tissue growth factor (CTGF) mRNA was used as the RNA trigger sequence, acting as a template for DNA toehold binding which in turn initiates strand exchange (Figure 2B). Since the antisense hybrid binds upstream on the RNA trigger, it was termed aH_{UP}. Similarly, the sense hybrid is referred to as sH_{DOWN}. Binding of the cognate hybrid pair to the trigger RNA positions the two RNA/DNA hybrid regions adjacent to one another in space. The close proximity of the trigger-bound cognate hybrids will induce strand exchange through progressive hybridization of the trigger-bound DNA strands to one another, forming a three-way junction with the RNA trigger, and leading to formation and release of a dsRNA product. Like the beacon-derived switch, this activatable RNA/DNA hybrid system could find use in instances where a cell population of interest can be distinguished by the high relative expression level of an endogenous RNA. However, this RNA/DNA hybrid system (and those that follow) could be of use in cases where conditional generation of a double-stranded RNA is desirable, which could take the form of an RNA interference substrate, saRNA, aptamer, or another functionally relevant dsRNA. In this instance, the dsRNA product was designed as a 25/27-mer DsiRNA.

Formation of the dsRNA product was visualized by non-denaturing PAGE. The initial aH_{UP}/sH_{DOWN} cognate pair did not induce strand exchange and dsRNA release when co-incubated with the CTGF trigger for 180 min (Figure 2C, "0 bp"). In the presence of the RNA trigger, a large fraction of the hybrid constructs appear to be stuck in an intermediate complex displaying slow electrophoretic mobility. Presumably, this observed band corresponds to a state in which both RNA/DNA hybrids are bound to the trigger through their respective toeholds, but strand exchange in not stimulated. Despite no observed dsRNA release from this system, the strand exchange reaction is predicted to be thermodynamically favored (Figure S2). In an attempt to provide a greater driving force for strand exchange, additional sets of cognate hybrids pairs were designed in which additional complementary DNA nucleotides were inserted between the toehold region and the RNA/DNA hybrid region of each hybrid construct. These complementary nts were inserted to essentially serve as a nucleation site for strand exchange between the cognate partners once bound to the RNA trigger. In total, four additional hybrid pairs were designed which contained between 1 and 4 additional bps to seed the strand exchange (Figure 2C).

Increasing the number of complementary DNA bps inserted immediately prior to the RNA/DNA hybrid regions resulted in increased DsiRNA release (Figure 2C,D). Insertion of at least 2 DNA bps was needed to observe significant increases in DsiRNA release in the presence of the trigger RNA, as compared to background in the absence of the trigger after three hours (Table S1). Insertion of 3 bps appears to be enough to achieve close to the maximal degree of product duplex release, as increasing to 4 inserted bps results in negligible further increases in DsiRNA release after 180 min. However, the gel electrophoresis experiments suggest that insertion of additional bps does seem to speed up the rate at which this plateau of apparent maximal possible product release is reached, as the +4 bp pair releases significantly more dsRNA after 30 min than the +3 bp system, and likewise the +3 bp system shows greater release than the +2 bp hybrid pair (Figure 2D and Table S2). Despite the +3 bp and +4 bp hybrid pairs eventually reaching a similar level of dsRNA release after three hours, their differences in the fraction of dsRNA released at early time points suggests that the initiation of strand exchange within the adjacent targeting system may be impeded by slow kinetics. Interestingly, despite these systems containing complementary DNA nts that could potentially serve as toeholds to promote strand exchange in the absence of the trigger RNA, increasing the number of inserted seed base pairs up to four did not result in significant differences in the degree of non-triggered dsRNA release when co-incubated over the longest duration examined (Table 1 and Table S1).

Table 1. Summary of observed dsRNA release from connective tissue growth factor (CTGF) trigger-inducible hybrid pairs. The average fraction of dsRNA release is reported in the presence and absence of CTGF trigger, at each of three time intervals examined. An efficiency score metric is determined for each hybrid pair at a given time point, with a larger score indicating better efficiency of conditional dsRNA release. The efficacy score takes into account both the fraction of dsRNA released and the signal-to-noise ratio. It is calculated as (fraction of triggered release) * (fraction triggered release/fraction non-triggered release). The hybrid pairing that yields greatest efficiency score at each of the three time intervals examined is bolded.

Hybrid Pair		Fraction dsRNA Released, 30 min			Fraction dsRNA Released, 90 min			Fraction dsRNA Released, 180 min		
Sense Hybrid	Antisense Hybrid	Non-Triggered	CTGF-Triggered	Efficiency Score	Non-Triggered	CTGF-Triggered	Efficiency Score	Non-Triggered	CTGF-Triggered	Efficiency Score
$sH_{DOWN.0bp}$	aH_{UR0bp}	0.07 ± 0.004	0.04 ± 0.01	0.03	0.10 ± 0.03	0.06 ± 0.01	0.04	0.07 ± 0.05	0.06 ± 0.03	0.05
$sH_{DOWN+1bp}$	aH_{UP+1bp}	0.09 ± 0.02	0.08 ± 0.02	0.07	0.08 ± 0.02	0.11 ± 0.06	0.15	0.08 ± 0.08	0.12 ± 0.04	0.20
$sH_{DOWN+2bp}$	aH_{UP+2bp}	0.05 ± 0.03	0.18 ± 0.10	0.62	0.05 ± 0.01	0.31 ± 0.16	1.79	0.06 ± 0.05	0.40 ± 0.03	2.69
$sH_{DOWN+3bp}$	aH_{UP+3bp}	0.08 ± 0.01	0.39 ± 0.13	2.01	0.06 ± 0.02	0.49 ± 0.08	4.00	0.06 ± 0.02	0.63 ± 0.01	6.29
$sH_{DOWN+4bp}$	aH_{UP+4bp}	0.07 ± 0.04	0.60 ± 0.05	5.28	0.06 ± 0.01	0.60 ± 0.06	5.63	0.04 ± 0.02	0.67 ± 0.04	10.77
$sH_{CTGF12/8}$	$aH_{CTGF-cpmt.12}$	0.05 ± 0.02	0.66 ± 0.06	9.5	0.06 ± 0.02	0.78 ± 0.06	9.7	0.07 ± 0.03	0.85 ± 0.04	10.0
$sH_{CTGF12/12}$	$aH_{CTGF-cpmt.12}$	0.05 ± 0.02	0.66 ± 0.05	8.1	0.09 ± 0.01	0.81 ± 0.05	7.6	0.11 ± 0.05	0.85 ± 0.04	6.8
$sH_{CTGF16/8}$	$aH_{CTGF-cpmt.12}$	0.04 ± 0.01	0.69 ± 0.09	13.3	0.05 ± 0.01	0.78 ± 0.06	12.0	0.07 ± 0.02	0.83 ± 0.04	10.5
$sH_{CTGF20/8}$	$aH_{CTGF-cpmt.12}$	**0.02 ± 0.01**	**0.66 ± 0.06**	**17.8**	**0.04 ± 0.01**	**0.83 ± 0.06**	**17.5**	**0.05 ± 0.02**	**0.79 ± 0.05**	**13.6**
$sH_{CTGF12/8}$	$aH_{CTGF-cpmt.16}$	0.04 ± 0.01	0.24 ± 0.08	1.5	0.05 ± 0.02	0.39 ± 0.04	3.0	0.09 ± 0.01	0.46 ± 0.07	2.4
$sH_{CTGF12/12}$	$aH_{CTGF-cpmt.16}$	0.06 ± 0.01	0.37 ± 0.09	2.5	0.09 ± 0.04	0.48 ± 0.02	2.6	0.11 ± 0.04	0.55 ± 0.03	2.8
$sH_{CTGF16/8}$	$aH_{CTGF-cpmt.16}$	0.05 ± 0.01	0.61 ± 0.13	7.5	0.05 ± 0.01	0.63 ± 0.02	7.2	0.07 ± 0.02	0.67 ± 0.03	6.5
$sH_{CTGF20/8}$	$aH_{CTGF-cpmt.16}$	0.03 ± 0.01	0.50 ± 0.06	9.4	0.04 ± 0.004	0.59 ± 0.05	9.7	0.05 ± 0.01	0.64 ± 0.02	9.2

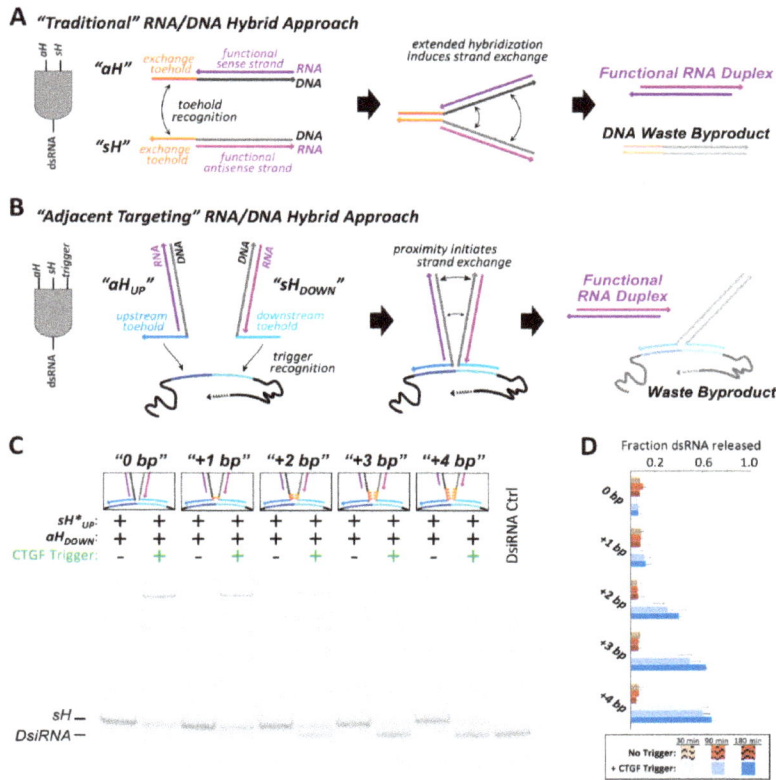

Figure 2. An RNA/DNA cognate pair system was designed to undergo conditional strand exchange by hybridizing to neighboring sites on an RNA trigger. (**A**) "Traditional" RNA/DNA hybrid pairs act as an 2-input *AND* gate. Hybridization between the single stranded toeholds of a sense hybrid (*sH*) and antisense hybrid (*aH*) initiates a thermodynamically driven strand exchange that generates a dsRNA duplex and DNA waste byproduct. (**B**) The "adjacent targeting" RNA/DNA hybrid system functions as a 3-input *AND* gate, requiring a hybrid pair as well as a specific RNA trigger sequence. The hybrid pair's respective toeholds bind to regions of the trigger that are immediately upstream and downstream from one another. Anchoring the cognate hybrids in close proximity leads to initiation of the thermodynamically favorable strand exchange reaction and dsRNA release. (**C**) Five different cognate pairs of adjacent targeting hybrids were analyzed by 12% acrylamide non-denaturing PAGE for their ability to release a DsiRNA product. Each sense hybrid and the DsiRNA control assembly contained a 3′ 6-carboxyfluorescein (6-FAM) labeled sense RNA strand for visualization. The pairs of constructs differ in the number of DNA nucleotides inserted between the single-strand toehold and the RNA/DNA hybrid duplex. These inserted nucleotides were complementary between cognate hybrids, resulting in either 0, +1, +2, +3 or +4 DNA bp that can seed the strand exchange (colored orange). The presence or absence of each component is indicated above each lane. The samples in the gel depicted were all incubated for 180 min at 37 °C. (**D**) Analysis of the fraction of dsRNA released by hybrid pairs in the presence and absence of the RNA trigger following 30, 90 or 180 min incubations at 37 °C. Error bars indicate standard deviation of three replicate experiments. Indication of statistical significance between samples is reported in the supporting information.

3.4. A Responsive Structural Element Can Act to Conditionally Induce Strand Exchange between RNA/DNA Hybrids

In an alternative approach for the implementation of conditional function within an RNA/DNA hybrid system, hybrid pairs were designed in which the accessibility of the toehold(s) needed to facilitate strand exchange was altered based on the presence or absence of a specific RNA trigger sequence. Although the adjacent targeting hybrid system described above performs its designed conditional function to release dsRNA, the fraction of dsRNA release for the best performing hybrid pair topped out at 0.67 after three hours. This second approach was pursued in an attempt to improve the efficiency of strand exchange and increase conditional dsRNA release. The "traditional" RNA/DNA hybrid methodology requiring the hybridization of complementary toeholds to one another for strand exchange serves as the basis of the conditional activation. We designate these single stranded toeholds as "exchange toeholds", since they are required to initiate strand exchange. To create a hybrid system responsive to conditional activation, a structured hairpin element was incorporated in the DNA strand immediately adjacent to the RNA/DNA hybrid duplex region of the sense hybrid (Figure 3A). This DNA hairpin ultimately controls the reassembly fate of the split functional RNA. In its initial folded state, the DNA hairpin is designed to sequester the entire length of the exchange toehold sequence within its helical stem, preventing the toehold from readily interacting with the complementary exchange toehold of the cognate antisense hybrid. The resulting hybrid pair initially exists in an "off" state that is unable to initiate strand exchange.

A new single stranded toehold, termed the "diagnostic toehold", is then implemented as a means to control the conditional activation of the hybrid by altering the accessibility of the exchange toehold imbedded within the DNA hairpin upon recognition of a specific RNA trigger sequence (Figure 3A). This single-stranded diagnostic toehold within the sense hybrid is positioned at the 5′ end of the DNA strand adjacent to the DNA hairpin (at the side opposite, the RNA/DNA hybrid region). By designing the sequence of the diagnostic toehold and the adjacent 5′ side of the DNA hairpin to be fully complementary to a region of an RNA trigger, hybridization of the trigger to the diagnostic toehold unzips the adjacent DNA hairpin and exposes the exchange toehold. Once the exchange toehold has been liberated, the complementary exchange toeholds of the hybrid pair can facilitate a strand exchange event and release a dsRNA output (Figure 3A). It is intended that this method of exchange toehold recognition, whereby the hybridization of complementary toeholds to one another forms a single duplex that can be directly extended by stacking additional DNA bps formed during RNA/DNA hybrid strand exchange, will exert a greater kinetic and/or thermodynamic drive than the three-way junction dependent method employed within the adjacent targeting system.

To illustrate the function of this "trigger-inducible" hybrid system, conditional hybrid constructs were designed to release a 25/27-mer DsiRNA when triggered by a fragment of the CTGF mRNA. The DNA strand of the sense hybrid was designed to contain a central hairpin with a 12 bp stem and 8 nt loop. This sense hybrid is referred to as *sH·*$_{CTGF.12/8}$, as the hybrid is designed to stimulate dsRNA release in the presence of CTGF ("^CTGF") and contains a DNA hairpin composed of a 12 bp stem and 8 nt loop ("12/8"). The exchange toehold within *sH·*$_{CTGF.12/8}$ is 12 nt in length and is initially completely sequestered within the DNA hairpin stem. The cognate partner hybrid is composed of an RNA/DNA hybrid duplex containing the DsiRNA antisense strand, with a 12 nt extension of the DNA strand at its 3′ end to encode the complementary exchange toehold. This hybrid is referred to as *aH·*$_{CTGF-cgnt.12}$ to reflect that it contains a 12 nt exchange toehold ("12") and is the cognate partner ("cgnt") to the CTGF-triggered *sH* hybrid ("^CTGF").

Non-denaturing PAGE and total nucleic acid staining was used to examine interactions occurring between the cognate hybrids, as well as between the hybrids and the trigger RNA (Figure 3B). While not quantitative, initial analysis using a nucleic acid stain allowed for surveillance of all molecular species and products. As expected, no changes to the hybrids' electrophoretic mobility is observed when incubated together at 37 °C in the absence of the trigger RNA, indicating that no interaction occurs between the hybrids and no dsRNA is released. Introduction of the RNA trigger activates *sH·CTGF.12/8* and induces the release of a dsRNA product when *aH·CTGF-cognt12* is also present. Higher migrating species are also observed when both hybrids are co-incubated with the trigger RNA. One of the high migrating bands corresponds to the expected waste product as indicated by similar migration of a control assembled from the RNA trigger and two DNA strands. An even slower migrating band is also observed and is likely to be a 5-molecule intermediate complex. Förster resonance energy transfer (FRET) experiments were performed to further verify the generation of the expected double stranded RNA product in the presence of the trigger molecule (Figure 3C). The cognate hybrids used for these FRET studies had a 3′ donor fluorophore on the RNA sense strand, and a 5′ acceptor fluorophore on the RNA antisense strand. In the absence of the RNA trigger, the FRET-labeled hybrid pair show no significant change in their emission spectrum after one hour at 37 °C. However, one hour after the introduction of the CTGF trigger a large decrease in donor emission (~515 nm) and increase in acceptor fluorescence (~565 nm) is observed, indicating formation of the DsiRNA duplex product.

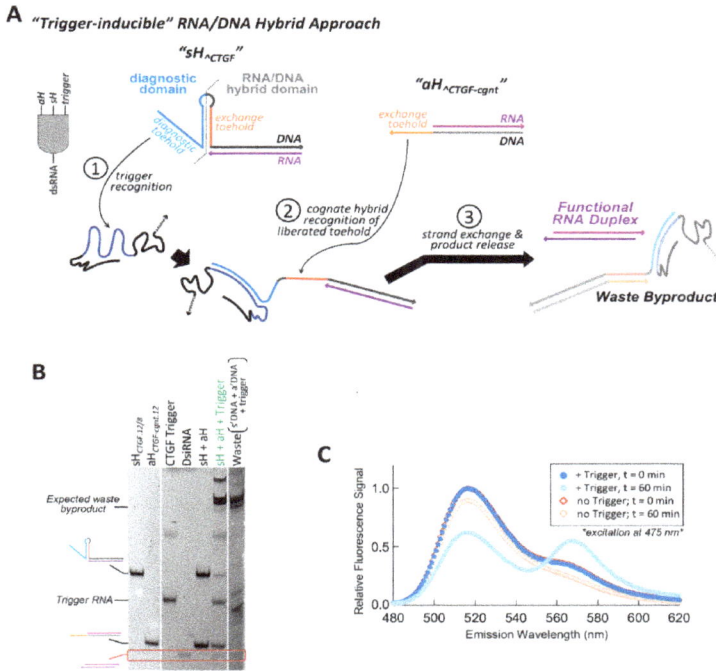

Figure 3. Incorporation of a structured responsive element can generate a trigger-inducible RNA/DNA hybrid system. (**A**) The inducible hybrid system functions as a three-input *AND* gate. The sense hybrid *sH·CTGF.12/8* contains a responsive DNA hairpin composed of a 12 bp stem and an 8 nt loop, and is flanked by an extended 5′ single strand that acts as a diagnostic toehold. Trigger hybridization to the diagnostic toehold progresses through the hairpin stem and unzips the hairpin (sequence regions colored blue). This liberates a previously sequestered toehold within *sH·CTGF.12/8* which can then hybridize with the complementary toehold of the cognate antisense hybrid, *aH·CTGF-cgnt.12*. Hybridization of these exchange toeholds (sequence regions colored orange) initiates strand exchange and releases a dsRNA product. (**B**) The function of this conditional system was assessed by 8% acrylamide non-denaturing PAGE and total staining with ethidium bromide.

DsiRNA release is observed when the sense and antisense hybrids are co-incubated in the presence of trigger (red box). Formation of the expected waste product is observed by comparison to a control assembly of the s′ and a′ DNA strands with the trigger molecule. All samples were incubated for 30 min at 37 °C. (C) Förster resonance energy transfer (FRET) analysis was performed as another method to verify conditional dsRNA formation. $sH \cdot_{CTGF.12/8}$ was assembled using a 3′ 6-carboxyfluorescein (6-FAM) (ex/em 495/520 nm) labeled sense RNA strand. $aH \cdot_{CTGF\text{-}cgnt.12}$ was assembled using a 5′-AlexaFluor546 (ex/em 555/570 nm) labeled antisense RNA strand. The hybrids were mixed and incubated at 37 °C for one hour in the presence or absence of the RNA trigger. Fluorescence emission spectra were recorded at t = 0 and t = 60 min using excitation at 475 nm.

3.5. Alteration of Structural Elements Was Explored as a Means to Optimize dsRNA Release from Cognate RNA/DNA Hybrids

Within the inducible hybrid system, the accessibility of one exchange toehold is impeded by being sequestered within a responsive DNA hairpin. This toehold becomes liberated upon opening of the hairpin in the presence of an RNA trigger and allows for a strand exchange to proceed. As such, altering the stability of this responsive hairpin structure, as well as the length and accessibility of the liberated exchange toehold once the hairpin is open, could potentially modulate the degree of strand exchange between a cognate hybrid pair. The initially characterized $sH \cdot_{CTGF.12/8}$ hybrid contained a 12 bp DNA hairpin stem capped by an 8 nt loop. Three additional CTGF-triggered sH hybrids were designed to investigate how changing the structure of the responsive DNA hairpin affects strand exchange and dsRNA release (Figure 4A). The first of these variants maintains a 12 bp DNA hairpin stem but expands the hairpin loop from 8 to 12 nts. This hybrid is denoted $sH \cdot_{CTGF.12/12}$. The two additional sH variants maintain the original 8 nt hairpin loop, but contain hairpin stems of 16 and 20 bps in length. These hybrids are named $sH \cdot_{CTGF.16/8}$ and $sH \cdot_{CTGF.20/8}$, respectively.

Each of the four $sH \cdot_{CTGF}$ constructs were assembled with a fluorescently labeled RNA sense strand to quantitatively examine their ability to liberate a dsRNA duplex following strand exchange with $aH \cdot_{CTGF\text{-}cgnt.12}$ in the presence and absence of the CTGF trigger RNA (Figure 4B). Interestingly, analysis using fluorescently labeled constructs revealed that the various $sH \cdot_{CTGF}$ /$aH \cdot_{CTGF\text{-}cgnt.12}$ hybrid pairs release a small fraction of dsRNA when incubated together in the absence of the trigger RNA. This was not originally observed in the initial qualitative experiments that utilized staining with ethidium bromide. The degree of non-triggered release among pairs of hybrid constructs was relatively minor after 30 min (~2–5% of signal) and was observed to marginally increase over time for each variant $sH \cdot_{CTGF}$ construct paired with $aH \cdot_{CTGF\text{-}cgnt.12}$ (Figure 4C). $sH \cdot_{CTGF.20/8}$, which is predicted to contain the most stable hairpin stem (Figure S3), exhibited the smallest degree of non-triggered DsiRNA release compared to other hybrids pairs after 30 min. Likewise, $sH \cdot_{CTGF.12/12}$ was predicted to have the weakest hairpin structure and displayed the greatest extent of non-triggered DsiRNA release after 30 min. This trend persists at longer time points; however, differences in non-triggered DsiRNA release among variant hybrids pairs were not all statistically significant, especially at longer time points (Table S3).

Structural changes to the responsive DNA hairpin of the $sH \cdot_{CTGF}$ hybrids resulted in negligible differences in trigger-induced dsRNA release between the four $sH \cdot_{CTGF}$/$aH \cdot_{CTGF\text{-}cgnt.12}$ pairs assayed (Figure 4C). However, these constructs did show a 12–18% improvement in conditional dsRNA release over the best performing adjacent targeting hybrid pair after three-hour incubations with the CTGF trigger (Table 1). The lack of differences in triggered DsiRNA release among the variant $sH \cdot_{CTGF}$ constructs was somewhat surprising based on the predicted change in free energy ($\Delta\Delta G$) between the unbound and CTGF trigger-bound states for each hybrid's responsive DNA element (Figure S3). However, it may be that the favorable change in free energy for each construct upon trigger binding is so great ($\Delta\Delta G < -25$ kcal mol^{-1} for each) that the comparatively small differences in $\Delta\Delta G$ between the various $sH \cdot_{CTGF}$ hybrids becomes inconsequential. Alternatively, differences in the $\Delta\Delta G$ of trigger binding could be offset by differences in steric accessibility of the newly liberated exchange toehold once the DNA hairpin has opened. Increasing the loop size or length of the hairpin stem increases the distance between the exchange toehold and the region bound by the RNA trigger once hybridized (Figure S3). This could in turn alter the accessibility of the liberated exchange toehold to the incoming cognate hybrid. The $sH\cdot_{CTGF.12/8}/aH\cdot_{CTGF-cgnt.12}$ hybrid pair has the shortest nucleotide distance between the region bound by the trigger and its exchange toehold, and time course FRET experiments indicate the observed rate constant of dsRNA release is slower for this hybrid pairing than for any of the other three $sH\cdot_{CTGF}$ hybrids paired with $aH\cdot_{CTGF-cgnt.12}$ (Figure S4).

Extending the length of the exchange toehold was also explored as a means to boost triggered dsRNA release within the CTGF-inducible hybrid system. A variant $aH\cdot_{CTGF-cgnt}$ hybrid was designed containing a 16 nt toehold and was termed $aH\cdot_{CTGF-cgnt.16}$. The toehold of $aH\cdot_{CTGF-cgnt.16}$ was designed to encode the same 12 nt sequence as the $aH\cdot_{CTGF-cgnt.12}$ toehold, with four additional nucleotides appended to the toehold's distal end. These four additional nucleotides are complementary to corresponding regions within $sH\cdot_{CTGF.16/8}$ and $sH\cdot_{CTGF.20/8}$ and result in complete pairing of the 16 nt exchange toeholds between these cognate hybrids. However, these four added nucleotides at the distal end of the $aH\cdot_{CTGF-cgnt.16}$ toehold do not have complementary sequences in $sH\cdot_{CTGF.12/8}$ and $sH\cdot_{CTGF.12/12}$ (Figure 4A), leaving the distal end of the $aH\cdot_{CTGF-cgnt.16}$ toehold unpaired.

Increasing the toehold length of the cognate antisense hybrid from 12 to 16 nucleotides was observed to have a negative impact on DsiRNA release when paired with any of the $sH\cdot_{CTGF}$ variants (Figure 4D,E). The use of $aH\cdot_{CTGF-cgnt.16}$ in place of $aH\cdot_{CTGF-cgnt.12}$ had a negligible effect on the degree of non-triggered release but presented a large significant impediment to CTGF-triggered release in nearly all instances (Table S4). The extent of diminished triggered-release was most pronounced when $aH\cdot_{CTGF-cgnt.16}$ was paired with $sH\cdot_{CTGF.12/8}$ and $sH\cdot_{CTGF.12/12}$, suggesting that having non-complementary nucleotides at the distal end of the $aH\cdot_{CTGF-cgnt.16}$ toehold interferes in some manner with the ability of the hybrids to promote the strand exchange reaction. As a way to compare the overall performance of the each conditionally-active hybrid pair, an "efficiency score" was determined for each time point examined. This efficiency score metric was calculated as the product of the fraction of triggered dsRNA release and the signal-to-noise ratio (triggered/non-triggered release). Larger scores indicated greater efficiency of conditional dsRNA release. Out of the eight pairs of CTGF-inducible hybrids and the five sets of adjacent-targeting hybrids, the $sH\cdot_{CTGF.20/8}/aH\cdot_{CTGF-cgnt.12}$ pairing displays the highest efficiency score for each time interval that was examined (Table 1).

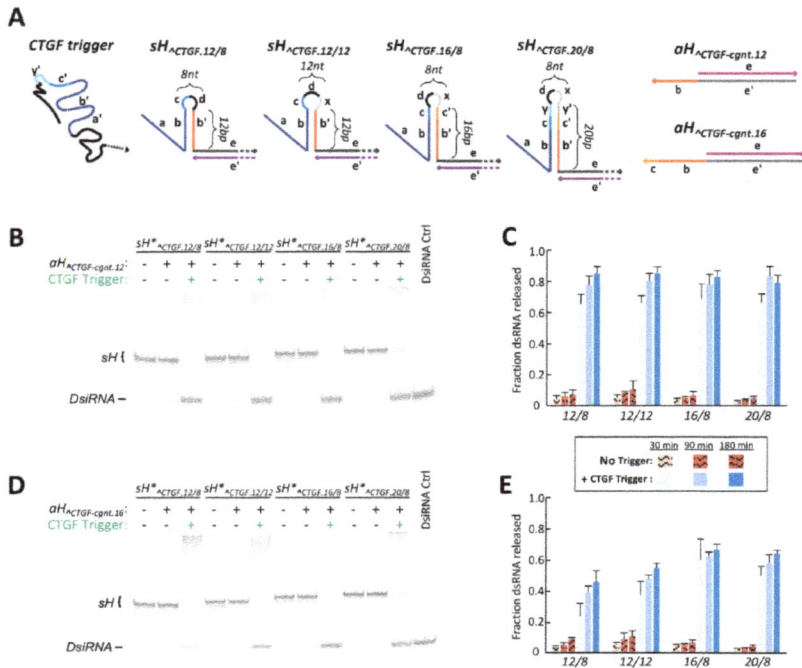

Figure 4. Effects of DNA structural alteration on the degree of trigger-inducible dsRNA release. (**A**) Four different sense hybrids that are responsive to the connective tissue growth factor (CTGF) trigger were designed, each having different features within the structured DNA hairpin. The hairpins differed in the size of their loop or the length of their stem. Two different cognate antisense hybrids were designed and differ in the length of their single-stranded toehold. Sequence regions are indicated by lowercase letters and different colors to convey sequence identity or sequence complementarity. (**B,D**) DsiRNA release in the presence and absence of trigger was assessed by 10% acrylamide non-denaturing PAGE for each sense hybrid paired with a cognate antisense hybrid exhibiting either (**B**) a 12 nt toehold ($aH_{\cdot CTGF\text{-}cgnt.12}$) or (**D**) a 16 nt toehold ($aH_{\cdot CTGF\text{-}cgnt.16}$). Each sense hybrid and the DsiRNA control contained a 3′ 6-carboxyfluorescein (6-FAM) labeled sense RNA strand for visualization and quantification. Gels in both (**B**) and (**D**) depict samples that were incubated for 30 min at 37 °C. (**C,E**) Analysis of the fraction of dsRNA released by the four sense hybrids paired with (**C**) $aH_{\cdot CTGF\text{-}cgnt.12}$ or (**E**) $aH_{\cdot CTGF\text{-}cgnt.16}$, in the presence and absence of the RNA trigger following 30, 90, or 180 min incubations at 37 °C. Error bars indicate standard deviation of three replicate experiments. Indication of statistical significance between samples is reported in the supporting information.

3.6. Redesigned Responsive Structural Elements Can Be Used to Inhibit Strand Exchange and Repress dsRNA Release

The concept and method of toehold sequestration used to impart conditional function within the trigger-inducible RNA/DNA hybrid system can be modified and redesigned to instead allow for the repression of strand exchange in the presence of a specific RNA trigger and thereby expands the degree of control over dsRNA release. In this embodiment, both exchange toeholds are initially free to undergo strand exchange, but one becomes sequestered into a DNA hairpin when interaction with an RNA trigger facilitates a structural rearrangement of that hybrid's responsive structural element (Figure 5A). Such a system would be of interest in situations where a cellular state of interest cannot be identified by the high expression of a particular RNA, but rather by a significant under expression of a specific RNA relative to the normal population.

Whereas the previously described inducible hybrid pairs contain a responsive hairpin element within the DNA strand of *sH* that is triggered by CTGF, the repressible hybrid pair contains a responsive DNA element within *aH* that is responsive to the KRAS mRNA-derived trigger. This new hybrid is termed "$aH \vee_{KRAS}$" to indicate that dsRNA release from the hybrid is negatively impacted by the KRAS trigger. In the absence of the cognate RNA trigger, the most stable DNA fold of $aH \vee_{KRAS}$ is that which results in a single stranded exchange toehold and a 14 bp DNA hairpin (Figure S5). When the trigger is present, however, it can bind to the 3′ diagnostic toehold present in $aH \vee_{KRAS}$ and proceed to unzip the 14 bp hairpin, as the trigger is complementary to the entire 3′ side of the hairpin stem. As the initial 14 bp hairpin can no longer form, a structural rearrangement can occur where the exchange toehold pairs to the 12 nts that compose the apical loop of the original hairpin. This new hairpin structure makes the exchange toehold inaccessible to the cognate hybrid and represses the ability for the hybrid pair to release a dsRNA duplex (Figure 5A).

The ability to repress hybrid strand exchange was examined for $aH \vee_{KRAS}$ with its cognate hybrid, "$sH \vee_{KRAS-cgnt}$", that contains a complementary 12 nt DNA exchange toehold extending from its RNA/DNA hybrid region. Analysis by non-denaturing PAGE at several time points illustrates that the cognate hybrids successfully undergo strand exchange and release dsRNA in the absence of the KRAS trigger (Figure 5B,C). However, when the KRAS trigger and $sH \vee_{KRAS-cgnt}$ hybrid are premixed and added simultaneously to $aH \vee_{KRAS}$, DsiRNA release is repressed more than 3-fold compared to in the absence of KRAS. A second context was also examined, where $aH \vee_{KRAS}$ was permitted to interact with the KRAS trigger for five minutes prior to the addition of the cognate $sH \vee_{KRAS-cgnt}$ hybrid. This scenario allowed the responsive DNA hairpin to rearrange and adopt its alternative "off"-state structure before the cognate exchange toehold was present in the reaction mix. In this context, DsiRNA release is reduced 12-fold after 30 min at 37 °C compared to in the absence of trigger, and maintains more than 7-fold repression after 3 h (Figure 5C). FRET experiments further illustrate that strand exchange occurs quickly and efficiently in the absence of the KRAS trigger, but is severely impeded upon introduction of KRAS (Figure S6).

To illustrate that repression of dsRNA release is dependent on the presence of a trigger RNA with a specific nucleotide sequence, additional non-cognate trigger molecules were co-incubated with the repressive hybrid pair. Neither of the non-cognate trigger molecules tested resulted in a reduction in dsRNA release (Figure S7). This same degree of trigger specificity is observed for the CTGF-inducible hybrid system, as the $sH \cdot_{CTGF.20/8}/aH \cdot_{CTGF-cgnt.12}$ hybrid pair are only observed to initiate dsRNA release in the presence of the CTGF trigger, and not when co-incubated with non-cognate trigger molecules (Figure S7). An orthogonal trigger-repressible system was also designed that is responsive to CTGF rather than KRAS, as a means to demonstrate versatility in accommodating various trigger sequence inputs, as well as an ability to position the response element at different locations within this generalized conditional system. In this system, the CTGF responsive DNA element was added to the sense hybrid rather than the antisense hybrid. Nonetheless, this cognate hybrid pair ($sH \vee_{CTGF}/aH \vee_{CTGF-cgnt}$) displays a repressed ability to generate dsRNA in the presence of the CTGF trigger, as intended (Figure S8).

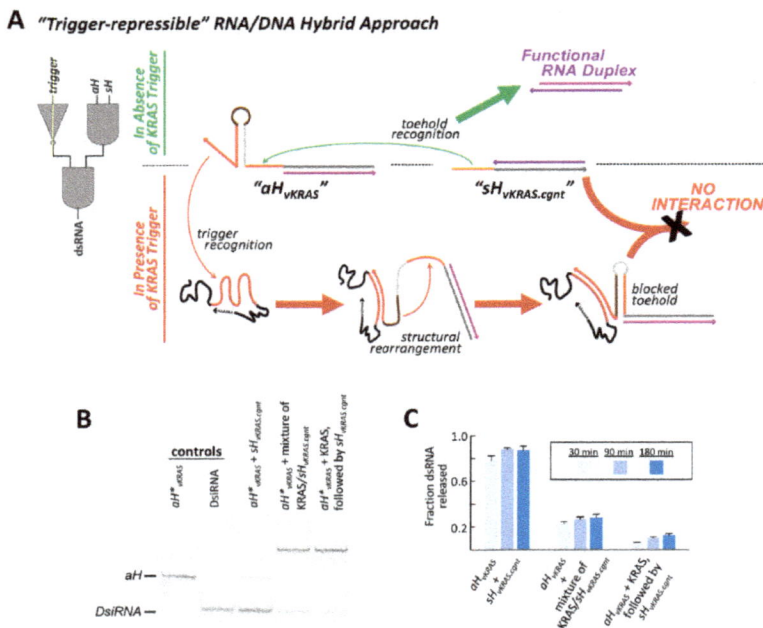

Figure 5. A redesign of the structured DNA responsive element allows for trigger-based conditional repression of dsRNA release. (**A**) A trigger-repressive hybrid system can be designed by combining a 2-input *AND* gate with a *NOT* gate. The antisense hybrid, $aH\vee_{KRAS}$, is designed to repress strand exchange in the presence of a trigger sequence derived from the Kirsten rat sarcoma proto-oncogene (KRAS) mRNA. If the trigger is absent, the exchange toehold of $aH\vee_{KRAS}$ (sequence region colored orange) is freely accessible and can promote dsRNA release. If the trigger is present, its hybridization to the diagnostic toehold of $aH\vee_{KRAS}$ (sequence region colored red) results in a structural rearrangement that blocks access to the exchange toehold and prevents interaction with the cognate sense hybrid, $sH\vee_{KRAS-cgnt}$. (**B**) The conditional function of this repressible system was assessed by 10% acrylamide non-denaturing PAGE. DsiRNA release from the $aH\vee_{KRAS}$ /$sH\vee_{KRAS-cgnt}$ pair was examined in three contexts: in the absence of the KRAS trigger (middle lane), when $sH\vee_{KRAS-cgnt}$ and the KRAS trigger are premixed and added simultaneously to $aH\vee_{KRAS}$ (2nd lane from right), or when $aH\vee_{KRAS}$ and the KRAS trigger are preincubated for 5 min prior to $sH\vee_{KRAS-cgnt}$ addition (right lane). The KRAS trigger was added in 3-fold excess in both cases. The depicted gel shows samples incubated for 180 min at 37 °C once all components are present. (**C**) Analysis of the fraction of dsRNA released from the KRAS repressible system following 30, 90, or 180 min incubations at 37 °C. Error bars indicate standard deviation of three replicate experiments. Indication of statistical significance between samples is reported in the supporting information.

3.7. Cognate Hybrids Pairs with Multiple Responsive Elements Allow for Multi-Trigger Regulation

Because the strand exchange reaction between cognate hybrid partners is dependent on the accessibility of a specific toehold sequence (exchange toehold) present on each of the two hybrids, it is possible to generate a system in which the accessibility of each toehold is under the control of a different RNA trigger sequence. In the case of the trigger-repressible hybrids, such as $aH\vee_{KRAS}$, the trigger RNA imparts no sequence constraints on the exchange toehold and allows the exchange toehold to be any sequence that permits proper folding. With this in mind, the exchange toehold of construct $aH\vee_{KRAS}$ was designed to be complementary to the exchange toehold of the $sH\cdot_{CTGF}$ hybrids characterized previously.

Hybrid construct $sH_{\text{·}CTGF.20/8}$ was partnered with $aH\vee_{KRAS}$ to generate a pair of conditional RNA/DNA hybrids whose function is dependent on the presence or absence of two RNA triggers, CTGF and KRAS (Figure 6A). The strand exchange reaction between these two hybrids is initially inhibited, as $sH_{\text{·}CTGF.20/8}$ initially exists in an "off" state and requires interaction with the CTGF trigger to promote strand exchange. $aH\vee_{KRAS}$ is initially in an active state; however, the exchange toehold of $aH\vee_{KRAS}$ becomes inaccessible upon interaction with the KRAS trigger. For efficient strand exchange to occur between this hybrid pair, the presence of the CTGF trigger is required, as well as the absence of the KRAS trigger.

The degree to which dsRNA could be conditionally released from this cognate hybrid pair was assessed by non-denaturing PAGE (Figure 6B) and FRET (Figure S9). In the absence of any trigger molecules, the $sH_{\text{·}CTGF.20/8}/aH\vee_{KRAS}$ hybrid pair releases very small amounts of dsRNA when co-incubated. The addition of the KRAS trigger to the hybrid pair reduces the degree of dsRNA release close to zero. However, if the CTGF trigger is added rather than the KRAS trigger, substantial release of dsRNA product occurs, as expected. Sequential addition of the KRAS trigger followed by the CTGF trigger to the hybrid pair results in very little dsRNA generation, suggesting that $aH\vee_{KRAS}$ inactivation by the KRAS trigger occurs relatively quickly. Additional characterization was performed to examine how differences in the relative concentration of the two triggers affect dsRNA release. Various ratios of CTGF and KRAS trigger molecules were premixed and added to the co-incubating $sH_{\text{·}CTGF.20/8}/aH\vee_{KRAS}$ hybrid pair. As might be expected, increasing the relative amount of CTGF trigger (activating) to KRAS trigger (deactivating) increases the extent of dsRNA release (Figure S10). When equal amounts of the KRAS and CTGF triggers are added to the hybrid pair, the degree of dsRNA release is about 60% of the maximum amount of dsRNA released when an excess of CTGF trigger is added to the hybrids in the absence of the KRAS trigger. However, when the ratio of CTGF/KRAS triggers is varied away from 1:1, induction/repression of dsRNA release disproportionately favors the trigger that is present in a greater amount, beyond what would be predicted based on the trigger stoichiometry (i.e.,: when a 3:2 ratio of KRAS/CTGF is present, the fraction of dsRNA is less than 40% of the maximal dsRNA released in the absence of any KRAS).

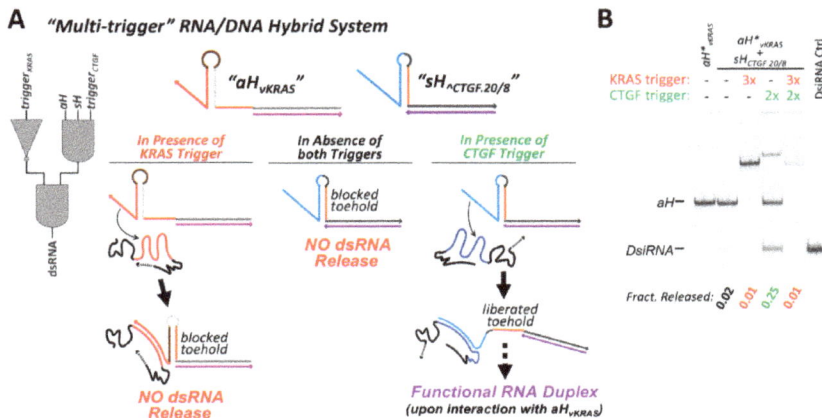

Figure 6. Multi-trigger systems can be composed in which each RNA/DNA hybrid contains a responsive DNA structural element. (**A**) A system comprising a 3-input *AND* gate and a *NOT* gate can be constructed by pairing $sH_{\text{·}CTGF.20/8}$ (activated by the connective tissue growth factor (CTGF) derived trigger) with $aH\vee_{KRAS}$ (repressed by the Kirsten rat sarcoma proto-oncogene (KRAS) mRNA derived trigger). Co-incubation of the two hybrids results in no interaction. Both hybrids and the CTGF trigger are required for dsRNA release, while the presence of the KRAS trigger will inhibit strand exchange.

(**B**) The multi-trigger system was assessed by 10% acrylamide non-denaturing PAGE. The fraction of DsiRNA released is indicated in the gel depicted, in the presence of indicated trigger combinations following 30 min incubation at 37 °C. The *sH* and *aH* hybrid were present at equimolar concentration, while the triggers were added at a 2-fold or 3-fold excess, as indicated. In samples when both triggers are present, they were added to premixed hybrids sequentially (KRAS followed by CTGF). The antisense hybrid and DsiRNA control in were assembled using a 5′-AlexaFluor546 labeled antisense RNA strand for the purpose of visualization and quantification.

3.8. Three-Strand RNA/DNA Hybrid Constructs Allow "Activated" Hybrids to Dissociate from Their Cognate Trigger

With both of the inducible and repressible conditional systems described above, the entirety of the hybrid construct containing the diagnostic toehold remains bound to the RNA trigger molecule following recognition and hybridization of the diagnostic toehold. However, one can imagine that there may be instances where the function of the conditional hybrid systems may benefit from allowing their RNA/DNA hybrid domains to freely diffuse away from their cognate trigger following hybridization through their diagnostic toehold/domain. A three-strand design approach was used to create an inducible hybrid that separates from the RNA trigger after hybridization. The design is based on that of the $sH_{\cdot CTGF.20/8}$ hybrid. The 8 nt hairpin loop is removed, splitting the 20 bp hairpin into a duplex that assembles from two distinct DNA strands. One DNA strand retained the 5′ diagnostic toehold, while the other maintained the RNA/DNA hybrid region (Figure S11). This new three-strand hybrid was termed "$sH_{\cdot CTGF.20split}$" and works in conjunction with $aH_{\cdot CTGF-cgnt.12}$. Analysis by non-denaturing PAGE illustrates that the three-piece hybrid $sH_{\cdot CTGF.20split}$ appears to function very similarly to that of $sH_{\cdot CTGF.20/8}$, although the three-piece hybrid seems to have a slight increase in its degree of non-triggered dsRNA release (Figure S11).

A similar approach was used to investigate a three-strand repressible hybrid construct based on the $aH{\vee}_{KRAS}$ hybrid. A nick was positioned within the 5′ strand of the DNA hairpin, aiming to maintain stable formation of the initial 14 bp hairpin and allow strand exchange in the absence of the KRAS trigger. Four different variants were designed to identify a nick position that retained the greatest conditional function. The function of the four variants partnered with $sH{\vee}_{KRAS-cgnt}$ was examined by non-denaturing PAGE (Figure S12). Each of the three-strand repressible systems tested show a diminished ability to promote desirable dsRNA release in the absence of the KRAS trigger compared to the original design. This may stem from the possibility that a larger fraction of the three-strand hybrids initially adopt their "off" state when assembled. However, some three-strand systems did retain their repressible function. "$aH{\vee}_{KRAS.nick14}$", where the nick was placed immediately below the hairpin loop and preserves the entire 14 bp stem, displayed the greatest degree of conditional function. Progressively moving the nick down the stem resulted in continued loss of the responsive function to the KRAS trigger.

4. Discussion

Due to their ability to easily store and recognize information at the level of their primary sequence, nucleic acids serve as an excellent material for the construction of logic elements and performance of molecular computing [39,40]. Nucleic acid strand displacement and conformational change driven by single-stranded toehold interactions have found use in a wide variety of applications [41], with many great successes achieved by utilizing these techniques to perform complex diagnostics both in the test-tube [42–44] and in-cell [25,30,45]. However, the development of conditional therapeutics using similar logic-based diagnostic elements has lagged behind. In part, this is likely due to an increase in design constraints associated with development of therapeutics, as both the input and output oligonucleotides need to adhere to predetermined sequences. In the case of logic-driven diagnostic systems, the reporter output often induces translation of fluorescent proteins or the release of fluorescent nucleic acid probes, each of which are governed by structural changes within strands whose primary sequences are largely malleable to fulfill structural requirements. In most cases, these diagnostic systems function by the opening of a structure, or releasing a single-stranded oligonucleotide, both of which tend to be easier functions to design than the generation and release of a double-stranded duplex required in the case of RNAi-based applications. These increases in design complexity are evident in the conditional systems presented above. Conditional release of a single-stranded oligonucleotide that is triggered by a single input can be regulated by using a small, simple bimolecular system, whereas the release of duplex RNA similarly governed by a single input required much more complexity.

For a conditional RNA-based therapeutic system to be considered for practical application, it likely needs to fulfill three essential criteria: first, the RNA components must be protected from potential ribonuclease degradation; second, the sequences of the trigger input and the oligonucleotide output must be functionally and structurally independent from one another, such that either can be altered without necessitating a change in the other; and, third, the systems must be cheap and easy to produce, meaning that the use of modified nucleotides should be kept to a minimum. Previous incarnations of conditional RNA therapeutic systems have struggled to simultaneously fulfill these criteria. An early scheme for conditional Dicer substrate generation by Masu et al. utilized a hairpin design similar to that of a molecular beacon, where trigger binding to the loop unpairs an adjacent stem and allows formation of the functional sense-antisense hybrid [27]. This construct was successful at separating the trigger and target sequence elements, and demonstrated conditional gene silencing in HeLa cells. However, the system required co-delivery of a completely single-stranded cognate antisense strand leaving it vulnerable to ribonuclease degradation, while the hairpin construct required significant 2′-O-methyl (2′-OMe) modification to prevent Dicer from processing the initial hairpin stem.

A similar approach was designed by Xie et al. whereby an antisense strand was pre-annealed to a chemically modified "protector" strand and this complex was co-delivered with a single-stranded sense RNA. The protector strand contained significant overhangs on each end and was completely complementary to a trigger RNA, allowing hybridization to the trigger to drive antisense release and siRNA formation [28]. This approach though fails to decouple trigger and output sequences, while suffering from the same ribonuclease susceptibility and extensive modification issues of the hairpin based approach. Hochrein et al. have constructed some of the most promising systems conceptually for the generation of a Dicer substrate siRNA. Their system utilizing stable small conditional RNAs both effectively separates trigger and output sequence requirements and provides nuclease protection of functional RNA regions through hybridization to 2′-OMe RNA [29], but again the significant use of chemically modified nucleotides can hinder efficient and cost effective synthesis. This use of extensive 2′-modification to not only protect regions of functional RNA, but also to prevent off-site Dicer cleavage is a reoccurring theme among these previous generations of conditional systems. Work by Kumar et al. explored a completely different approach by genetically expressing a pri-miRNA-like conditional construct from a plasmid [26]. While the system was able to perform conditional silencing in cell culture, this system required the delivery of modified oligonucleotide triggers to induce activation and processing by Drosha. Additionally, the need to prepare and express constructs from a plasmid brings additional complications.

The original incarnation of split-function RNA/DNA hybrid pairs from Afonin et al. lacks a true diagnostic component that would be able to sense an RNA biomarker but was able to successfully separate the output oligonucleotide sequences from the toehold regions that controlled the conditional strand–exchange reaction. Importantly, they were able to demonstrate that formation of an RNA/DNA hybrid duplex both protected the RNA from degradation and prevented the initial duplexes from being processed by Dicer [31]. As described above, these "traditional" RNA/DNA hybrid duplexes served as a jumping off point for the development of a new generation of conditional RNA-based therapeutics, by inserting or appending new diagnostic components able to respond to the presence of RNA biomarkers. These new systems fulfill our outlined criteria for conditional therapeutics by combining the nuclease resistance of the RNA/DNA hybrid technology with independent diagnostic components. In addition, the earlier generations of conditional systems described above each functioned as Boolean *AND* gates, releasing a double stranded RNAi substrate based on the presence of a single RNA trigger. We demonstrate in this current work that these next-generation RNA/DNA hybrid systems can be based on *AND* gates, *NOT* gates and combinatorial systems able to sense multiple biomarker triggers. Furthermore, although the proof of principle demonstrations we have presented here release a single dsRNA duplex from each RNA/DNA hybrid system, it should be possible to release multiple different dsRNA products in a conditional fashion from a single pair of conditional RNA/DNA hybrids, as this has been demonstrated previously with the "traditional" hybrid approach [37]. As an assembled collection, this suite of nucleic acid logic systems represents a robust toolkit for the conditional generation of nucleic acid species that encode a specific sequence and function while serving as a foundation for future therapeutic development.

Supplementary Materials: The following are available online at http://www.mdpi.com/2079-4991/9/4/615/s1: a detailed description of conditional RNA/DNA hybrid construct design, all RNA and DNA sequences used in this study, Figure S1: free energy calculations pertaining to the beacon-derived switch, Figure S2: Free energy calculations pertaining to the +0 bp adjacent targeting hybrid system, Figure S3: Free energy calculations pertaining to the responsive DNA hairpin elements of the $sH_{\cdot CTGF}$ hybrids, Figure S4: Time course FRET experiments of variant $sH_{\cdot CTGF}$ hybrids paired with $aH_{\cdot CTGF.cgnt12}$, Figure S5: Free energy calculations pertaining to the responsive DNA hairpin element of $aH\vee_{KRAS}$, Figure S6: Time course FRET experiments of $aH\vee_{KRAS}$ paired with $sH\vee_{KRAS.cgnt}$, Figure S7: PAGE analysis of conditional dsRNA in presence of non-cognate trigger molecules, Figure S8: PAGE analysis of a CTGF-repressible conditional hybrid pair, Figure S9: Time course FRET experiments of a multi-input conditional hybrid pair, Figure S10: Effect of trigger titration on dsRNA release from a multi-input conditional hybrid pair, Figure S11: PAGE analysis of a 3-piece trigger-inducible RNA/DNA hybrid, Figure S12: PAGE analysis of a 3-piece trigger-repressible RNA/DNA hybrid, Table S1: Statistical significance of dsRNA release from a single adjacent-targeting hybrid pair at various timepoints, Table S2: Statistical significance of dsRNA release between various adjacent-targeting hybrid pairs at a single time point, Table S3: Statistical significance of dsRNA release among various trigger-inducible $sH_{\cdot CTGF}$ hybrids pairs at a single time point, Table S4: Statistical significance of dsRNA released for a given $sH_{\cdot CTGF}$ when paired with $aH_{\cdot CTGF\text{-}cgnt.12}$ versus $aH_{\cdot CTGF\text{-}cgnt.16}$, Table S5: Statistical significance of dsRNA release from a single CTGF trigger-inducible hybrid pair at various timepoints, Table S6: Statistical significance of dsRNA released for the KRAS-repressible hybrid pair.

Author Contributions: P.Z. and B.A.S. conceived the research. P.Z. and E.B. designed the methodology of conditional systems, designed sequences and performed computational analysis. P.Z., H.H., M.V. and N.D. performed experiments and analyzed experimental results. P.Z. and B.A.S. wrote the manuscript. P.Z., E.B., H.H., M.V. and B.A.S. provided comments and edits on manuscript drafts.

Funding: This work was supported in whole or in part with Federal funds from the Frederick National Laboratory for Cancer Research, National Institutes of Health, under Contract No. HHSN261200800001E. This research was supported (in part) by the Intramural Research Program of the NIH, National Cancer Institute, Center for Cancer Research. Additionally, this work was in part supported by the LDER grant from the Frederick National Laboratory for Cancer Research.

Acknowledgments: The content of this publication does not necessarily reflect the views or policies of the Department of Health and Human Services, nor does mention of trade names, commercial products, or organizations imply endorsement by the U.S. Government.

Conflicts of Interest: The authors declare no conflict of interest.

References

1. McClorey, G.; Wood, M.J. An overview of the clinical application of antisense oligonucleotides for RNA-targeting therapies. *Curr. Opin. Pharmacol.* **2015**, *24*, 52–58. [CrossRef] [PubMed]

2. Bobbin, M.L.; Rossi, J.J. RNA Interference (RNAi)-Based Therapeutics: Delivering on the Promise? *Annu. Rev. Pharmacol. Toxicol.* **2016**, *56*, 103–122. [CrossRef]

3. Zheng, L.; Wang, L.; Gan, J.; Zhang, H. RNA activation: Promise as a new weapon against cancer. *Cancer Lett.* **2014**, *355*, 18–24. [CrossRef]

4. Dias, N.; Stein, C.A. Antisense oligonucleotides: Basic concepts and mechanisms. *Mol. Cancer Ther.* **2002**, *1*, 347–355. [PubMed]

5. Giles, R.V.; Spiller, D.G.; Green, J.A.; Clark, R.E.; Tidd, D.M. Optimization of antisense oligodeoxynucleotide structure for targeting bcr-abl mRNA. *Blood* **1995**, *86*, 744–754. [PubMed]

6. Roberts, J.; Palma, E.; Sazani, P.; Orum, H.; Cho, M.; Kole, R. Efficient and persistent splice switching by systemically delivered LNA oligonucleotides in mice. *Mol. Ther.* **2006**, *14*, 471–475. [CrossRef]

7. Fire, A.; Xu, S.; Montgomery, M.K.; Kostas, S.A.; Driver, S.E.; Mello, C.C. Potent and specific genetic interference by double-stranded RNA in Caenorhabditis elegans. *Nature* **1998**, *391*, 806–811. [CrossRef]

8. Reynolds, A.; Leake, D.; Boese, Q.; Scaringe, S.; Marshall, W.S.; Khvorova, A. Rational siRNA design for RNA interference. *Nat. Biotechnol.* **2004**, *22*, 326–330. [CrossRef]

9. Kim, D.H.; Behlke, M.A.; Rose, S.D.; Chang, M.S.; Choi, S.; Rossi, J.J. Synthetic dsRNA Dicer substrates enhance RNAi potency and efficacy. *Nat. Biotechnol.* **2005**, *23*, 222–226. [CrossRef] [PubMed]

10. Bramsen, J.B.; Laursen, M.B.; Damgaard, C.K.; Lena, S.W.; Babu, B.R.; Wengel, J.; Kjems, J. Improved silencing properties using small internally segmented interfering RNAs. *Nucleic Acids Res.* **2007**, *35*, 5886–5897. [CrossRef]

11. Chang, C.I.; Yoo, J.W.; Hong, S.W.; Lee, S.E.; Kang, H.S.; Sun, X.; Rogoff, H.A.; Ban, C.; Kim, S.; Li, C.J.; et al. Asymmetric shorter-duplex siRNA structures trigger efficient gene silencing with reduced nonspecific effects. *Mol. Ther.* **2009**, *17*, 725–732. [CrossRef]

12. Janowski, B.A.; Younger, S.T.; Hardy, D.B.; Ram, R.; Huffman, K.E.; Corey, D.R. Activating gene expression in mammalian cells with promoter-targeted duplex RNAs. *Nat. Chem. Biol.* **2007**, *3*, 166–173. [CrossRef]

13. Ma, L.; Reinhardt, F.; Pan, E.; Soutschek, J.; Bhat, B.; Marcusson, E.G.; Teruya-Feldstein, J.; Bell, G.W.; Weinberg, R.A. Therapeutic silencing of miR-10b inhibits metastasis in a mouse mammary tumor model. *Nat. Biotechnol.* **2010**, *28*, 341–347. [CrossRef]

14. Kotula, J.W.; Pratico, E.D.; Ming, X.; Nakagawa, O.; Juliano, R.L.; Sullenger, B.A. Aptamer-mediated delivery of splice-switching oligonucleotides to the nuclei of cancer cells. *Nucleic Acid Ther.* **2012**, *22*, 187–195. [CrossRef]

15. Song, E.; Zhu, P.; Lee, S.K.; Chowdhury, D.; Kussman, S.; Dykxhoorn, D.M.; Feng, Y.; Palliser, D.; Weiner, D.B.; Shankar, P.; et al. Antibody mediated in vivo delivery of small interfering RNAs via cell-surface receptors. *Nat. Biotechnol.* **2005**, *23*, 709–717. [CrossRef]

16. Rozema, D.B.; Lewis, D.L.; Wakefield, D.H.; Wong, S.C.; Klein, J.J.; Roesch, P.L.; Bertin, S.L.; Reppen, T.W.; Chu, Q.; Blokhin, A.V.; et al. Dynamic PolyConjugates for targeted in vivo delivery of siRNA to hepatocytes. *Proc. Natl. Acad. Sci. USA* **2007**, *104*, 12982–12987. [CrossRef] [PubMed]

17. Rychahou, P.; Haque, F.; Shu, Y.; Zaytseva, Y.; Weiss, H.L.; Lee, E.Y.; Mustain, W.; Valentino, J.; Guo, P.; Evers, B.M. Delivery of RNA nanoparticles into colorectal cancer metastases following systemic administration. *ACS Nano* **2015**, *9*, 1108–1116. [CrossRef]

18. Jiang, W.G.; Watkins, G.; Fodstad, O.; Douglas-Jones, A.; Mokbel, K.; Mansel, R.E. Differential expression of the CCN family members Cyr61, CTGF and Nov in human breast cancer. *Endocr. Relat. Cancer* **2004**, *11*, 781–791. [CrossRef]

19. Lapointe, J.; Li, C.; Higgins, J.P.; van de Rijn, M.; Bair, E.; Montgomery, K.; Ferrari, M.; Egevad, L.; Rayford, W.; Bergerheim, U.; et al. Gene expression profiling identifies clinically relevant subtypes of prostate cancer. *Proc. Natl. Acad. Sci. USA* **2004**, *101*, 811–816. [CrossRef] [PubMed]

20. Croce, C.M. Causes and consequences of microRNA dysregulation in cancer. *Nat. Rev. Genet.* **2009**, *10*, 704–714. [CrossRef]

21. Tyagi, S.; Kramer, F.R. Molecular beacons: Probes that fluoresce upon hybridization. *Nat. Biotechnol.* **1996**, *14*, 303–308. [CrossRef]

22. Zhang, P.; Beck, T.; Tan, W. Design of a Molecular Beacon DNA Probe with Two Fluorophores. *Angew. Chem. Int. Ed. Engl.* **2001**, *40*, 402–405. [CrossRef]

23. Penchovsky, R. Computational design of allosteric ribozymes as molecular biosensors. *Biotechnol. Adv.* **2014**, *32*, 1015–1027. [CrossRef]

24. Kennedy, A.B.; Liang, J.C.; Smolke, C.D. A versatile cis-blocking and trans-activation strategy for ribozyme characterization. *Nucleic Acids Res.* **2013**, *41*, e41. [CrossRef]

25. Groves, B.; Chen, Y.J.; Zurla, C.; Pochekailov, S.; Kirschman, J.L.; Santangelo, P.J.; Seelig, G. Computing in mammalian cells with nucleic acid strand exchange. *Nat. Nanotechnol.* **2016**, *11*, 287–294. [CrossRef]

26. Kumar, D.; Kim, S.H.; Yokobayashi, Y. Combinatorially inducible RNA interference triggered by chemically modified oligonucleotides. *J. Am. Chem. Soc.* **2011**, *133*, 2783–2788. [CrossRef]

27. Masu, H.; Narita, A.; Tokunaga, T.; Ohashi, M.; Aoyama, Y.; Sando, S. An activatable siRNA probe: Trigger-RNA-dependent activation of RNAi function. *Angew. Chem. Int. Ed. Engl.* **2009**, *48*, 9481–9483. [CrossRef]

28. Xie, Z.; Liu, S.J.; Bleris, L.; Benenson, Y. Logic integration of mRNA signals by an RNAi-based molecular computer. *Nucleic Acids Res.* **2010**, *38*, 2692–2701. [CrossRef]
29. Hochrein, L.M.; Schwarzkopf, M.; Shahgholi, M.; Yin, P.; Pierce, N.A. Conditional Dicer substrate formation via shape and sequence transduction with small conditional RNAs. *J. Am. Chem. Soc.* **2013**, *135*, 17322–17330. [CrossRef]
30. Hemphill, J.; Deiters, A. DNA computation in mammalian cells: microRNA logic operations. *J. Am. Chem. Soc.* **2013**, *135*, 10512–10518. [CrossRef]
31. Afonin, K.A.; Viard, M.; Martins, A.N.; Lockett, S.J.; Maciag, A.E.; Freed, E.O.; Heldman, E.; Jaeger, L.; Blumenthal, R.; Shapiro, B.A. Activation of different split functionalities on re-association of RNA-DNA hybrids. *Nat. Nanotechnol.* **2013**, *8*, 296–304. [CrossRef]
32. Bindewald, E.; Afonin, K.A.; Viard, M.; Zakrevsky, P.; Kim, T.; Shapiro, B.A. Multistrand Structure Prediction of Nucleic Acid Assemblies and Design of RNA Switches. *Nano Lett.* **2016**, *16*, 1726–1735. [CrossRef]
33. Zakrevsky, P.; Parlea, L.; Viard, M.; Bindewald, E.; Afonin, K.A.; Shapiro, B.A. Preparation of a Conditional RNA Switch. *Methods Mol. Biol.* **2017**, *1632*, 303–324. [CrossRef]
34. Ke, W.; Hong, E.; Saito, R.F.; Rangel, M.C.; Wang, J.; Viard, M.; Richardson, M.; Khisamutdinov, E.F.; Panigaj, M.; Dokholyan, N.V.; et al. RNA-DNA fibers and polygons with controlled immunorecognition activate RNAi, FRET and transcriptional regulation of NF-kappaB in human cells. *Nucleic Acids Res.* **2018**. [CrossRef]
35. Zakrevsky, P.; Bindewald, E.; Shapiro, B.A. RNA toehold interactins initiate conditional gene silencing. *DNA RNA Nanotechnol.* **2016**, *3*, 11–13. [CrossRef]
36. Lu, J.S.; Bindewald, E.; Kasprzak, W.; Shapiro, B.A. RiboSketch: Versatile Visualization of Multi-stranded RNA and DNA Secondary Structure. *Bioinformatics* **2018**, *34*, 4297–4299. [CrossRef]
37. Afonin, K.A.; Desai, R.; Viard, M.; Kireeva, M.L.; Bindewald, E.; Case, C.L.; Maciag, A.E.; Kasprzak, W.K.; Kim, T.; Sappe, A.; et al. Co-transcriptional production of RNA-DNA hybrids for simultaneous release of multiple split functionalities. *Nucleic Acids Res.* **2014**, *42*, 2085–2097. [CrossRef]
38. Afonin, K.A.; Viard, M.; Tedbury, P.; Bindewald, E.; Parlea, L.; Howington, M.; Valdman, M.; Johns-Boehme, A.; Brainerd, C.; Freed, E.O.; et al. The Use of Minimal RNA Toeholds to Trigger the Activation of Multiple Functionalities. *Nano Lett.* **2016**, *16*, 1746–1753. [CrossRef]
39. Abels, S.G.; Khisamutdinov, E.F. Nucleic Acid Computing and its Potential to Transform Silicon-Based Technology. *DNA RNA Nanotechnol.* **2015**, *2*, 13–22. [CrossRef]
40. Wu, C.; Wan, S.; Hou, W.; Zhang, L.; Xu, J.; Cui, C.; Wang, Y.; Hu, J.; Tan, W. A survey of advancements in nucleic acid-based logic gates and computing for applications in biotechnology and biomedicine. *Chem. Commun.* **2015**, *51*, 3723–3734. [CrossRef]
41. Zhang, D.Y.; Seelig, G. Dynamic DNA nanotechnology using strand-displacement reactions. *Nat. Chem.* **2011**, *3*, 103–113. [CrossRef]
42. Seelig, G.; Soloveichik, D.; Zhang, D.Y.; Winfree, E. Enzyme-free nucleic acid logic circuits. *Science* **2006**, *314*, 1585–1588. [CrossRef]
43. Li, W.; Yang, Y.; Yan, H.; Liu, Y. Three-input majority logic gate and multiple input logic circuit based on DNA strand displacement. *Nano Lett.* **2013**, *13*, 2980–2988. [CrossRef] [PubMed]
44. Brown, C.W., 3rd; Lakin, M.R.; Stefanovic, D.; Graves, S.W. Catalytic molecular logic devices by DNAzyme displacement. *Chembiochem* **2014**, *15*, 950–954. [CrossRef] [PubMed]
45. Green, A.A.; Kim, J.; Ma, D.; Silver, P.A.; Collins, J.J.; Yin, P. Complex cellular logic computation using ribocomputing devices. *Nature* **2017**, *548*, 117–121. [CrossRef] [PubMed]

nanomaterials

MDPI

Review

Smart-Responsive Nucleic Acid Nanoparticles (NANPs) with the Potential to Modulate Immune Behavior

Morgan Chandler and Kirill A. Afonin *

Nanoscale Science Program, Department of Chemistry, University of North Carolina at Charlotte, Charlotte, NC 28223, USA; mchand11@uncc.edu
* Correspondence: kafonin@uncc.edu; Tel.: +1-704-687-0685

Received: 14 March 2019; Accepted: 8 April 2019; Published: 12 April 2019

Abstract: Nucleic acids are programmable and biocompatible polymers that have beneficial uses in nanotechnology with broad applications in biosensing and therapeutics. In some cases, however, the development of the latter has been impeded by the unknown immunostimulatory properties of nucleic acid-based materials, as well as a lack of functional dynamicity due to stagnant structural design. Recent research advancements have explored these obstacles in tandem via the assembly of three-dimensional, planar, and fibrous cognate nucleic acid-based nanoparticles, called NANPs, for the conditional activation of embedded and otherwise quiescent functions. Furthermore, a library of the most representative NANPs was extensively analyzed in human peripheral blood mononuclear cells (PBMCs), and the links between the programmable architectural and physicochemical parameters of NANPs and their immunomodulatory properties have been established. This overview will cover the recent development of design principles that allow for fine-tuning of both the physicochemical and immunostimulatory properties of dynamic NANPs and discuss the potential impacts of these novel strategies.

Keywords: nucleic acid nanoparticles; NANPs; immunostimulation; dynamic; conditionally activated; RNA interference; RNA nanotechnology

1. Introduction

Nanotechnology has been integrated into many aspects of modern life [1] by providing a means of additional control over the unique physicochemical properties of functional moieties—including size, surface charge, and hydrophobicity, as well as their precise incorporation—and making them useful for biomedical applications. The ability to fine-tune these properties subsequently allows for the improved efficacy of therapeutic treatments and has implications for the future of personalized medicine [2–5]. Nucleic acids, including both DNA and RNA, represent a branch of biopolymers which additionally offer a biocompatible and programmable therapeutic approach. Beyond their traditionally known roles as passive carriers of genetic information, DNA and RNA have emerged as building materials for versatile biological drugs, called therapeutic nucleic acids (TNAs), which can take advantage of cellular pathways for the sensing, targeting, and silencing of a broad spectrum of various diseases, including asthma, cystic fibrosis, viral infections, and cancers [6,7]. TNAs are a diverse class of biomacromolecules that include antisense oligonucleotides, triplex-forming oligodeoxyribonucleotides, immunostimulatory oligos, catalytic oligos, inhibitory DNAs, interfering RNAs, and aptamers, which differ by composition, secondary structure, and mechanism of action [8]. Each TNA class may include multiple subtypes. For example, RNA interference (RNAi) inducers include siRNAs, miRNAs, and shRNAs, to name just a few. The great potential of RNAi technologies became apparent from the recent inspiring example of the very first siRNA therapeutic agent, patisiran (trade name ONPATTRO®), which was developed against polyneuropathy in patients with hereditary transthyretin-mediated amyloidosis and approved by the U.S. Food and Drug Administration (FDA) in 2018 [9], in addition to the many studies showing the versatility and modularity of TNAs [10–12]. Usually, TNAs affect the flow of genetic information, mimic antibodies, or stimulate the immune system, causing either the suppression of disease-specific genes or the stimulation of gene expression in response to an antigen. Besides specific siRNAs, several other promising therapeutically potent classes of nucleic acids, such as antisense oligonucleotides, aptamers, DNA decoys, and ribozymes, are also under consideration [13]. Currently, three additional successful examples of TNAs are approved for therapeutic uses: fomivirsen (brand name Vitravene™), an antisense oligonucleotide designed against cytomegalovirus retinitis which was also the very first nucleic acid-based therapy approved by the FDA [14]; mipomersen (trade name KYNAMRO®), an antisense oligonucleotide approved for the treatment of homozygous familial hypercholesterolemia [15]; and pegaptanib (brand name MACUGEN®), an aptamer approved for the therapy of age-related macular degeneration [16].

If several different TNAs are simultaneously chosen for the same treatment and they must target the same cells, the optimal route for their controlled co-administration would be through the introduction of nucleic acid-based nanoparticles, or NANPs [17–27]. This strategy has resulted in multiple new methods for the design and assembly of nano-TNAs, in which nucleic acids also serve as building blocks to assemble programmable scaffolds with well-defined properties. To achieve this, biomedical sciences benefit not only from the natural roles of nucleic acids, but also from their known ability to form both canonical Watson–Crick (e.g., G–C and A–U (or T for DNA)) and non-canonical base pairings [28]. Non-canonical base pairs are mostly characteristic of RNAs and include 12 basic geometric families [28], thus leading to a diverse set of oligonucleotide structures, called motifs, that can fold into complexes with a precise 3D shape. The existing RNA motifs [25,29–35] can be rationally combined to promote the assembly of various NANPs that can be further decorated with therapeutic domains and employed as drug delivery platforms [36]. Such a novel use of RNA as a starting material in bottom-up assembly has helped to establish the burgeoning field of RNA nanotechnology, which also utilizes the biological functions of nucleic acids to address specific biomedical challenges [21,33,36]. Programmable NANPs guarantee precise control over versatile functionalization with different moieties, such as aptamers, fluorescent dyes, and proteins, and their simultaneous delivery with numerous siRNAs targeting different biological pathways [20–23,25,31,37]. NANPs' ability to successfully combat diseases at their source has already been confirmed by multiple animal studies [25,38–43]. Additionally, the design principles being developed in RNA nanotechnology address fundamental problems relevant to the biophysics of RNA co-transcriptional folding [24,44–47], structure–activity relationships [48], and their physicochemical interactions with other classes of biomolecules (e.g., lipids [41,49,50], proteins [51,52]) or inorganic materials [53,54].

Though TNAs in general and NANP-based nano-TNAs in particular are strong candidates in nanotherapeutics, their transition into a clinical setting has been hindered by a lack of general knowledge about their immunostimulatory properties [55], while their statically designed structures have posed limits to the conditional activation or deactivation of preprogrammed biological functions. To overcome these obstacles, recent efforts, as described in this review, have been focused on a new platform of design principles for assembling dynamic nucleic acid assemblies with a controlled and fine-tunable immune response (Figure 1).

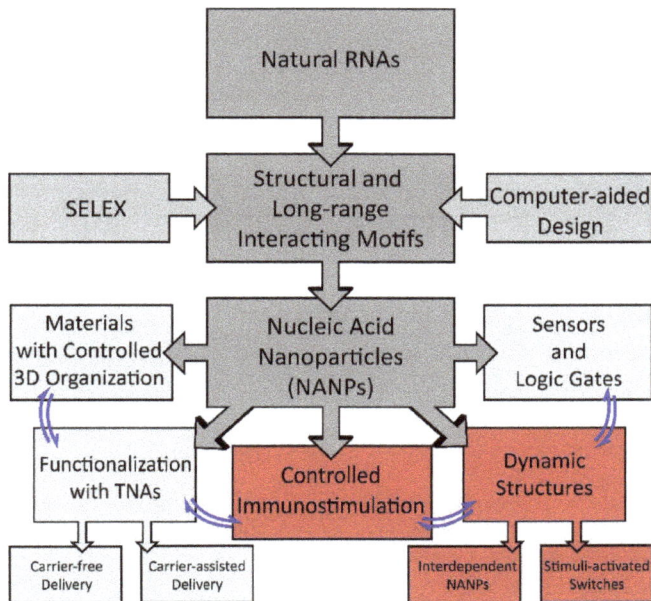

Figure 1. Schematic of the development of nucleic acid-based nanoparticle (NANP) technologies and their current applications. Structural motifs inspired by structure–activity studies of natural RNAs, computer-assisted design, or determined by SELEX (systematic evolution of ligands by exponential enrichment) are used for the rational design of NANPs. NANPs can then be programmed for multiple applications for materials organization, sensors, and dynamic structures, or can be functionalized with therapeutic nucleic acids (TNAs) or used to elicit a controlled immune response.

2. Dynamic Shape-Switching and Functional Activation with NANPs

To take full advantage of the programmability of nucleic acids, therapeutic NANPs could be conditionally activated inside human cells for the release of preprogrammed functionalities which would offer higher targeting specificity, thereby potentially reducing off-target effects. The "dynamicity" preprogrammed in NANPs' behavior defines their ability to be activated in response to various stimuli. By interacting with, for example, a target strand or environmental variable of choice as a diagnostic step, switching NANPs can be designed to release therapeutics only when these interactions occur [56,57]. In the absence of the predetermined intracellular trigger mRNA, characteristic only for the diseased cells, these switching NANPs are not in active therapeutic conformations (Figure 2). The single-stranded bait sequences made of RNA (or chemically modified nucleotides) that deactivate switching NANPs are computationally designed to provide a thermodynamically more favorable binding to the trigger mRNA than to NANP strands [56,57]. The interactions between the switching NANPs and trigger mRNA removes the bait strand and consequently exposes shRNA-like hairpins that become the next most thermodynamically stable fold of the remaining NANP. The human enzyme Dicer can now recognize and cleave these refolded structures and load the RISC with the "guide" strand which, in turn, will activate RNAi. RNAi activation results in the suppression of targeted anti-apoptotic genes, thus inducing the programmed death of the diseased cell.

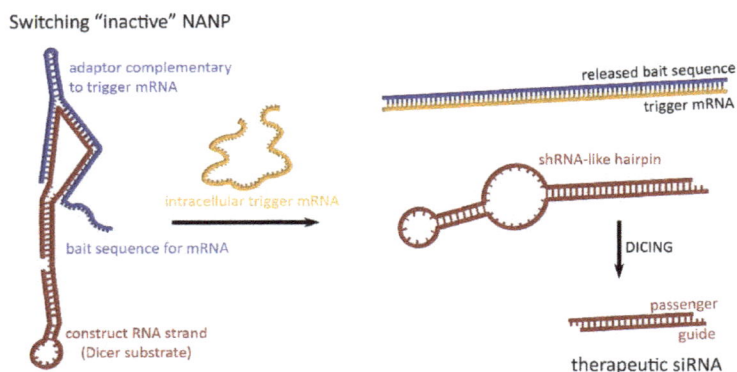

Figure 2. Dynamic NANPs exemplified by two-stranded RNA switch. The presence of trigger mRNA promotes strand rearrangements in an otherwise inactive switch, thus allowing for the release of therapeutic siRNA.

Besides activation from environmental stimuli, RNA/DNA assemblies can also be designed to interact with their cognate counterparts for their activation [40]. Hybrid structures composed of both RNA and DNA (or chemical analogs [58]) can be used to synchronize the activation of functionalities embedded into the hybrid structures [40,59]. Thus, for RNAi activation, two halves of a Dicer substrate (DS) RNA [60] can be split between two hybrid duplexes which undergo strand displacement when both are present to form a complete duplex for processing and subsequent gene silencing. The process of reassociation is thermodynamically driven and can be initiated via complementary ssDNA [40,61,62] or ssRNA [26] toehold interactions. The use of hybrids also offers some preclinical benefits, such as controlled rates of reassociation, significantly reduced degradation in human blood serum, and the possibility to chemically introduce additional functionalities into the DNA strands without affecting the function of the released RNAs. Besides RNAi inducers, split aptamers have also been tested which, when presented as separate halves, are non-functional. However, when the halves of the split aptamer are brought together during the reassociation, the completed aptamer regains its function [40,63,64]. The development of split fluorescent aptamers, such as malachite green [65], Spinach [63], and Broccoli [66], has produced tools which are especially useful when applied as a validating output for dynamic RNA nanotechnology [63,64,66–68], in the visualization of NANP assemblies [45,69], and for logic gating [70].

Another design scheme, which evolved from the approach using RNA/DNA hybrids, utilizes NANPs which are designed to completely reassociate with one another and thus activate embedded functionalities. An example of such conditional activation has been shown via interdependent NANPs in the shape of three-dimensional cubes which interact with their cognate "anti-cubes" under isothermal conditions to trigger a change in the shape and successive activation of attached functionalities [71]. Cubes are composed of six strands of RNA and/or DNA, while the anti-cube strands are designed as the reverse complements of the cube sequences (Figure 3A). Only two NANPs—the cube and anti-cube—are required for activation. Upon interaction, the two cubes that each have a total of nine unpaired bases per corner undergo a thermodynamically driven switch into six double-stranded duplexes. Moieties which are split across the cube and anti-cube complement strands are then brought together for activation of functionality.

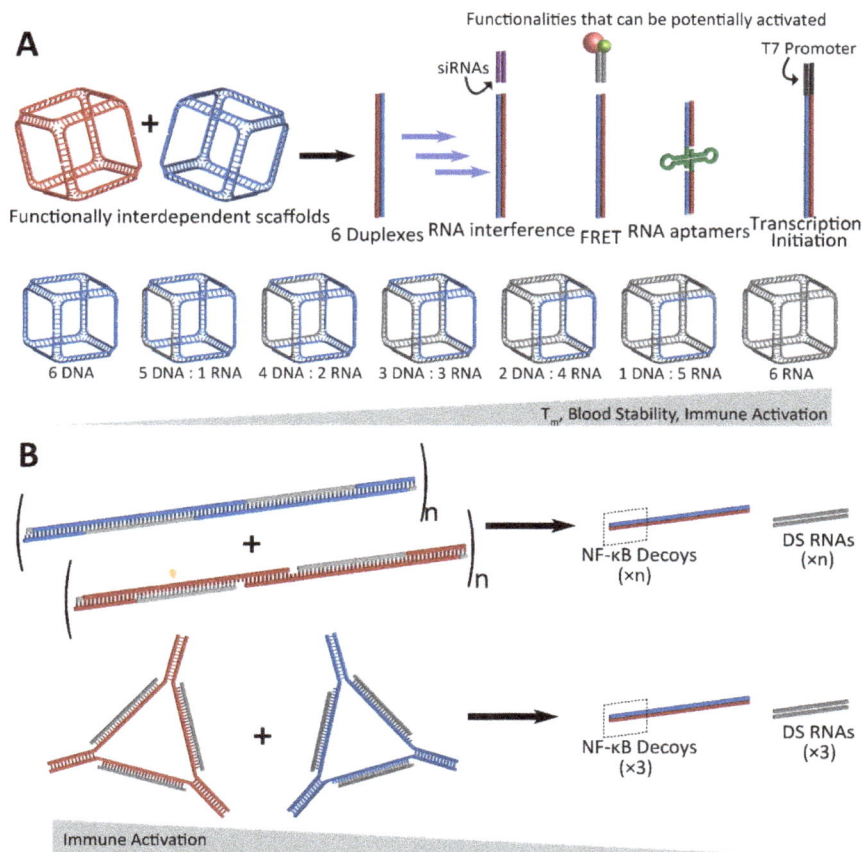

Figure 3. Conceptual representation of interdependent shape-switching NANPs programmed to activate multiple functionalities upon their isothermal reassociation. (**A**) Hexameric NANP cubes and complementary anti-cubes reassociate to drive the formation of six double-stranded duplexes. Functionalities which can potentially be added to the NANPs are split between the two functionally interdependent cubes to become active upon reassociation, resulting in the completion of siRNAs for gene silencing, Förster resonance energy transfer (FRET) pairs for visualization, aptamers, or T7 RNA polymerase promoter sequences for the co-transcriptional assembly of RNA NANPs. The cubes can be composed of varying ratios of DNA and/or RNA, with melting temperature, blood stability, and immune activation increasing with RNA composition. (**B**) Hybrid DNA/RNA fibers (top) and polygons (bottom) reassociate to form DNA duplexes containing NF-κB decoys, as well as Dicer substrate (DS) RNA duplexes for gene silencing. The NF-κB decoy binds to and prevents NF-κB from entering the nucleus, thereby stopping it from producing pro-inflammatory cytokines. As a result, immune activation decreases as the hybrid structures reassociate.

2.1. Activation of RNA Interference, FRET, RNA Aptamers, and Transcription Initiation

Conditional activation of the functionalities upon reassociation of the cube and anti-cube has been demonstrated with a variety of methods which also establish the wide range of applications for this technology [71]. A set of cognate cubes were decorated with split DS RNAs against multiple different genes (*BCL2*, *PLK1*, and green fluoroscent proteins—*GFP*). DS RNAs were labeled with fluorophores chosen to undergo a Förster resonance energy transfer (FRET) for real-time intracellular analysis. Upon reassociation, the formation of DS RNA duplexes could be further processed in a cellular environment, through dicing, for the release of functional siRNAs. Cognate functionally interdependent NANPs carrying DS RNAs against GFP [60] showed a significant knockdown in the fluorescence of human breast cancer cells expressing enhanced GFP [71]. Also, the downregulation of *BCL2* and *PLK1* genes [27], which have been shown to induce apoptosis, was confirmed by a significant decrease in the viability of human cervical and prostate cancer cells.

The reassociation of the cube and anti-cube and the consecutive formation of RNA fibers was also demonstrated with the activation of a split aptamer. The RNA aptamer Broccoli [72], which binds the chemically synthesized fluorophore DFHBI-1T to mimic the natural function of GFP, was split into two separate strands termed Broc and Coli [66,68]. With both parts of the aptamer attached to cognate RNA cube strands, the Broc cube and Coli anti-cube reassociated in the presence of DFHBI-1T, allowing for the interaction to be traced via fluorescence in real time [71].

The reassociation of the DNA cubes and anti-cubes carrying split T7 RNA polymerase promoters was shown to form templates for the co-transcriptional assembly of complete RNA cubes in vitro. Anti-cubes carrying a complete T7 promoter can be used for the RNA cube's co-transcriptional assembly, thus providing a template for the future intracellular production of RNA nanoparticles that can be activated upon interaction of functionally interdependent NANPs [71].

Lastly, hybrid RNA/DNA fibers and polygons can interact with their cognate structures for reassociation, resulting in the release of DS RNAs as well as double-stranded DNAs carrying NF-κB decoys (Figure 3B) [73]. NF-κB (nuclear factor kappa-light-chain-enhancer of activated B cells) is expressed in most animal cells and generally remains in an inactive state in the cytoplasm until it can be activated for translocation to the nucleus, where it is then involved in the production of pro-inflammatory cytokines. With the release of NF-κB decoys [74,75], which can bind and retain NF-κB in the cytoplasm, the hybrid fibers and polygons dynamically modulate the immune response in addition to RNAi activation. Further, this strategy takes advantage of all strands included in the assembly to utilize them in functional roles, leaving no static byproducts [73].

2.2. Fine-Tunable Properties

Besides the activation of multiple functionalities, an important feature of the rational design of the cube/anti-cube NANP system is the ability to fine-tune the physicochemical and immunological properties of NANPs simply by adjusting the ratios of DNA and RNA strands in their composition [71]. With an increasing number of RNA strands, both the thermodynamic stability and the reassociation time of cubes increased. The immune response to these NANPs was also assessed via the measurement of the activation and secretion of interferon (IFN)α and a panel of pro-inflammatory cytokines and chemokines (IL-1β, TNFα, IL-8, and MIP-1α). NANPs composed of all RNA strands were the most potent stimulators of an immune response, indicating that they may serve an application as a vaccine adjuvant. By adjusting the nucleic acid composition, the immune response to nanoparticles can be fine-tuned in such a way that assemblies could be utilized for drug delivery or immunotherapy.

3. Immunostimulatory Properties of NANPs

It is becoming apparent that interactions between NANPs and the immune system must be defined to permit the successful translation of this technology to the clinic. Foreign nucleic acids can produce a robust and severe response in immune cells. Pro-inflammatory cytokines and type I interferons are characteristic of nucleic acid sensing in immune cells, and in animals, it can produce anywhere from minor inflammation to severe cytokine storms. Therefore, to address fundamental questions regarding the immune recognition of these novel materials in a timely fashion, the relation of features such as the size, shape, composition, and physicochemical properties of various polygonal NANPs [76] to the activation of immune responses in human microglia-like cells (hμglia or hHμ) was examined [48] using a series of assembled RNA, DNA, and RNA/DNA hybrid NANPs. A set of several polygons designed based on the versatile tetra-U helix linking motif [76] was assembled by using a ubiquitous set of strands (both RNA and DNA) so that their immunostimulatory effects—when composed of all RNA, RNA with a DNA center, all DNA, and DNA with a RNA center—could be characterized (Figure 4A). The engineered NANPs were designed to assume various shapes and sizes as well as variations in their content (RNA vs. DNA) and physicochemical stabilities while having minimal differences in their sequences. The measured biomarkers of a pro-inflammatory response were cytokines and type I interferons. Using modern machine learning techniques, the quantitative structure–activity relationship (QSAR) models were developed to successfully predict and engineer NANPs able to stimulate an intended immune response, or lack thereof [48].

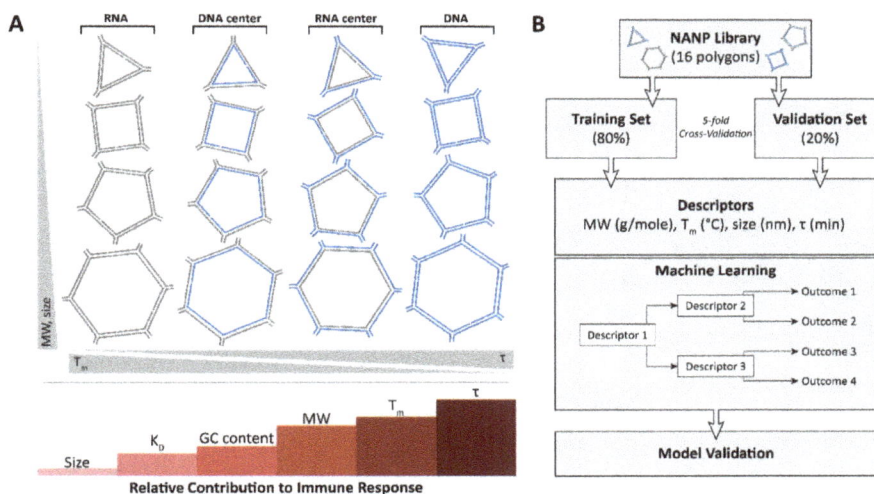

Figure 4. The very first application of quantitative structure–activity relationship (QSAR) modeling for NANP immunostimulation uses measured physical and chemical properties as inputs to predict a given NANP's ability to generate a pro-inflammatory response. (**A**) A library of 16 NANPs which incorporate the same sequences was used to determine the physicochemical properties contributing to immune response. From the study, size (diameter) was determined to contribute the least to immune response, followed by K_D, GC content, molecular weight, T_m and, finally, stability with the highest contribution into the random forest model. (**B**) A schematic showing the QSAR modeling approach which uses the 16 NANP library.

This very first application of QSAR modeling for NANPS studied measured physical and chemical properties as inputs to predict a given NANP's ability to generate a pro-inflammatory response (Figure 4B). Importantly, this QSAR model [48] can be used to more intelligently design nucleic acid-based pharmaceuticals to reduce detrimental immune responses, stimulate desired protective immune responses, and increase their intended activity. This work is instrumental in bridging the rapidly narrowing gap between basic research on NANPs and advanced pharmaceuticals containing these novel materials.

Following these findings, the very first systematic investigation of NANP recognition by immune cells using primary human peripheral blood mononuclear cells (PBMCs) from a cohort of more than 100 healthy human donors was recently designed and executed [77]. Despite expectations [78], the researchers did not find a strong, uniform immune response for all NANPs. Instead, the tests found varying and specific responses from different immune cells, depending on each NANP's shape and formulation. It was discovered that all NANPs used without a delivery carrier were immunoquiescent, and that type I and III IFNs are key cytokines triggered by NANPs after their internalization by phagocytic cells. It was shown that overall immunostimulation relies on the NANPs' shape, type of connectivity, and composition. Importantly, plasmacytoid dendritic cells were identified as the primary interferon producers among all PBMCs treated with NANPs, and it was demonstrated that scavenger receptor-mediated uptake and endosomal toll-like receptor (TLR) signaling are essential for NANPs' recognition. In particular, TLR 7 (Figure 5) was identified as a key player in immunostimulation triggered by NANPs that was observed both in model HEK-Blue TLR 7 cells [77] and in human PBMCs [79], as well as being later confirmed by extensive mechanistic studies [79]. All immunological studies strongly suggest that the further understanding of how particular NANPs can trigger the immune response is required to open up possibilities in a new field where NANPs can be used as vaccine adjuvants.

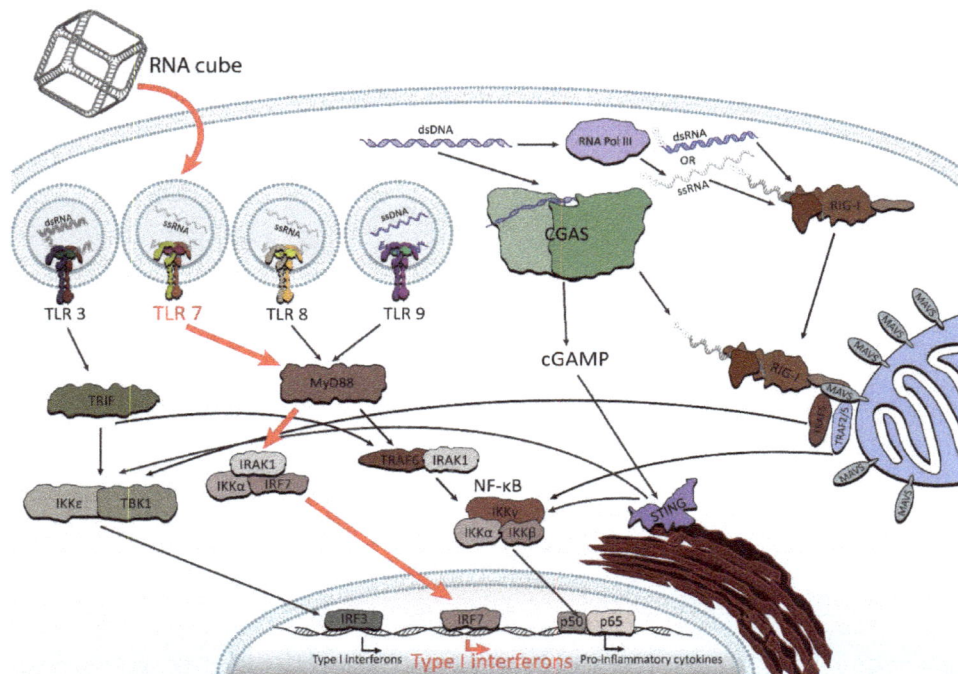

Figure 5. Abridged illustration detailing the pathways involved in the endosomal and cytosolic sensing of smaller therapeutically relevant nucleic acids. NANPs have been observed to enter through the endosomal pathway (when complexed with Lipofectamine 2000) and trigger the toll-like receptors (TLRs). The pathway shown in red (for TLR 7) demonstrates the identified route for RNA cube recognition. For the purpose of this review, the figure shows all individual TLRs (related to NA sensing) in separate endosomes in order to better highlight the particular pathway of NANPs' recognition.

Importantly, the following work has also demonstrated that the functionalization of NANPs with RNAi inducers completely changes their known immunorecognition [80]. However, the possibility to control the magnitude and specificity of the immunostimulatory response by varying the design parameters and functional moieties in each NANP has also been revealed (Figure 6). Additionally, through using RNA fibrous structures which have been shown to have minimal recognition by the immune system, the delivery of multiple modalities, such as siRNAs and fluorophores, can be coordinated with minimal immunorecognition. Relying on HIV-like (~180°) kissing loop interactions, dumbbell-shaped hairpins are the modular building blocks for these assemblies and allow for simple customizability [80].

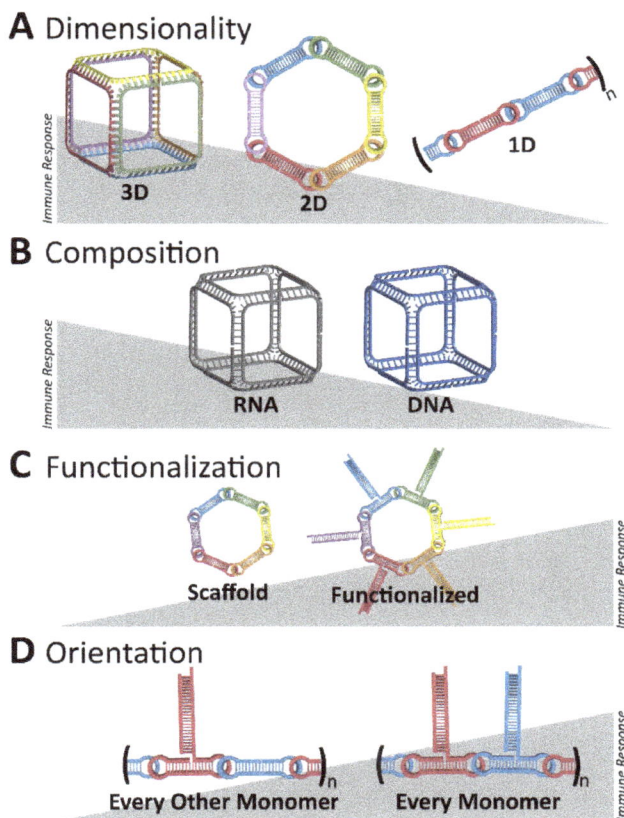

Figure 6. Initial overall immunostimulatory trends observed with variations in NANP designs. (**A**) Three-dimensional RNA cubes are potent immunostimulants, followed by two-dimensional RNA rings and one-dimensional RNA fibers. (**B**) Between structures of the same dimensionality, RNA NANPs are more immunostimulatory than DNA NANPs. (**C**) A NANP scaffold becomes more immunostimulatory if it is functionalized, for example, with TNAs. (**D**) For RNA fibers which are less immunostimulatory in terms of dimensionality, functionalization at every monomer results in a greater immune response than if the fibers are functionalized at every other monomer. Different colors emphasize the architecture of NANPs consisting of multiple strands (A,C,D) or changes in their composition (B) such as DNA vs. RNA.

4. Conclusions

In conclusion, the responsive behaviors of NANPs can be determined by the specific design principles for individual NANPs, their assembly type, compositions, physicochemical properties, and the presence of any additional functionalities in their structure. The use of NANPs for the controlled design of dynamic stimuli-responsive systems presents multiple advantages: (i) NANPs can be programmed to gather multiple different functionalities for their simultaneous delivery to cells [17,25,37,42,61]; (ii) RNA NANPs and their chemical analogs can be co-transcriptionally assembled [23,24,45,46]; (iii) the relatively inexpensive production of NANPs with a high batch-to-batch consistency [22] enables their economic industrial scale production; (iv) the thermal and chemical stabilities of NANPs can be fine-tuned by introducing chemically modified nucleotides or DNA strands into their composition [45,71,77]; (v) carrier-free NANPs avoid nonspecific cell penetration due to their negative charge [77] and, thus, can be used for extracellular tasks; and (vi) the immunological properties of NANPs are controllable and can potentially be predicted [48,77]. The ability to fine-tune immunostimulation and the conditional activation of functionalities are important facets in the rational design strategies of programmable biopolymers. Stimulation of the controlled production of IFNs and pro-inflammatory cytokines by human immune cells is instrumental for immunotherapies and vaccine use when activation of the immune response is necessary [77]. The excessive and uncontrolled induction of cytokines, however, may promote tissue necrosis and become harmful [81]. The innovative development of NANP-based tools that can be used for communication with the human immune system can be employed in achieving the desirable activation of the immune system for vaccines and cancer immunotherapies, while immunoquiescent NANPs can be used as nano-TNAs and for dynamic NANP construction.

Predicting and controlling the immune responses triggered by NANPs, especially when the magnitudes of those responses may vary between donors, remains a substantial challenge for the furtherment of nano-TNAs. In addition, due to their negative charges which inhibit free passage across the cell membrane, a carrier or targeting moieties are needed for delivery. While there is a plethora of nanoparticles formulated for nucleic acid delivery, the effects of the carrier on the immunostimulatory properties of NANPs remain unclear. Furthermore, the healthy host immune tolerance to nucleic acids and the generation of antibodies against DNA and RNA after exposure to NANPs is an important safety topic which, so far, has not been addressed. Thus, assessing the immunogenicity of NANPs is a future logical direction in this field.

The activation of RNA interference, FRET, RNA aptamers, transcription, and other responses upon the reassociation of cognate NANPs at isothermal conditions exhibits the potential for dynamic constructs to operate only in the presence of a target. Additionally, the degree to which the properties of TNAs contribute to the elicitation of an immune response will allow for their intentional activation or evasion, depending on whether immunotherapy or drug delivery is desired. By optimizing conditional activation and immunostimulation, NANPs can be better designed for future clinical applications.

Author Contributions: M.C. and K.A.A. co-wrote the manuscript.

Funding: Research reported in this publication was supported by the National Institute of General Medical Sciences of the National Institutes of Health under Award Number R01GM120487. The content is solely the responsibility of the authors and does not necessarily represent the official views of the National Institutes of Health.

Acknowledgments: The authors would like to thank Justin Halman for contributing Figure 5 of this review.

Conflicts of Interest: The authors declare no conflict of interest.

References

1. Vance, M.E.; Kuiken, T.; Vejerano, E.P.; McGinnis, S.P.; Hochella, M.F., Jr.; Rejeski, D.; Hull, M.S. Nanotechnology in the real world: Redeveloping the nanomaterial consumer products inventory. *Beilstein J. Nanotechnol.* **2015**, *6*, 1769–1780. [CrossRef] [PubMed]

2. Dobrovolskaia, M.A. Pre-clinical immunotoxity studies of nanotechnology-formulated drugs: Challenges, considerations and strategy. *J. Control. Release* **2015**, *220*, 571–583. [CrossRef] [PubMed]

3. Dobrovolskaia, M.A. Self-assembled DNA/RNA nanoparticles as a new generation of therapeutic nucleic acids: Immunological compatibility and other translational considerations. *DNA RNA Nanotechnol.* **2016**, *3*, 1–10. [CrossRef]

4. Dobrovolskaia, M.A.; McNeil, S.E. Immunological properties of engineered nanomaterials. *Nat. Nanotechnol.* **2007**, *2*, 469–478. [CrossRef] [PubMed]

5. Hall, J.B.; Dobrovolskaia, M.A.; Patri, A.K.; McNeil, S.E. Characterization of nanoparticles for therapeutics. *Nanomedicine* **2007**, *2*, 789–803. [CrossRef]

6. Alvarez-Salas, L.M. Nucleic acids as therapeutic agents. *Curr. Top. Med. Chem.* **2008**, *8*, 1379–1404. [CrossRef]

7. Catuogno, S.; Esposito, C.L.; Condorelli, G.; de Franciscis, V. Nucleic acids delivering nucleic acids. *Adv. Drug Deliv. Rev.* **2018**, *134*, 79–93. [CrossRef]

8. Sridharan, K.; Gogtay, N.J. Therapeutic nucleic acids: Current clinical status. *Br. J. Clin. Pharmacol.* **2016**, *82*, 659–672. [CrossRef] [PubMed]

9. Adams, D.; Gonzalez-Duarte, A.; O'Riordan, W.D.; Yang, C.C.; Ueda, M.; Kristen, A.V.; Tournev, I.; Schmidt, H.H.; Coelho, T.; Berk, J.L.; et al. Patisiran, an RNAi Therapeutic, for Hereditary Transthyretin Amyloidosis. *N. Engl. J. Med.* **2018**, *379*, 11–21. [CrossRef]

10. Grijalvo, S.; Alagia, A.; Jorge, A.F.; Eritja, R. Covalent Strategies for Targeting Messenger and Non-Coding RNAs: An Updated Review on siRNA, miRNA and antimiR Conjugates. *Genes* **2018**, *9*, 74. [CrossRef] [PubMed]

11. Sasaki, S.; Guo, S. Nucleic Acid Therapies for Cystic Fibrosis. *Nucleic Acid Ther.* **2018**, *28*, 1–9. [CrossRef]

12. Nimjee, S.M.; White, R.R.; Becker, R.C.; Sullenger, B.A. Aptamers as Therapeutics. *Annu. Rev. Pharmacol. Toxicol.* **2017**, *57*, 61–79. [CrossRef]

13. Burnett, J.C.; Rossi, J.J. RNA-based therapeutics: Current progress and future prospects. *Chem. Biol.* **2012**, *19*, 60–71. [CrossRef] [PubMed]

14. Holmlund, J.T.; Monia, B.P.; Kwoh, T.J.; Dorr, F.A. Toward antisense oligonucleotide therapy for cancer: ISIS compounds in clinical development. *Curr. Opin. Mol. Ther.* **1999**, *1*, 372–385. [PubMed]

15. Sehgal, A.; Vaishnaw, A.; Fitzgerald, K. Liver as a target for oligonucleotide therapeutics. *J. Hepatol.* **2013**, *59*, 1354–1359. [CrossRef]

16. Gilboa, E.; McNamara, J.n.; Pastor, F. Use of oligonucleotide aptamer ligands to modulate the function of immune receptors. *Clin. Cancer Res.* **2013**, *19*, 1054–1062. [CrossRef]

17. Li, H.; Lee, T.; Dziubla, T.; Pi, F.; Guo, S.; Xu, J.; Li, C.; Haque, F.; Liang, X.J.; Guo, P. RNA as a stable polymer to build controllable and defined nanostructures for material and biomedical applications. *Nano Today* **2015**, *10*, 631–655. [CrossRef] [PubMed]

18. Shu, Y.; Pi, F.; Sharma, A.; Rajabi, M.; Haque, F.; Shu, D.; Leggas, M.; Evers, B.M.; Guo, P. Stable RNA nanoparticles as potential new generation drugs for cancer therapy. *Adv. Drug Deliv. Rev.* **2014**, *66*, 74–89. [CrossRef]

19. Shu, Y.; Haque, F.; Shu, D.; Li, W.; Zhu, Z.; Kotb, M.; Lyubchenko, Y.; Guo, P. Fabrication of 14 different RNA nanoparticles for specific tumor targeting without accumulation in normal organs. *RNA* **2013**, *19*, 767–777. [CrossRef]

20. Shukla, G.C.; Haque, F.; Tor, Y.; Wilhelmsson, L.M.; Toulmé, J.-J.; Isambert, H.; Guo, P.; Rossi, J.J.; Tenenbaum, S.A.; Shapiro, B.A. A boost for the emerging field of RNA nanotechnology. *ACS Nano* **2011**, *5*, 3405–3418. [CrossRef]

21. Guo, P. The emerging field of RNA nanotechnology. *Nat. Nanotechnol.* **2010**, *5*, 833–842. [CrossRef] [PubMed]

22. Afonin, K.A.; Grabow, W.W.; Walker, F.M.; Bindewald, E.; Dobrovolskaia, M.A.; Shapiro, B.A.; Jaeger, L. Design and self-assembly of siRNA-functionalized RNA nanoparticles for use in automated nanomedicine. *Nat. Protoc.* **2011**, *6*, 2022–2034. [CrossRef]

23. Afonin, K.A.; Kasprzak, W.K.; Bindewald, E.; Kireeva, M.; Viard, M.; Kashlev, M.; Shapiro, B.A. In Silico Design and Enzymatic Synthesis of Functional RNA Nanoparticles. *Acc. Chem. Res.* **2014**, *47*, 1731–1741. [CrossRef]

24. Afonin, K.A.; Kireeva, M.; Grabow, W.W.; Kashlev, M.; Jaeger, L.; Shapiro, B.A. Co-transcriptional assembly of chemically modified RNA nanoparticles functionalized with siRNAs. *Nano Lett.* **2012**, *12*, 5192–5195. [CrossRef]

25. Afonin, K.A.; Viard, M.; Koyfman, A.Y.; Martins, A.N.; Kasprzak, W.K.; Panigaj, M.; Desai, R.; Santhanam, A.; Grabow, W.W.; Jaeger, L.; et al. Multifunctional RNA nanoparticles. *Nano Lett.* **2014**, *14*, 5662–5671. [CrossRef]

26. Parlea, L.; Puri, A.; Kasprzak, W.; Bindewald, E.; Zakrevsky, P.; Satterwhite, E.; Joseph, K.; Afonin, K.A.; Shapiro, B.A. Cellular Delivery of RNA Nanoparticles. *ACS Comb. Sci.* **2016**, *18*, 527–547. [CrossRef]

27. Stewart, J.M.; Viard, M.; Subramanian, H.K.; Roark, B.K.; Afonin, K.A.; Franco, E. Programmable RNA microstructures for coordinated delivery of siRNAs. *Nanoscale* **2016**, *8*, 17542–17550. [CrossRef]

28. Leontis, N.B.; Stombaugh, J.; Westhof, E. The non-Watson-Crick base pairs and their associated isostericity matrices. *Nucleic Acids Res.* **2002**, *30*, 3497–3531. [CrossRef] [PubMed]

29. Geary, C.; Chworos, A.; Verzemnieks, E.; Voss, N.R.; Jaeger, L. Composing RNA Nanostructures from a Syntax of RNA Structural Modules. *Nano Lett.* **2017**, *17*, 7095–7101. [CrossRef] [PubMed]

30. Sarver, M.; Zirbel, C.L.; Stombaugh, J.; Mokdad, A.; Leontis, N.B. FR3D: Finding local and composite recurrent structural motifs in RNA 3D structures. *J. Math. Biol.* **2008**, *56*, 215–252. [CrossRef]

31. Hoiberg, H.C.; Sparvath, S.M.; Andersen, V.L.; Kjems, J.; Andersen, E.S. An RNA origami octahedron with intrinsic siRNAs for potent gene knockdown. *Biotechnol. J.* **2018**, *14*, e1700634. [CrossRef] [PubMed]

32. Chworos, A.; Severcan, I.; Koyfman, A.Y.; Weinkam, P.; Oroudjev, E.; Hansma, H.G.; Jaeger, L. Building programmable jigsaw puzzles with RNA. *Science* **2004**, *306*, 2068–2072. [CrossRef]

33. Leontis, N.; Sweeney, B.; Haque, F.; Guo, P. Conference Scene: Advances in RNA nanotechnology promise to transform medicine. *Nanomedicine* **2013**, *8*, 1051–1054. [CrossRef]

34. Dibrov, S.M.; McLean, J.; Parsons, J.; Hermann, T. Self-assembling RNA square. *Proc. Natl. Acad. Sci. USA* **2011**, *108*, 6405–6408. [CrossRef] [PubMed]

35. Sajja, S.; Chandler, M.; Fedorov, D.; Kasprzak, W.K.; Lushnikov, A.; Viard, M.; Shah, A.; Dang, D.; Dahl, J.; Worku, B.; et al. Dynamic Behavior of RNA Nanoparticles Analyzed by AFM on a Mica/Air Interface. *Langmuir* **2018**, *34*, 15099–15108. [CrossRef] [PubMed]

36. Jasinski, D.; Haque, F.; Binzel, D.W.; Guo, P. Advancement of the Emerging Field of RNA Nanotechnology. *Acs Nano* **2017**, *11*, 1142–1164. [CrossRef] [PubMed]

37. Shu, D.; Shu, Y.; Haque, F.; Abdelmawla, S.; Guo, P. Thermodynamically stable RNA three-way junction for constructing multifunctional nanoparticles for delivery of therapeutics. *Nat. Nanotechnol.* **2011**, *6*, 658–667. [CrossRef]

38. Rychahou, P.; Haque, F.; Shu, Y.; Zaytseva, Y.; Weiss, H.L.; Lee, E.Y.; Mustain, W.; Valentino, J.; Guo, P.; Evers, B.M. Delivery of RNA nanoparticles into colorectal cancer metastases following systemic administration. *ACS Nano* **2015**, *9*, 1108–1116. [CrossRef]

39. Feng, L.; Li, S.K.; Liu, H.; Liu, C.Y.; LaSance, K.; Haque, F.; Shu, D.; Guo, P. Ocular delivery of pRNA nanoparticles: Distribution and clearance after subconjunctival injection. *Pharm. Res.* **2014**, *31*, 1046–1058. [CrossRef]

40. Afonin, K.A.; Viard, M.; Martins, A.N.; Lockett, S.J.; Maciag, A.E.; Freed, E.O.; Heldman, E.; Jaeger, L.; Blumenthal, R.; Shapiro, B.A. Activation of different split functionalities upon re-association of RNA-DNA hybrids. *Nat. Nanotechnol.* **2013**, *8*, 296–304. [CrossRef]

41. Kim, T.; Afonin, K.A.; Viard, M.; Koyfman, A.Y.; Sparks, S.; Heldman, E.; Grinberg, S.; Linder, C.; Blumenthal, R.P.; Shapiro, B.A. In Silico, In Vitro, and In Vivo Studies Indicate the Potential Use of Bolaamphiphiles for Therapeutic siRNAs Delivery. *Mol. Ther. Nucleic Acids* **2013**, *2*, e80. [CrossRef]

42. Lee, H.; Lytton-Jean, A.K.; Chen, Y.; Love, K.T.; Park, A.I.; Karagiannis, E.D.; Sehgal, A.; Querbes, W.; Zurenko, C.S.; Jayaraman, M.; et al. Molecularly self-assembled nucleic acid nanoparticles for targeted in vivo siRNA delivery. *Nat. Nanotechnol.* **2012**, *7*, 389–393. [CrossRef] [PubMed]

43. Binzel, D.W.; Shu, Y.; Li, H.; Sun, M.; Zhang, Q.; Shu, D.; Guo, B.; Guo, P. Specific Delivery of MiRNA for High Efficient Inhibition of Prostate Cancer by RNA Nanotechnology. *Mol. Ther.* **2016**, *24*, 1267–1277. [CrossRef]

44. Geary, C.; Rothemund, P.W.; Andersen, E.S. RNA nanostructures. A single-stranded architecture for cotranscriptional folding of RNA nanostructures. *Science* **2014**, *345*, 799–804. [CrossRef]

45. Afonin, K.A.; Bindewald, E.; Yaghoubian, A.J.; Voss, N.; Jacovetty, E.; Shapiro, B.A.; Jaeger, L. In vitro assembly of cubic RNA-based scaffolds designed in silico. *Nat. Nanotechnol.* **2010**, *5*, 676–682. [CrossRef]

46. Afonin, K.A.; Lin, Y.P.; Calkins, E.R.; Jaeger, L. Attenuation of loop-receptor interactions with pseudoknot formation. *Nucleic Acids Res.* **2012**, *40*, 2168–2180. [CrossRef]

47. Lee, J.B.; Hong, J.; Bonner, D.K.; Poon, Z.; Hammond, P.T. Self-assembled RNA interference microsponges for efficient siRNA delivery. *Nat. Mater.* **2012**, *11*, 316–322. [CrossRef] [PubMed]

48. Johnson, M.B.; Halman, J.R.; Satterwhite, E.; Zakharov, A.V.; Bui, M.N.; Benkato, K.; Goldsworthy, V.; Kim, T.; Hong, E.; Dobrovolskaia, M.A.; et al. Programmable Nucleic Acid Based Polygons with Controlled Neuroimmunomodulatory Properties for Predictive QSAR Modeling. *Small* **2017**, *13*, 42. [CrossRef]

49. Gupta, K.; Afonin, K.A.; Viard, M.; Herrero, V.; Kasprzak, W.; Kagiampakis, I.; Kim, T.; Koyfman, A.Y.; Puri, A.; Stepler, M.; et al. Bolaamphiphiles as carriers for siRNA delivery: From chemical syntheses to practical applications. *J. Control. Release* **2015**, *213*, 142–151. [CrossRef] [PubMed]

50. Gupta, K.; Mattingly, S.J.; Knipp, R.J.; Afonin, K.A.; Viard, M.; Bergman, J.T.; Stepler, M.; Nantz, M.H.; Puri, A.; Shapiro, B.A. Oxime ether lipids containing hydroxylated head groups are more superior siRNA delivery agents than their nonhydroxylated counterparts. *Nanomedicine* **2015**, *10*, 2805–2818. [CrossRef] [PubMed]

51. Osada, E.; Suzuki, Y.; Hidaka, K.; Ohno, H.; Sugiyama, H.; Endo, M.; Saito, H. Engineering RNA-protein complexes with different shapes for imaging and therapeutic applications. *ACS Nano* **2014**, *8*, 8130–8140. [CrossRef] [PubMed]

52. Schwarz-Schilling, M.; Dupin, A.; Chizzolini, F.; Krishnan, S.; Mansy, S.S.; Simmel, F.C. Optimized Assembly of a Multifunctional RNA-Protein Nanostructure in a Cell-Free Gene Expression System. *Nano Lett.* **2018**, *18*, 2650–2657. [CrossRef]

53. Cruz-Acuna, M.; Halman, J.R.; Afonin, K.A.; Dobson, J.; Rinaldi, C. Magnetic nanoparticles loaded with functional RNA nanoparticles. *Nanoscale* **2018**, *10*, 17761–17770. [CrossRef] [PubMed]

54. Juneja, R.; Lyles, Z.; Vadarevu, H.; Afonin, K.A.; Vivero-Escoto, J.L. Multimodal Polysilsesquioxane Nanoparticles for Combinatorial Therapy and Gene Delivery in Triple-Negative Breast Cancer. *ACS Appl. Mater. Interfaces* **2019**, *11*, 12308–12320. [CrossRef]

55. Dobrovolskaia, M.A.; McNeil, S.E. Immunological and hematological toxicities challenging clinical translation of nucleic acid-based therapeutics. *Expert Opin. Biol. Ther.* **2015**, *15*, 1023–1048. [CrossRef] [PubMed]

56. Bindewald, E.; Afonin, K.A.; Viard, M.; Zakrevsky, P.; Kim, T.; Shapiro, B.A. Multistrand Structure Prediction of Nucleic Acid Assemblies and Design of RNA Switches. *Nano Lett.* **2016**, *16*, 1726–1735. [CrossRef]

57. Zakrevsky, P.; Parlea, L.; Viard, M.; Bindewald, E.; Afonin, K.A.; Shapiro, B.A. Preparation of a Conditional RNA Switch. *Methods Mol. Biol.* **2017**, *1632*, 303–324.

58. Groves, B.; Chen, Y.J.; Zurla, C.; Pochekailov, S.; Kirschman, J.L.; Santangelo, P.J.; Seelig, G. Computing in mammalian cells with nucleic acid strand exchange. *Nat. Nanotechnol.* **2016**, *11*, 287–294. [CrossRef] [PubMed]

59. Afonin, K.A.; Lindsay, B.; Shapiro, B.A. Engineered RNA Nanodesigns for Applications in RNA Nanotechnology. *RNA Nanotechnol.* **2013**. [CrossRef]

60. Rose, S.D.; Kim, D.H.; Amarzguioui, M.; Heidel, J.D.; Collingwood, M.A.; Davis, M.E.; Rossi, J.J.; Behlke, M.A. Functional polarity is introduced by Dicer processing of short substrate RNAs. *Nucleic Acids Res.* **2005**, *33*, 4140–4156. [CrossRef] [PubMed]

61. Afonin, K.A.; Viard, M.; Kagiampakis, I.; Case, C.L.; Dobrovolskaia, M.A.; Hofmann, J.; Vrzak, A.; Kireeva, M.; Kasprzak, W.K.; KewalRamani, V.N.; et al. Triggering of RNA Interference with RNA-RNA, RNA-DNA, and DNA-RNA Nanoparticles. *ACS Nano* **2015**, *9*, 251–259. [CrossRef] [PubMed]

62. Martins, A.N.; Ke, W.; Jawahar, V.; Striplin, M.; Striplin, C.; Freed, E.O.; Afonin, K.A. Intracellular Reassociation of RNA-DNA Hybrids that Activates RNAi in HIV-Infected Cells. *Methods Mol. Biol.* **2017**, *1632*, 269–283. [PubMed]

63. Rogers, T.A.; Andrews, G.E.; Jaeger, L.; Grabow, W.W. Fluorescent monitoring of RNA assembly and processing using the split-spinach aptamer. *ACS Synth. Biol.* **2015**, *4*, 162–166. [CrossRef] [PubMed]

64. Afonin, K.A.; Desai, R.; Viard, M.; Kireeva, M.L.; Bindewald, E.; Case, C.L.; Maciag, A.E.; Kasprzak, W.K.; Kim, T.; Sappe, A.; et al. Co-transcriptional production of RNA-DNA hybrids for simultaneous release of multiple split functionalities. *Nucleic Acids Res.* **2014**, *42*, 2085–2097. [CrossRef] [PubMed]

65. Kolpashchikov, D.M. Binary malachite green aptamer for fluorescent detection of nucleic acids. *J. Am. Chem. Soc.* **2005**, *127*, 12442–12443. [CrossRef]

66. Sajja, S.; Chandler, M.; Striplin, C.D.; Afonin, K.A. Activation of Split RNA Aptamers: Experiments Demonstrating the Enzymatic Synthesis of Short RNAs and Their Assembly As Observed by Fluorescent Response. *J. Chem. Educ.* **2018**, *95*, 1861–1866. [CrossRef]

67. Chandler, M.; Ke, W.; Halman, J.R.; Panigaj, M.; Afonin, K.A. *Nanooncology: Engineering Nanomaterials for Cancer Therapy and Diagnosis*; Gonçalves, G., Tobias, G., Eds.; Springer International Publishing: Cham, Switzerland, 2018; pp. 365–385.

68. Chandler, M.; Lyalina, T.; Halman, J.; Rackley, L.; Lee, L.; Dang, D.; Ke, W.; Sajja, S.; Woods, S.; Acharya, S.; et al. Broccoli Fluorets: Split Aptamers as a User-Friendly Fluorescent Toolkit for Dynamic RNA Nanotechnology. *Molecules* **2018**, *23*, 3178. [CrossRef]

69. O'Hara, J.M.; Marashi, D.; Morton, S.; Jaeger, L.; Grabow, W.W. Optimization of the Split-Spinach Aptamer for Monitoring Nanoparticle Assembly Involving Multiple Contiguous RNAs. *Nanomaterials* **2019**, *9*, 378. [CrossRef]

70. Goldsworthy, V.; LaForce, G.; Abels, S.; Khisamutdinov, E.F. Fluorogenic RNA Aptamers: A Nano-platform for Fabrication of Simple and Combinatorial Logic Gates. *Nanomaterials* **2018**, *8*, 984. [CrossRef]

71. Halman, J.R.; Satterwhite, E.; Roark, B.; Chandler, M.; Viard, M.; Ivanina, A.; Bindewald, E.; Kasprzak, W.K.; Panigaj, M.; Bui, M.N.; et al. Functionally-interdependent shape-switching nanoparticles with controllable properties. *Nucleic Acids Res.* **2017**, *45*, 2210–2220. [CrossRef] [PubMed]

72. Filonov, G.S.; Moon, J.D.; Svensen, N.; Jaffrey, S.R. Broccoli: Rapid Selection of an RNA Mimic of Green Fluorescent Protein by Fluorescence-Based Selection and Directed Evolution. *J. Am. Chem. Soc.* **2014**, *136*, 16299–16308. [CrossRef]

73. Ke, W.; Hong, E.; Saito, R.F.; Rangel, M.C.; Wang, J.; Viard, M.; Richardson, M.; Khisamutdinov, E.F.; Panigaj, M.; Dokholyan, N.V.; et al. RNA–DNA fibers and polygons with controlled immunorecognition activate RNAi, FRET and transcriptional regulation of NF-κB in human cells. *Nucleic Acids Res.* **2018**, *47*, 1350–1361. [CrossRef] [PubMed]

74. Kim, K.H.; Lee, E.S.; Cha, S.H.; Park, J.H.; Park, J.S.; Chang, Y.C.; Park, K.K. Transcriptional regulation of NF-kappaB by ring type decoy oligodeoxynucleotide in an animal model of nephropathy. *Exp. Mol. Pathol.* **2009**, *86*, 114–120. [CrossRef] [PubMed]

75. Porciani, D.; Tedeschi, L.; Marchetti, L.; Citti, L.; Piazza, V.; Beltram, F.; Signore, G. Aptamer-Mediated Codelivery of Doxorubicin and NF-κB Decoy Enhances Chemosensitivity of Pancreatic Tumor Cells. *Mol. Ther. Nucleic Acids* **2015**, *4*, e235. [CrossRef] [PubMed]

76. Bui, M.N.; Brittany Johnson, M.; Viard, M.; Satterwhite, E.; Martins, A.N.; Li, Z.; Marriott, I.; Afonin, K.A.; Khisamutdinov, E.F. Versatile RNA tetra-U helix linking motif as a toolkit for nucleic acid nanotechnology. *Nanomedicine* **2017**, *13*, 1137–1146. [CrossRef]

77. Hong, E.; Halman, J.R.; Shah, A.B.; Khisamutdinov, E.F.; Dobrovolskaia, M.A.; Afonin, K.A. Structure and Composition Define Immunorecognition of Nucleic Acid Nanoparticles. *Nano Lett.* **2018**, *18*, 4309–4321. [CrossRef]

78. Surana, S.; Shenoy, A.R.; Krishnan, Y. Designing DNA nanodevices for compatibility with the immune system of higher organisms. *Nat. Nanotechnol.* **2015**, *10*, 741–747. [CrossRef]

79. Hong, E.; Halman, J.R.; Shah, A.; Cedrone, E.; Truong, N.; Afonin, K.A.; Dobrovolskaia, M.A. Toll-Like Receptor-Mediated Recognition of Nucleic Acid Nanoparticles (NANPs) in Human Primary Blood Cells. *Molecules* **2019**, *24*, 1094. [CrossRef] [PubMed]

80. Rackley, L.; Stewart, J.M.; Salotti, J.; Krokhotin, A.; Shah, A.; Halman, J.R.; Juneja, R.; Smollett, J.; Lee, L.; Roark, K.; et al. RNA Fibers as Optimized Nanoscaffolds for siRNA Coordination and Reduced Immunological Recognition. *Adv. Funct. Mater.* **2018**, *28*, 1805959. [CrossRef]

81. Kondo, S.; Sauder, D.N. Tumor necrosis factor (TNF) receptor type 1 (p55) is a main mediator for TNF-alpha-induced skin inflammation. *Eur. J. Immunol.* **1997**, *27*, 1713–1718. [CrossRef] [PubMed]

MDPI

Article

Suicide Gene Therapy By Amphiphilic Copolymer Nanocarrier for Spinal Cord Tumor

So-Jung Gwak [1,2] and Jeoung Soo Lee [1,*]

[1] Department of Bioengineering, Drug Design, Development and Delivery (4D) Laboratory, Clemson University, Clemson, SC 29634-0905, USA; plus38317@gmail.com
[2] Department of Chemical Engineering, Wonkwang University, 460, Iksandae-ro, Iksan, Jeonbuk 54538, Korea
* Correspondence: ljspia@clemson.edu; Tel.: +1-864-656-3212; Fax: +1-864-656-4466

Received: 11 February 2019; Accepted: 30 March 2019; Published: 8 April 2019

Abstract: Spinal cord tumors (SCT) are uncommon neoplasms characterized by irregular growth of tissue inside the spinal cord that can result in non-mechanical back pain. Current treatments for SCT include surgery, radiation therapy, and chemotherapy, but these conventional therapies have many limitations. Suicide gene therapy using plasmid encoding herpes simplex virus-thymidine kinase (pHSV-TK) and ganciclovir (GCV) has been an alternative approach to overcome the limitations of current therapies. However, there is a need to develop a carrier that can deliver both pHSV-TK and GCV for improving therapeutic efficacy. Our group developed a cationic, amphiphilic copolymer, poly (lactide-co-glycolide) -graft-polyethylenimine (PgP), and demonstrated its efficacy as a drug and gene carrier in both cell culture studies and animal models. In this study, we evaluated PgP as a gene carrier and demonstrate that PgP can efficiently deliver reporter genes, pGFP in rat glioma (C6) cells in vitro, and pβ-gal in a rat T5 SCT model in vivo. We also show that PgP/pHSV-TK with GCV treatment showed significantly higher anticancer activity in C6 cells compared to PgP/pHSV-TK without GCV treatment. Finally, we demonstrate that PgP/pHSV-TK with GCV treatment increases the suicide effect and apoptosis of tumor cells and reduces tumor size in a rat T5 SCT model.

Keywords: suicide gene therapy; non-viral gene delivery; ganciclovir; spinal cord tumor

1. Introduction

Intramedullary spinal cord tumors (IMSCT) constitute 8 to 10% of primary spinal cord tumors [1] and are uncommon neoplasms characterized by irregular growth of tissue found inside the spinal cord. These tumors cause significant neurologic morbidity and mortality [2], such as pressure on sensitive tissues and reduced function, resulting in pain, sensory changes, and motor deficits. Current treatments for spinal cord tumor include surgical therapy, radiotherapy, and chemotherapy [3–5]. However, these conventional therapies have limitations such as tumor cell survival that leads to high recurrence rate, as well as infiltration of tumors into the spinal cord [6,7]. Radiation therapy is limited by dose-related toxicity to the normal spinal cord and surrounding tissues, and systemic toxicity [8]. Chemotherapy has also been used to treat SCT [9], but it has dose-limiting toxicity and side effects such as sensory neuropathies and gastrointestinal disturbances after systemic administration [10,11].

Gene therapy has been investigated as an alternative approach to overcome the limitations of current therapies for SCTs [12,13]. Among gene therapies, suicide gene therapy has received considerable attention in the field of cancer gene therapy. Suicide gene therapy is based on the gene-directed enzyme prodrug strategy, which requires suicide genes and prodrugs. In gene-directed enzyme prodrug therapy, a viral or non-viral vector is used to transfect tumor cells with a gene encoding a specific exogenous enzyme. A prodrug is then administered that can be only activated by the specific enzyme expressed in the tumor cells. The Herpes simplex virus thymidine kinase (HSV-TK) gene is the most frequently used suicide gene with the anticancer prodrug, ganciclovir

(GCV) [13–16]. Won et al. reported that reducible poly (oligo-D-arginine) (rPOA) could deliver plasmid encoding herpes simplex virus-thymidine kinase (pHSV-TK)and demonstrated that locally injected rPOA/pHSV-TK with systemically administered GCV significantly reduced tumor volume and improved locomotor function compared to the GCV only group and naked pHSV-TK with GCV-treated group [12,13]. One of the drawbacks of this pHSV-TK and GCV combination therapy is the GCV concentration in tumor cells. GCV has a very low oral bioavailability (~5%) and short plasma half-life (~2–6 h), so GCV has to be injected daily (5 mg/kg) to maintain the therapeutic plasma concentration. To improve GCV concentration in plasma, several delivery systems have been studied, including liposomes [17,18] and silicone pellets [19]. Kajiwara et al. reported that they achieved longer circulation of liposome encapsulated GCV (PEG-GCV-lipo) in blood and the area under the curve (AUC) of liposome encapsulated GCV (PEG-GCV-lipo) was 36-fold and 32-fold higher compared to GCV solution after intravenous and intraperitoneal injection in mice, respectively [18]. They reported that PEG-GCV-lipo was three times more effective than GCV solution in inhibiting tumor growth in a mouse KB xenografts model. Therefore, there is a need to develop a carrier that can efficiently deliver both pHSV-TK and GCV to improve the efficacy of this suicide gene therapy.

Our group developed a cationic, amphiphilic copolymer, poly (lactide-co-glycolide)-graft-polyethylenimine (PgP), as a drug and nucleic acid delivery carrier [20–22]. We reported its ability to efficiently deliver plasmid DNA (pDNA) and small interfering RNA (siRNA) in various cell lines and primary neurons in vitro, as well as in the normal rat spinal cord [20]. We showed that PgP loaded with a fluorescent, hydrophobic dye (DiR; 1,1'-Dioctadecyl-3,3,3',3'-Tetramethylindotricarbocyanine Iodide) is retained within the injured spinal cord for up to five days after intraspinal injection. Injection of PgP complexed with siRNA targeting Ras homolog gene family, member A (RhoA) in the injured rat spinal cord resulted in RhoA knockdown, reduced astrogliosis and necrotic cavity formation, and increased axonal sparing/regeneration [21]. Finally, we reported that a hydrophobic drug, rolipram, can be encapsulated in the hydrophobic core of PgP, increasing its aqueous solubility seven times compared to that in water alone [22].

Our long-term goal is to develop PgP as a platform technology for delivery of therapeutic drugs and nucleic acids to spinal cord tumors. In this study, we investigated PgP as a suicide gene carrier. We show that PgP can efficiently deliver pDNA encoding green fluorescence protein (pGFP) in rat glioblastoma (C6) cells in vitro and deliver pDNA encoding β-galactosidase protein (pβ-gal) in a rat spinal cord tumor model in vivo. We demonstrate that PgP/pHSV-TK with GCV achieves a suicide effect in C6 cells in vitro. Finally, we show the suicide effect of PgP/pHSV-TK with GCV treatment in a rat T5 spinal cord tumor model.

2. Material and Methods

2.1. Plasmid Amplification and Purification

Plasmids encoding the Monster Green Fluorescent Protein (phMGFP Vector, pGFP, Promega, Madison, WI, USA), beta-galactosidase (pSV40-βGal, βGal, Promega), and herpes simplex virus-thymidine kinase (pHSV-TK, Invivogen, San Diego, CA, USA) were transformed into *Escherichia coli* DH5α (Life Technologies, Grand Island, NY, USA) and amplified in Lysogeny broth (LB) medium with ampicillin at 37 °C overnight with shaking at 250 rpm. The amplified plasmids were purified by the Endorsee Maxi plasmid purification kit (Qiagen, Valencia, CA, USA). The quantity of plasmid was determined by the absorbance at 260 nm and the quality of plasmid was determined by the ratio of 260/280 nm using BioTek Take 3 microplate reader (BioTek, Synergy HT, Winooski, VT, USA).

2.2. Particle Size and Surface Charge of PgP/pDNA Polyplex

PgP was synthesized using PLGA (lactide:glycolide 75:25, 25 kDa, 120 µmole, Durect Corporation, Pelham, AL, USA) and branched polyethylenimine (bPEI, MW: 25 kDa, 100 µmol, Sigma, Milwaukee,

WI, USA) as previously published by our laboratory [22]. PgP/pGFP polyplexes were prepared by mixing PgP and pGFP at various N/P (number of nitrogen atoms in PgP/number of phosphorus atoms in DNA) ratio and incubated for 30 min at 37 °C. Polyplex size distribution was measured using dynamic laser light scattering and zeta potential was measured electrophoretically by Zeta PALS (Brookhaven Instruments Corp., Holtsville, NY, USA). The mean diameter and zeta potential of polyplexes were measured in triplicate.

2.3. Transfection Efficiency and Cytotoxicity of PgP/pDNA Complex in 10% Serum Condition

To evaluate PgP as a non-viral gene carrier, we first measured the transfection efficiency and cytotoxicity of polyplexes formed with PgP and reporter gene, *pGFP*, in rat glioma (C6) cells in vitro. C6 cells were maintained in DMEM/F12 supplemented with 10% FBS and 1% penicillin/streptomycin at 37 °C in 5% CO_2. For transfection, the cells were seeded at a density 1×10^5 cells/well in a 12-well plate and cultured overnight. The polyplexes of PgP/pGFP (2 μg pGFP) were prepared at various N/P ratios ranging from 15/1 to 60/1 and incubated for 30 min at 37 °C. Polyplexes (2 μg pDNA/well) were added to the cells in the media containing 10% FBS and incubated at 37 °C. At 24 h post-transfection, the media containing polyplexes were removed and the cells were washed and replenished with fresh media and then cultured for an additional 24 h. GFP-expressing cells were counted by easyCyte flow cytometer (Millipore, Darmstadt, Germany) and transfection efficiency is expressed as % transfected cells according to the following equation:

$$\% \text{ transfection efficiency} = (\text{number of GFP-positive cells/number of total cells}) \times 100$$

Transfected cells were imaged using an inverted fluorescent microscope (Zeiss Axiovert 200, Göttingen, Deutschland). The bPEI/pGFP (N/P ratio 5/1) complex was used as a positive control.

Cytotoxicity was analyzed in parallel experiments using the MTT assay. At 48 h post-transfection, the medium was removed and cells were washed with PBS, and incubated with 1 mL of fresh medium containing 3-(4,5-dimethylthiazol-2-yl)-2,5-diphenyltetrazolium bromide (MTT, 2 mg/mL in PBS, Sigma, Milwaukee, WI, USA). After incubation for 4 h at 37 °C, the MTT solution was removed and 1.5 mL of dimethylsulfoxide (DMSO) was added to dissolve the formazan crystals formed by the live cells. Absorbance (A) was measured at 570 nm. The cell viability (%) was calculated compared to non-transfected control according to the following equation:

$$\text{Cell viability (\%)} = (A_{570(\text{sample})}/A_{570(\text{control})}) \times 100\%$$

2.4. Characterization of PgP/pDNA Polyplexes

2.4.1. Stability of PgP/pDNA Polyplex

To evaluate the stable complex formation of the PgP/pDNA polyplex, PgP/pGFP complexes were prepared at various N/P ratios in deionized water and incubated for 30 min at 37 °C. The complexes were loaded on a 1% (*w/v*) agarose gel and electrophoresed for 60 min at 80 V. The gel was stained with ethidium bromide for 30 min and then imaged on an ultraviolet (UV) illuminator (GELDOC-IT2 imager, UVP, Waltham, MA, USA) to visualize the retention of complexes and migration of naked pDNA.

2.4.2. Heparin Competition Assay

The stability of the PgP/pGFP polyplexes was evaluated using a heparin competition assay. Briefly, PgP/pGFP at an N/P ratio of 60/1 was prepared as described above. Ten microliter aliquots of complex solutions were placed into tubes. Then heparin, a negatively charged polysaccharide, was added at 0–40 heparin/pDNA, *w/w* ratio and the polyplex/heparin solutions were incubated at 37 °C for 30 min. The samples were immediately analyzed by 1% agarose gel electrophoresis as described above.

2.4.3. Stability of PgP/pDNA in Serum

To measure the stability of the PgP/pGFP polyplexes in serum, PgP/pGFP polyplexes were prepared at a N/P ratio of 60/1 and then incubated in the media containing 10% FBS at 37 °C. Naked pDNA incubated in the media containing10% FBS was used for comparison. At 30 min, and 1, 3, 6, 24, and 72 h post-incubation, the samples were evaluated by 1% agarose gel electrophoresis as described above.

2.5. Long-Term Storage Stability of PgP/pDNA Polyplexes

To evaluate stability during long-term storage, PgP/pGFP polyplexes at an N/P ratio of 60/1 were selected based on the results from in vitro transfection and cytotoxicity. PgP/pGFP polyplexes at an N/P ratio of 60/1 (2 μg pGFP) were prepared and incubated at 4 °C for 6 months. The stability of stored PgP/pGFP polyplexes was evaluated at predetermined time points (6 h, 1 day, 3 days, 1 week, and 1, 3, 4, 5, and 6 months) using 1% agarose gel electrophoresis and transfection efficiency in C6 cells as described above. Freshly prepared PgP/pGFP polyplexes were used as a control. GFP-positive transfected cells were imaged using an inverted fluorescent microscope (Zeiss Axiovert 200, Göttingen, Germany).

2.6. Suicide Effects of PgP/pHSV-TK Polyplex and GCV Treatment In Vitro

To evaluate the suicide effect of PgP/pHSV-TK polyplex with GCV prodrug, C6 cells were seeded in 12-well plates at 1×10^5 cells/well and cultured for 24 h. PgP/pHSV-TK polyplexes (2 μg, pHSV-TK) were prepared at an N/P ratio of 60/1, added to the cells cultured in DMEM/F12 containing 10% FBS, and incubated at 37 °C. At 24 h post-transfection, the media were relaced with fresh media containing 2 different GCV concentrations (50 and 100 μg/mL). At 1 and 4 days post-GCV treatment, the anti-cancer efficacy of PgP/pHSV-TK with GCV was analyzed via the MTT assay as described above. PgP/pHSV-TK at the N/P ratio of 60/1 only, GCV (100 μg/mL) only, and bPEI/pHSV-TK at an N/P ratio of 5/1 with GCV (100 μg/mL) were used for comparison. We used PgP/pGFP at an N/P ratio of 60/1 as a control to eliminate the cytotoxicity of the polyplex itself.

2.7. Generation of Spinal Cord Tumor Model

All surgical procedures and postoperative care were conducted according to National Institute of Health (NIH) guidelines for the care and use of laboratory animals (NIH publication No. 86–23, revised 1996) and under the supervision of the Clemson University Institutional Animal Care and Use Committee. Sprague Dawley rats (250–300 g, male) were anesthetized using isoflurane gas and laminectomy was performed at the T5 spinous process using an orthopedic bone cutter. C6 glioma cells (1.0×10^6 cells in 3 μL PBS) were injected into the T5 position using a Hamilton syringe (26 G) (Hamilton, Bonaduz, Switzerland) [12,13,23]. As a control, PBS was injected into the T5 spinal cord. After C6 cell injection, the paraspinal muscles and the skin were closed with 3–0 silk suture. After surgery, animals were warmed with a heating blanket for recovery. To verify the tumor formation, animals were euthanized 12 days post-tumor cell injection by CO_2 overdose, and spinal cord tissue from the region surrounding the injection site was explanted. Samples were embedded in Tissue-Tek® O.C.T compound (Sakura Finetek Inc., Torrance, CA, USA) on liquid nitrogen, sectioned, and stained with hematoxylin and eosin (H&E).

2.8. Transfection Efficiency of PgP/pβ-Gal in a Rat Spinal Cord Tumor Model In Vivo

To evaluate PgP as a gene carrier in vivo, we used pβ-gal instead of pGFP for transfection to avoid potential interference from tissue autofluorescence. Animals were injected with C6 glioma cells (1.0×10^6 cells in 3 μL PBS) at T5 as described above. At 5 days post-injection, the rats were randomly assigned to one of three experimental groups: (1) PgP/ pβ-gal polyplexes (*n* = 4 rats/group), (2) bPEI/pβ-gal complex, and (3) naked pβ-gal. PgP/ pβ-gal polyplexes (20 μL, 10 μg pβ-gal) at an N/P ratio of 60/1 were injected into the spinal cord tumor using a Hamilton syringe (26 G). bPEI/pβ-gal complex (20 μL, 10 μg pβ-gal) at an N/P of ratio 5/1 and naked pβ-gal (20 μL, 10 μg pβ-gal) were injected for comparison. After injection, the paraspinal muscles and the skin were closed with 3-0 silk suture. Seven days after treatment, rats were sacrificed by cardiac perfusion with 4% para-formaldehyde (PFA) in saline. The retrieved spinal cords were embedded in Tissue-Tek® O.C.T. compound and sectioned into 10-μm thickness using a cryostat (Lecia CM 1950, Leica, Wetzlar, Germany) for histological analyses. The sections were fixed with 4% PFA, washed two times in PBS (pH 7.4) for 5 min, stained using X-gal staining solution (Invitrogen, NY, USA) overnight, washed in distilled water, and then counter stained with eosin.

2.9. Suicide Effect of PgP/pHSV-TK Polyplexes with GCV in a Rat Spinal Cord Tumor In Vivo

To evaluate the suicide effect of PgP /pHSV-TK polyplexes with GCV in vivo, rats were randomly assigned to one of three experimental groups (*n* = 4/group): (1) PgP/pHSV-TK polyplexes with GCV, (2) PgP/pHSV-TK polyplexes without GCV, and (3) saline injection (Untreated) at 5 days after C6 cell injection, as described above. PgP/pHSV-TK polyplexes (10 μg pHSV-TK, 20 μL) at an N/P ratio of 60/1 were prepared and injected in the tumor lesion by Hamilton syringe (26 G) and GCV (40 mg/kg) was administered by intraperitoneal (i.p.) injection every day for 10 days [13]. PgP/pHSV-TK polyplexes (10 μg pHSV-TK, 20 μL) at an N/P ratio of 60/1 without GCV treatment were used for comparison and saline-injected animals were used as the untreated control. At 10 days post-GCV treatment, the rats were sacrificed by cardiac perfusion with 4% PFA in saline. The spinal cords were explanted, embedded, and sectioned as described above. To measure the tumor size, 16 sections (4 sections/rat, 4 rats/group) were stained with H&E, imaged. Quantitative measurements of tumor area were analyzed by Image J. The percent tumor area was calculated according to the following equation:

$$\% \text{ Tumor area} = (\text{Tumor area}/\text{Total area of spinal cord}) \times 100$$

To evaluate the suicide effect by PgP/pHSV-TK polyplex (N/P ratio of 60/1) with GCV, the terminal deoxynucleotidyl transferase dUTP nick end labeling (TUNEL) assay was performed to identify apoptotic cells. Briefly, the sections were stained by using the ApopTag Plus Fluorescein In situ Apoptosis Detection kit (Chemicon International, Temecula, CA, USA) and nuclei were counterstained by 4′,6-diamidino-2-phenylindole (DAPI) and digitally imaged using an inverted epifluorescence microscope (Zeiss Axiovert 200, Göttingen, Germany). The untreated spinal cord tumor group and the PgP/pHSV-TK polyplex (N/P ratio of 60/1) without GCV group were used for comparison. We also evaluated the suicide effect of the PgP/pHSV-TK polyplex (N/P ratio of 60/1) with GCV on expression of Bcl-2-associatied X protein (Bax), a pro-apoptotic protein. Briefly, sections were stained using antibody against Bax (1:200, sc-23959, Santa Cruz Biotechnology, Dallas, TX, USA), followed by Cy3-conjugated anti-rabbit IgG (1:200, ab97035, Abcam, Cambridge, MA, USA). The stained sections were counterstained with DAPI and imaged using an inverted epifluorescence microscope.

2.10. Statistical Analysis

Quantitative data are presented as the mean ± standard deviation. Statistical analysis was performed by one-way ANOVA with the least significant difference (LSD) method used for *post-hoc* comparisons between subgroups. A *p*-value less than 0.05 was considered significant.

3. Results

3.1. Characterization of PgP/pDNA Polyplexes

The particle size and zeta potential of PgP/pDNA polyplexes at various N/P ratios were evaluated. The mean size of the PgP/pDNA polyplexes at all N/P ratios was about 143.4 nm with narrow polydispersity. The negatively charged pDNA was completely neutralized by positively charged PgP and the surface charge of PgP/pDNA polyplexes at N/P ratio of 15/1 or above was positive (Table 1).

Table 1. Mean particle size (PS), zeta potential (ZP), and polydispersity index (PDI) of PgP/pDNA polyplexes at various N/P ratios.

N/P ratio	15	30	45	60
Particle Size (nm)	141.2 ± 3.8	148.5 ± 3.8	138.0 ± 3.2	145.7 ± 1.5
PDI	0.17 ± 0.01	0.16 ± 0.01	0.20 ± 0.01	0.17 ± 0.01
Zeta potential (mV)	34.4 ± 0.2	41.3 ± 2.5	41.5 ± 0.7	41.5 ± 0.3

3.2. Transfection Efficiency and Cytotoxicity of PgP/pDNA Polyplexes in 10% Serum Condition In Vitro

Transfection efficiency of PgP/pGFP polyplexes was evaluated in C6 cells in media containing 10% serum. GFP expression increased with increasing polyplex N/P ratio and the highest transfection efficiency (69%) was observed at an N/P ratio of 60/1. In contrast, the transfection efficiency of bPEI/pDNA polyplexes at an N/P ratio of 5/1 was approximately 1% (Figure 1A). Cell viability was modestly lower for bPEI/pGFP and PgP/pGFP at all N/P ratios, but the difference was only statistically significant at N/P ratios of 50/1 and 60/1 (Figure 1B). Figure 1C shows representative images of GFP-positive cells after transfection with PgP/pGFP polyplexes at various N/P ratios. Based on the high transfection efficiency with low cytotoxicity, we performed all subsequent experiments using PgP/pDNA at an N/P ratio of 60/1.

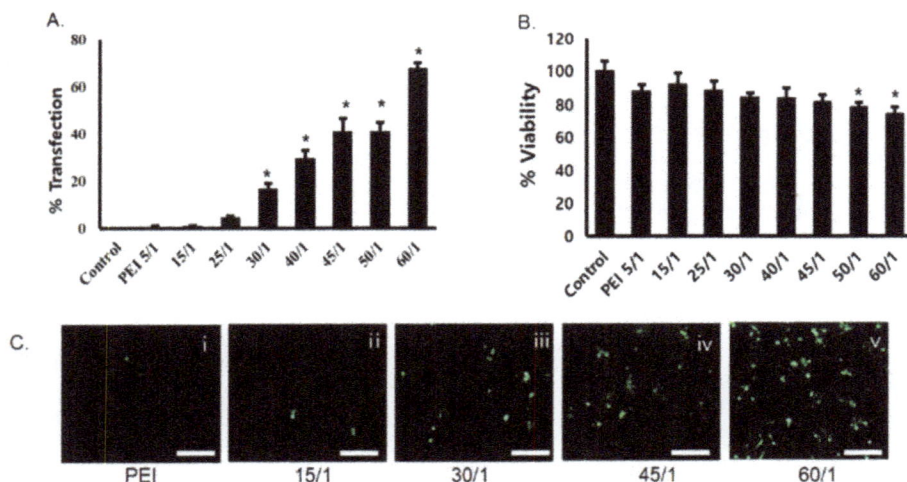

Figure 1. Transfection efficiency and cytotoxicity of PgP/pGFP polyplexes: (**A**) percent transfection efficiency and (**B**) percent cell viability after transfection of PgP/pGFP polyplexes in C6 cells in media containing 10% serum. Data represent the mean ± SD. (**C**) Representative images of GFP-positive cells after transfection with (i) bPEI/pGFP at N/P ratio of 5/1 and (ii–v) PgP/pGFP polyplexes at N/P ratios of 15/1, 30/1, 45/1 and 60/1. Scale bars indicate 100 μm.

3.3. Stability of PgP/pDNA Polyplex

The ability of PgP to condense pGFP was evaluated by gel retardation assay. Complete retardation of electrophoretic mobility was achieved at N/P ratios of 15/1 or above (Figure 2A). The stability of PgP/pGFP polyplexes was also evaluated by heparin competition assay. PgP/pDNA polyplexes at an N/P ratio of 60/1 that selected for highest transfection efficiency were prepared and incubated in solutions with varying heparin concentrations. PgP/pDNA polyplexes were stable in the presence of up to 6 *w/w* heparin/pDNA ratio and completely dissociated at ratios of 10 or higher (Figure 2B). The integrity of PgP/pDNA polyplexes at the N/P ratio of 60/1 in the presence of serum was evaluated after incubation in media containing 10% FBS. Naked pDNA incubated in the presence of serum was used for comparison. Naked pDNA was degraded by nucleases in the serum and undetectable after 3 h incubation (Figure 2Ci), whereas PgP/pDNA polyplexes were stable up to 24 h incubation (Figure 2Cii).

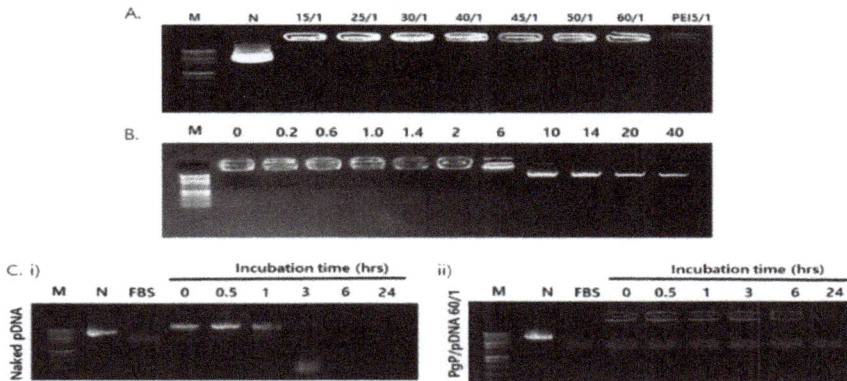

Figure 2. Analysis of polyplex stability. (**A**) Gel retardation assay of PgP/pDNA polyplexes at varying N/P ratios. Molecular weight marker (M, lane 1), naked pDNA (N, lane 2), PgP/pDNA polyplexes prepared at N/P ratios of 15/1, 25/1, 30/1, 40/1, 45/1. 50/1, and 60/1 (lanes 3–8) and bPEI/pDNA polyplex at N/P ratio of 5/1 (lane 9). (**B**) Heparin competition assay. PgP/pDNA polyplexes (2 μg pDNA) at N/P ratio of 60/1 were incubated with solutions containing heparin at varying concentration (0–40 heparin/pDNA, *w/w* ratio) at 37 °C for 30 min. M: Molecular marker. (**C**) Naked pDNA and PgP/pDNA polyplexes at N/P ratio 60/1 at various time points during incubation in 10% serum-containing media. (i) Naked pDNA and (ii) PgP/pDNA polyplex at N/P ratio 60/1, N is naked, untreated pDNA control.

3.4. Long-Term Storage Stability of PgP/pGFP Polyplexes

To evaluate long-term stability, PgP/pGFP polyplexes at N/P ratio 60/1 were stored at 4 °C for six months. Gel electrophoresis analysis showed that PgP/pGFP polyplexes were stable up to six months at 4 °C (Figure 3A). Transfection efficiency of PgP/pGFP polyplexes stored at 4 °C for up to four months was not significantly different compared to freshly prepared PgP/pGFP polyplexes, but significantly decreased after six months' storage (Figure 3B). Figure 3C shows representative images of GFP-positive cells after transfection with PgP/pGFP polyplexes stored at 4 °C.

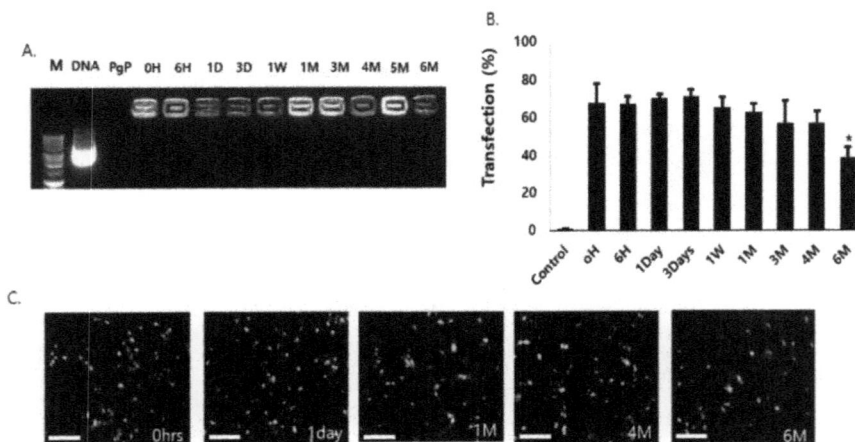

Figure 3. Long-term stability of PgP/pGFP polyplexes (2 µg pGFP, N/P ratio 60/1) at 4 °C. (**A**) Gel retardation assay by 1% agarose gel electrophoresis of PgP/pGFP polyplexes. M: Molecular weight marker (lane1), DNA: naked pDNA (lane 2), PgP only (lane 3), lanes 4–13: Polyplexes at pre-determined time points (0, 6 h, 1, 3, 7 days, 1, 3, 4, 5 and 6 months, respectively) during storage at 4 °C. (**B**) Transfection efficacy of polyplexes stored at 4 °C in C6 cells in media containing 10% serum at pre-determined time points. Data represent the mean ± SD. * $p < 0.05$ compared with freshly prepared polyplexes. (**C**) Representative images of GFP-positive cells at 2 days post-transfection with PgP/pGFP polyplexes stored 4 °C. Scale bars indicate 100 µm.

3.5. Suicide Effect of PgP/pHSV-TK Polyplex with GSV Treatment In Vitro

The suicide effect of PgP/pHSV-TK (N/P ratio of 60/1, 2 µg pHSV-TK) polyplexes with two different GCV doses (50 and 100 µg/mL) was evaluated at 1 and 4 days after GCV treatment in C6 cells. PgP/pGFP polyplexes (N/P ratio of 60/1, 2 µg pGFP) were used as a control instead of untreated cells to eliminate the cytotoxicity caused by the polyplex itself. PgP/pHSV-TK only or GCV (100 µg/mL) only, and PEI/pHSV-TK (N/P ratio of 5/1, 2 µg pHSV-TK) were used for comparison. PgP/pHSV-TK only or GCV only (100 µg/mL) did not show significantly different anti-cancer efficacy compared to PgP/pGFP polyplexes at both 1 and 4 days after treatment (Figure 4). The anti-cancer efficacy of PgP/pHSV-TK polyplexes with both GCV doses of 50 and 100 µg/mL was significantly higher than that of PgP/pGFP polyplexes at both 1 and 4 days post-treatment. We also observed that the anti-cancer efficacy of PgP/pHSV-TK polyplexes was significantly higher than that of bPEI/pGFP polyplexes (both with 100 µg/mL GCV) at 1 and 4 days post-treatment.

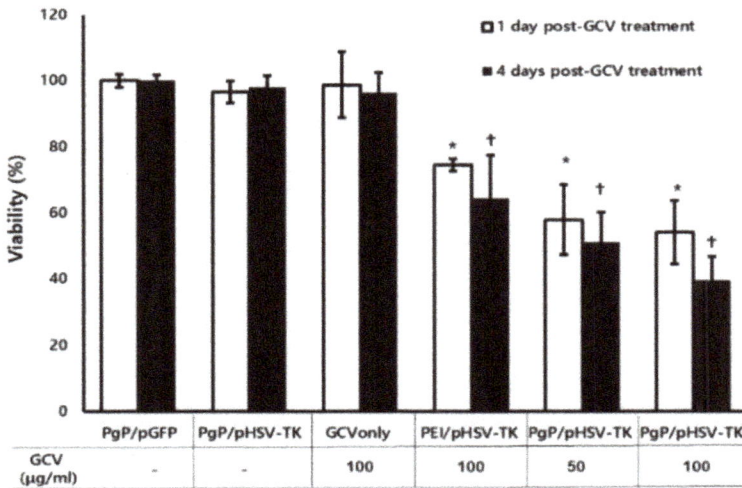

Figure 4. The suicide effect of PgP/pHSV-TK polyplexes and GCV. C6 cells were transfected with PgP/pHSV-TK (2 μg pHSV-TK) and then treated with GCV (50 and 100 μg/mL). At 1 and 4 days post-GCV treatment, anti-cancer efficacy was analyzed by MTT assay. PgP/pGFP (N/P of 60/1), PgP/pTK (N/P ratio of 60/1), GCV only (100 μg/mL), bPEI/pTK (N/P ratio of 5/1) with GCV (100 μg/mL) were used for comparison. Data represent the mean ± SD. *: $p < 0.05$ compared to PgP/pGFP at 1 day post-GCV treatment and +: $p < 0.05$ compared to PgP/pGFP at 4 days post-GCV treatment.

3.6. Transfection Efficiency of PgP/pβ-Gal in a Rat Spinal Cord Tumor Model In Vivo

To verify spinal cord tumor formation, rats were sacrificed at 12 days post-injection of C6 cells, tissues sectioned, and stained with H&E. The tumor masses were observed by high-density cell growth in between T5 and T6 in C6 cell-injected animals (Figure 5B,D), compared to normal spinal cord (Figure 5A,C). To evaluate the efficacy of PgP as a gene carrier in vivo, PgP/pβ-gal polyplexes (N/P 60/1, 10 μg) were intratumorally injected 5 days after C6 cell injection. Seven days later, the transfection efficiency of PgP/pβ-gal polyplexes in the spinal cord tumor was evaluated by x-gal staining. Figure 6 shows representative images of β-Gal-positive cells stained in blue within the spinal cord tumor area. We observed that the x-gal positively stained area in animals receiving PgP/pβ-gal polyplexes was substantially larger compared to animals receiving bPEI/pβ-gal polyplexes or naked pβ-gal (Figure 6).

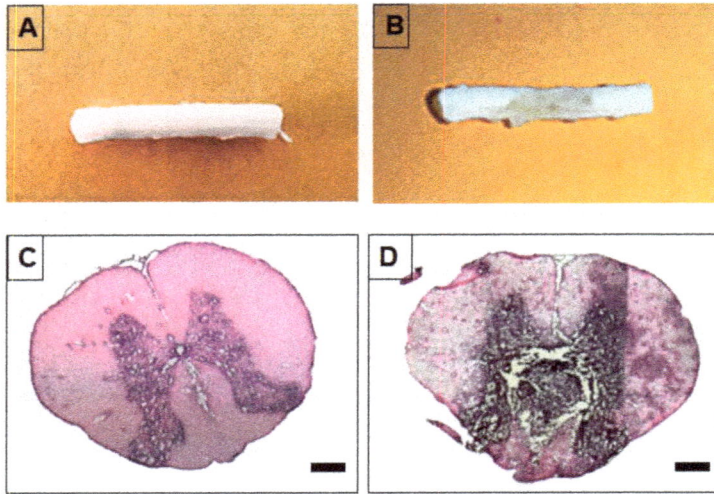

Figure 5. Generation of a rat T5 spinal cord tumor model at 12 days post-injection of C6 cells. (**A,B**) Representative images of isolated spinal cord and (**C,D**) H&E stained spinal cord section. (**A,C**) Normal spinal cord and (**B,D**) spinal cord with C6 cell-derived tumor. Scale bars indicate 500 μm.

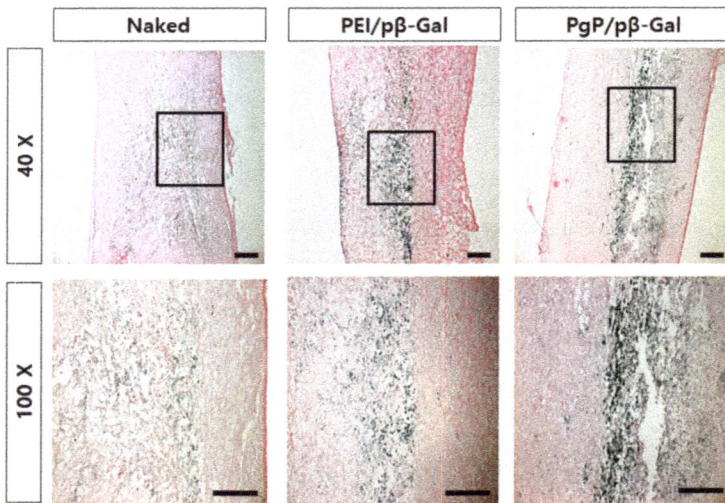

Figure 6. Representative images of β-Galactosidase-positive cells at 7 days post-injection of PgP/pβ-Gal polyplexes (N/P of 60/1) in rat T5 spinal cord tumor in vivo. (left) Naked pβ-Gal, (middle) bPEI/pβ-Gal polyplexes at N/P of 5/1, and (right) PgP/pβ-Gal polyplexes at N/P of 60/1. Original magnification: (**Top**) 40× and (**Bottom**) 100×. Scale bars indicate 100 μm.

3.7. Suicide Effect of PgP/pHSV-TK Polyplexes with GCV in a Rat Spinal Cord Tumor In Vivo

To evaluate the suicide effect of PgP/pHSV-TK with GCV treatment, PgP/pHSV-TK polyplexes at an N/P ratio 60/1 (10 μg pHSV-TK) were injected into spinal cord tumors at 5 days post-injection of C6 cells and then received i.p. injection of 40 mg GCV/kg for 10 days. Figure 7A shows representative images of H&E stained spinal cord tumor sections from various animal groups. The percent tumor area in animals treated with PgP/pHSV-TK polyplexes and GCV injection was significantly lower than that of groups receiving PgP/pHSV-TK polyplexes alone and the untreated SCT group (Figure 7B). We also evaluated the effect of PgP/pHSV-TK with GCV treatment on apoptosis using the TUNEL assay and immunohistochemistry (IHC) for expression of the pro-apoptotic protein, Bax. The number of TUNEL-positive cells in spinal cord tumors was substantially higher in animals receiving PgP/pHSV-TK polyplexes with GCV injection than both the PgP/pHSV-TK polyplexes only group and the untreated spinal cord tumor group (Figure 8A). We also observed that Bax expression was highly upregulated in animals injected with PgP/pHSV-TK polyplexes plus GCV compared to both PgP/pHSV-TK polyplexes only and untreated spinal cord tumor groups (Figure 8B).

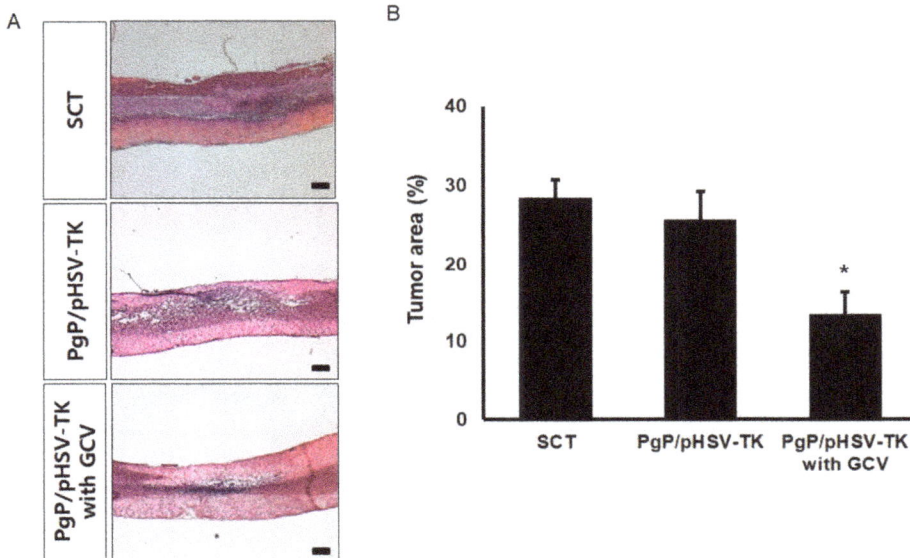

Figure 7. Histological analysis of the suicide effect on tumor size at 10 days post-intratumoral injection of PgP/pHSV-TK (10 μg pHSV-TK, N/P ratio of 60/1) polyplexes with GCV (40 mg/kg, intraperitoneal injection). (**A**) Representative images of H&E stained longitudinal spinal cord sections. Scale bars indicate 500 μm. (**B**) The percent tumor area of spinal cord tumors. The % tumor area was measured and averaged from 16 different sections of spinal cords from each group (4 sections/rat, 4 rats/ group). * $p < 0.05$ compared with untreated SCT group.

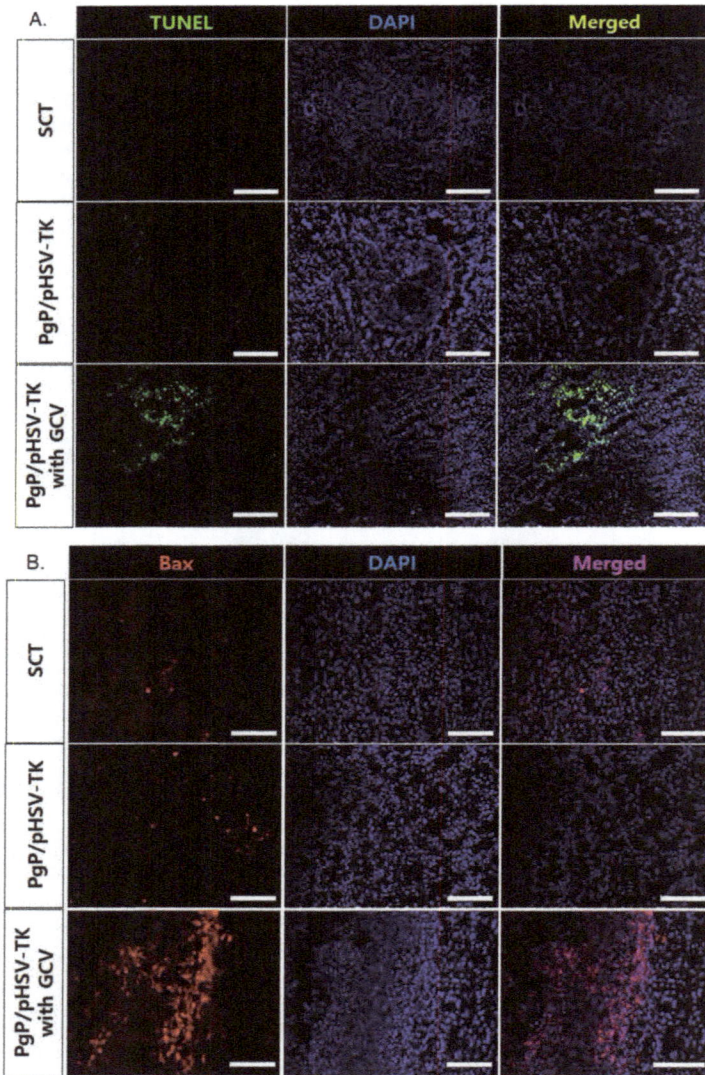

Figure 8. Representative images of (**A**) TUNEL+ cells (green), (**B**) Bax+ cells (red), in spinal cord tumor at 10 days post-intratumoral injection of PgP/pHSV-TK (10 µg pHSV-TK, N/P ratio of 60/1) polyplexes with GCV (40 mg/kg, intraperitoneal injection). Cell nuclei were counter stained with DAPI (blue). Scale bars indicate 100 µm.

4. Discussion

Intramedullary spinal cord tumor (IMSCT) is an uncommon neoplasm that causes significant neurologic morbidity and mortality [2]. Surgical therapy, radiotherapy, and chemotherapy are currently the most common treatments [3–5]. In addition to these current treatments, gene therapy has been explored as a possible alternative approach to overcome some of the limitations associated with current therapies for SCTs [12,13].

A cationic, amphiphilic copolymer, PgP, has been developed by our group and has been shown to perform as an efficient drug and non-viral gene carrier in vitro as well as in *in vivo* animal models [21,22,24]. As a block copolymer that forms polymeric micelles in aqueous solution with a hydrophobic core and hydrophilic, cationic shell, PgP has the potential to serve as a combinatorial carrier for the simultaneous delivery of hydrophobic drugs and anionic therapeutic nucleic acids. Our long-term goal is to use PgP as a carrier for GCV and pHSV-TK co-delivery. In this study, we evaluated PgP as a gene carrier in glioma (C6) cells in vitro using reporter genes such as pGFP and a rat spinal cord tumor model in vivo using reporter gen, pβ-Gal, and therapeutic gene, pHSV-TK. We first evaluated the particle size and surface charge of PgP/pGFP polyplexes at various N/P ratios and found that PgP can form stable complexes with pDNA 138–148.5 nm in size and positive surface charge at all N/P ratios (Table 1). This size range and surface charge are suitable for intracellular uptake by endocytosis and avoidance of rapid clearance by the reticuloendothelial system (RES) in systemic delivery [25,26].

Barriers for systemic gene therapy in vivo are polyplex stability and degradation by nucleases in the blood stream. To model the stability of PgP/pGFP in the blood stream, we conducted a heparin competition study using negatively charged polysaccharide, heparin. PgP/pGFP at an N/P ratio of 60/1 showed high stability and heparin was only able to dissociate pGFP from PgP at 10 *w/w* ratio of heparin/pGFP and higher (Figure 1B). We also observed that PgP can efficiently protect pDNA from nucleases in serum for up to 24 h, whereas pDNA without PgP was completely degraded by serum nucleases within 3 h. These results show that PgP may be a promising non-viral gene carrier for systemic gene therapy *in vivo*. In the long-term storage stability study, we observed that the transfection efficiency of PgP/pGFP polyplexes in C6 (rat glioma) cells was maintained up to four months at 4 °C and this result is consistent with the long term stability of PgP/pGFP and PgP/siRhoA observed in B35 (neuroblastoma) cells in our laboratory [21,24]. This demonstrates that PgP can form a stable complex with both pDNA and siRNA and preserve the bioactivity of nucleic acids during long-term storage, an important challenge for the clinical translation of non-viral vectors.

To evaluate PgP as a pHSV-TK gene carrier, we administered two doses of GCV (50 and 100 μg/mL) after PgP/pHSV-TK transfection in C6 cells in vitro. The PgP/pHSV-TK transfected group with both GCV doses showed significantly higher anti-cancer efficacy compared to control groups including GCV only, PgP/pGFP only, and PgP/pHSV-TK without GCV treatment at both one and four days. The lack of a significant effect of GCV dose on the anti-cancer efficacy of PgP/pHSV-TK is consistent with results obtained for rPOA/HSV-TK transfection and GCV treatment by Won et al. [13]. We observed that the PgP/pHSV-TK transfection with GCV (100 μg/mL) treatment group showed significantly higher cell cytotoxicity compared to bPEI/pHSV-TK transfection with GCV (100 μg/mL) treatment group one day post-GCV treatment. In vivo β-Gal expression was substantially higher in animals receiving PgP/pβ-Gal polyplexes compared with bPEI/pβ-Gal polyplexes in the rat spinal cord tumor model. bPEI is known as an effective transfection reagent, being widely studied both in vitro and in vivo due to its high transfection efficacy and the ability of the proton sponge effect to facilitate endosomal escape [27,28]. However, poor transfection efficiency in the presence of serum and cytotoxicity are limitations to clinical application [29].

Finally, we evaluated the suicide effect of PgP/pHSV-TK with GCV in the rat spinal cord tumor model and observed that the percent tumor area in animals treated with PgP/pHSV-TK and GCV was significantly smaller than that of those receiving PgP/pHSV-TK polyplexes without GCV (Figure 7). This result was further confirmed by the presence of more TUNEL+ cells as well as increased expression of the pro-apoptotic protein, Bax, in the PgP/pHSV-TK polyplexes with GCV treatment group compared with the PgP/pHSV-TK without GCV group. These data demonstrate that PgP can serve as an effective pHSV-TK carrier to activate the prodrug GCV for the treatment of spinal cord tumors.

5. Conclusions

In this study, we demonstrated that the cationic amphiphilic copolymer PgP can be a gene carrier due to the stability of polyplexes in the presence of negatively charged serum proteins and the ability to protect pDNA from nucleases in the serum, PgP/pDNA polyplexes maintain bioactivity for transfection after storage at 4 °C for up to 4 months—an important feature for commercial and clinical application. We demonstrated that PgP can efficiently deliver pHSV-TK in C6 cells and PgP/pHSV-TK in combination with GCV showed significantly higher suicide effect on C6 cells compared to various control groups in vitro. Finally, we demonstrated that PgP/pHSV-TK with GCV treatment increases the suicide effect and apoptosis of tumor cells and reduces tumor size in a rat T5 spinal cord tumor model compared with PgP/pHSV-TK without GCV treatment. In the future, we will evaluate the potential of PgP as a GCV and pHSV-TK co-delivery carrier to improve the bioavailability and half-life of GCV as well as increase the survival rate after treatment of PgP/pHSV-TK with PgP-GCV in rat T5 spinal cord tumor model.

Author Contributions: S.-J.G. and J.S.L. designed all the experiments and S.-J.G., performed the experiments. S.-J.G. and J.S.L. analyzed and reviewed the data, contributed to manuscript preparation, and approved the final version of the manuscript.

Funding: This research was funded by the National Institute of General Medical Sciences (NIGMS) under Grant No. 5P20GM103444-07 and partly supported by the South Carolina Spinal Cord Injury Fund under award number SCIRF #2017 B-01.

Acknowledgments: The authors would like to thank to Christian Macks for his assistance with in vivo animal study. The authors would like to thank Godley-Snell Research Center for animal care. The authors thank Ken Webb, Bioengineering, Clemson University for his careful review and editing of the manuscript.

Conflicts of Interest: The authors declare no conflict of interest.

References

1. Adams, H.; Avendano, J.; Raza, S.M.; Gokaslan, Z.L.; Jallo, G.I.; Quinones-Hinojosa, A. Prognostic factors and survival in primary malignant astrocytomas of the spinal cord: A population-based analysis from 1973 to 2007. *Spine* **2012**, *37*, E727–E735. [PubMed]

2. Mechtler, L.L.; Nandigam, K. Spinal cord tumors: New views and future directions. *Neurol. Clin.* **2013**, *31*, 241–268. [PubMed]

3. Parsa, A.T.; Lee, J.; Parney, I.F.; Weinstein, P.; McCormick, P.C.; Ames, C. Spinal cord and intradural-extraparenchymal spinal tumors: Current best care practices and strategies. *J. Neurooncol.* **2004**, *69*, 291–318.

4. Witham, T.F.; Khavkin, Y.A.; Gallia, G.L.; Wolinsky, J.P.; Gokaslan, Z.L. Surgery insight: Current management of epidural spinal cord compression from metastatic spine disease. *Nat. Clin. Pract. Neurol.* **2006**, *2*, 87–94, quiz 116.

5. Bowers, D.C.; Weprin, B.E. Intramedullary spinal cord tumors. *Curr. Treat. Options Neurol.* **2003**, *5*, 207–212. [PubMed]

6. Legler, J.M.; Ries, L.A.; Smith, M.A.; Warren, J.L.; Heineman, E.F.; Kaplan, R.S.; Linet, M.S. Cancer surveillance series [corrected]: Brain and other central nervous system cancers: Recent trends in incidence and mortality. *J. Natl. Cancer Inst.* **1999**, *91*, 1382–1390. [CrossRef] [PubMed]

7. Alemany, R.; Gomez-Manzano, C.; Balague, C.; Yung, W.K.; Curiel, D.T.; Kyritsis, A.P.; Fueyo, J. Gene therapy for gliomas: Molecular targets, adenoviral vectors, and oncolytic adenoviruses. *Exp. Cell Res.* **1999**, *252*, 1–12. [PubMed]

8. Werner-Wasik, M.; Yu, X.; Marks, L.B.; Schultheiss, T.E. Normal-tissue toxicities of thoracic radiation therapy: Esophagus, lung, and spinal cord as organs at risk. *Hematol. Oncol. Clin. N. Am.* **2004**, *18*, 131–160, x–xi. [CrossRef]

9. Balmaceda, C. Chemotherapy for intramedullary spinal cord tumors. *J. Neurooncol.* **2000**, *47*, 293–307. [CrossRef] [PubMed]

10. Hagiwara, H.; Sunada, Y. Mechanism of taxane neurotoxicity. *Breast Cancer* **2004**, *11*, 82–85. [CrossRef]

11. Poirier, V.J.; Hershey, A.E.; Burgess, K.E.; Phillips, B.; Turek, M.M.; Forrest, L.J.; Beaver, L.; Vail, D.M. Efficacy and toxicity of paclitaxel (taxol) for the treatment of canine malignant tumors. *J. Vet. Intern. Med.* **2004**, *18*, 219–222. [CrossRef] [PubMed]

12. Pennant, W.A.; An, S.; Gwak, S.J.; Choi, S.; Banh, D.T.; Nguyen, A.B.; Song, H.Y.; Ha, Y.; Park, J.S. Local non-viral gene delivery of apoptin delays the onset of paresis in an experimental model of intramedullary spinal cord tumor. *Spinal Cord* **2014**, *52*, 3–8. [CrossRef] [PubMed]

13. Won, Y.W.; Kim, K.M.; An, S.S.; Lee, M.; Ha, Y.; Kim, Y.H. Suicide gene therapy using reducible poly (oligo-d-arginine) for the treatment of spinal cord tumors. *Biomaterials* **2011**, *32*, 9766–9775. [CrossRef] [PubMed]

14. Hattori, Y.; Maitani, Y. Folate-linked nanoparticle-mediated suicide gene therapy in human prostate cancer and nasopharyngeal cancer with herpes simplex virus thymidine kinase. *Cancer Gene Ther.* **2005**, *12*, 796–809. [CrossRef]

15. Garcia-Rodriguez, L.; Abate-Daga, D.; Rojas, A.; Gonzalez, J.R.; Fillat, C. E-cadherin contributes to the bystander effect of tk/gcv suicide therapy and enhances its antitumoral activity in pancreatic cancer models. *Gene Ther.* **2011**, *18*, 73–81. [CrossRef]

16. Pu, K.; Li, S.Y.; Gao, Y.; Ma, L.; Ma, W.; Liu, Y. Bystander effect in suicide gene therapy using immortalized neural stem cells transduced with herpes simplex virus thymidine kinase gene on medulloblastoma regression. *Brain Res.* **2011**, *1369*, 245–252. [CrossRef]

17. Engelmann, C.; Panis, Y.; Bolard, J.; Diquet, B.; Fabre, M.; Nagy, H.; Soubrane, O.; Houssin, D.; Klatzmann, D. Liposomal encapsulation of ganciclovir enhances the efficacy of herpes simplex virus type 1 thymidine kinase suicide gene therapy against hepatic tumors in rats. *Hum. Gene Ther.* **1999**, *10*, 1545–1551. [CrossRef]

18. Kajiwara, E.; Kawano, K.; Hattori, Y.; Fukushima, M.; Hayashi, K.; Maitani, Y. Long-circulating liposome-encapsulated ganciclovir enhances the efficacy of HSV-TK suicide gene therapy. *J. Control. Release* **2007**, *120*, 104–110. [CrossRef] [PubMed]

19. Miura, F.; Moriuchi, S.; Maeda, M.; Sano, A.; Maruno, M.; Tsanaclis, A.M.; Marino, R., Jr.; Glorioso, J.C.; Yoshimine, T. Sustained release of low-dose ganciclovir from a silicone formulation prolonged the survival of rats with gliosarcomas under herpes simplex virus thymidine kinase suicide gene therapy. *Gene Ther.* **2002**, *9*, 1653–1658. [CrossRef]

20. Jeon, O.; Yang, H.S.; Lee, T.J.; Kim, B.S. Heparin-conjugated polyethylenimine for gene delivery. *J. Control. Release* **2008**, *132*, 236–242. [CrossRef]

21. Gwak, S.J.; Macks, C.; Jeong, D.U.; Kindy, M.; Lynn, M.; Webb, K.; Lee, J.S. Rhoa knockdown by cationic amphiphilic copolymer/siRhoA polyplexes enhances axonal regeneration in rat spinal cord injury model. *Biomaterials* **2017**, *121*, 155–166. [CrossRef] [PubMed]

22. Macks, C.; Gwak, S.J.; Lynn, M.; Lee, J.S. Rolipram-loaded polymeric micelle nanoparticle reduces secondary injury after rat compression spinal cord injury. *J. Neurotrauma* **2018**, *35*, 582–592. [CrossRef] [PubMed]

23. Gwak, S.J.; An, S.S.; Yang, M.S.; Joe, E.; Kim, D.H.; Yoon, D.H.; Kim, K.N.; Ha, Y. Effect of combined bevacizumab and temozolomide treatment on intramedullary spinal cord tumor. *Spine* **2014**, *39*, E65–E73. [CrossRef] [PubMed]

24. Gwak, S.J.; Macks, C.; Bae, S.; Cecil, N.; Lee, J.S. Physicochemical stability and transfection efficiency of cationic amphiphilic copolymer/pdna polyplexes for spinal cord injury repair. *Sci. Rep.* **2017**, *7*, 11247. [CrossRef]

25. Li, S.D.; Huang, L. Stealth nanoparticles: High density but sheddable peg is a key for tumor targeting. *J. Control. Release* **2010**, *145*, 178–181. [CrossRef] [PubMed]

26. Guo, S.; Huang, L. Nanoparticles escaping res and endosome: Challenges for sirna delivery for cancer therapy. *J. Nanomater.* **2011**, *2011*, 11. [CrossRef]

27. Abdallah, B.; Hassan, A.; Benoist, C.; Goula, D.; Behr, J.P.; Demeneix, B.A. A powerful nonviral vector for in vivo gene transfer into the adult mammalian brain: Polyethylenimine. *Hum. Gene Ther.* **1996**, *7*, 1947–1954. [CrossRef]

28. Godbey, W.T.; Wu, K.K.; Mikos, A.G. Size matters: Molecular weight affects the efficiency of poly(ethylenimine) as a gene delivery vehicle. *J. Biomed. Mater. Res.* **1999**, *45*, 268–275. [CrossRef]

29. Cao, D.; Qin, L.; Huang, H.; Feng, M.; Pan, S.; Chen, J. Transfection activity and the mechanism of pdna-complexes based on the hybrid of low-generation pamam and branched PEI-1.8k. *Mol. Biosyst.* **2013**, *9*, 3175–3186. [CrossRef]

Review

Vectors for Glioblastoma Gene Therapy: Viral & Non-Viral Delivery Strategies

Breanne Caffery [1], Jeoung Soo Lee [1] and Angela A. Alexander-Bryant [1,2,*]

[1]   Drug Design, Development, and Delivery (4D) Laboratory, Clemson University, Clemson, SC 29634, USA;
      bhourig@g.clemson.edu (B.C.); ljspia@clemson.edu (J.S.L.)
[2]   Nanobiotechnology Laboratory, Department of Bioengineering, Clemson University, Clemson,
      SC 29634, USA
*   Correspondence: angelaa@clemson.edu; Tel.: +1-(864)-656-5232

Received: 13 November 2018; Accepted: 3 January 2019; Published: 16 January 2019

Abstract: Glioblastoma multiforme is the most common and aggressive primary brain tumor. Even with aggressive treatment including surgical resection, radiation, and chemotherapy, patient outcomes remain poor, with five-year survival rates at only 10%. Barriers to treatment include inefficient drug delivery across the blood brain barrier and development of drug resistance. Because gliomas occur due to sequential acquisition of genetic alterations, gene therapy represents a promising alternative to overcome limitations of conventional therapy. Gene or nucleic acid carriers must be used to deliver these therapies successfully into tumor tissue and have been extensively studied. Viral vectors have been evaluated in clinical trials for glioblastoma gene therapy but have not achieved FDA approval due to issues with viral delivery, inefficient tumor penetration, and limited efficacy. Non-viral vectors have been explored for delivery of glioma gene therapy and have shown promise as gene vectors for glioma treatment in preclinical studies and a few non-polymeric vectors have entered clinical trials. In this review, delivery systems including viral, non-polymeric, and polymeric vectors that have been used in glioblastoma multiforme (GBM) gene therapy are discussed. Additionally, advances in glioblastoma gene therapy using viral and non-polymeric vectors in clinical trials and emerging polymeric vectors for glioma gene therapy are discussed.

Keywords: glioblastoma multiforme; gene therapy; viral vector; non-viral vector; gene delivery; siRNA

1. Introduction

Glioblastoma multiforme (GBM) is a type of glioma that arises from astrocytes, defined by the World Health Organization (WHO) as a grade IV glioma [1,2]. GBM is not only the most common malignant primary brain tumor, but also the most aggressive of malignant tumors, with recurrence in nearly all patients [3]. GBM affects about three people out of every 100,000 per year, accounting for approximately 15% of primary brain tumors, and 80% of malignant primary brain tumors. GBM is about two times more common in whites than in blacks, and 1.5 times more common in men than in women, with an average age at onset of 64 [4].

High-grade gliomas are typically located in undesirable locations in the cerebral hemisphere and are classified as diffuse gliomas due to their high rate of infiltration into surrounding brain tissue. These factors allow for persistent tumor growth and lessen the chance of remission, with progression to grade III or IV gliomas likely even in most low-grade diffuse gliomas [2,5]. The current standard of care for treating GBM includes surgical resection of the tumor, radiation therapy, and chemotherapy via temozolomide (TMZ) [6]. Carmustine (BCNU, Gliadel™) wafers have been used as local adjuvant therapy in combination with systemic TMZ since its approval; however, its use has been limited due to observed toxicities and ambiguity of overall survival benefit [7,8]. Additionally, bevacizumab, a monoclonal antibody that inhibits vascular endothelial growth factor (VEGF), is used for the treatment of recurrent glioblastomas [9]. However, even with aggressive treatment, survival rates remain between 12 and 15 months, and the 3-year survival rate is less than 16% [1,10]. GBM remains an essentially incurable disease, resulting in a patient death rate of greater than 95% within five years of diagnosis [11]. Consequently, there is a clear need for advancement in treatment strategies to improve outcomes for patients with GBM. Gene therapy may provide a viable alternative to conventional treatments towards combating cancer progression in GBM.

This review discusses gene expression in GBM, the limitations to conventional therapy, and current approaches to circumvent these barriers using gene therapy. Advances in gene delivery systems will be reviewed, highlighting viral and non-viral vectors used for GBM gene therapy. Trials bringing these therapies closer to clinical approval to date will be discussed, as well as preclinical studies, particularly using polymeric nanoparticles, which have shown promise as future vectors for delivery of gene therapy in GBM patients.

2. Gene Expression in GBM

Glioma occurs due to sequential acquisition of genetic alterations, causing a transformation from benign to malignant tissue [12]. Glioblastoma can occur in four clinical subtypes including classical, proneural, mesenchymal and neural GBM [13]. Classical, or primary GBM arises de novo, and occurs in about 95% of cases, only requiring about 3–6 months to develop [1]. Proneural or secondary GBM arises as a recurrence from a previous anaplastic or low-grade astrocytoma, usually requiring 10–15 years to develop [14]. Classical GBM can be identified by chromosome 7 amplification paired with chromosome 10 loss, as well as by increased expression of the epidermal growth factor receptor (EGFR) and mutations in phosphatase and tensin homologue (PTEN) [1,13]. In a study conducted by Verhaak et al., point or vIII EGFR mutations were found in over half of GBM cases analyzed [13]. EGFR overexpression, observed in 97% of patients with classical GBM, causes a reduction in apoptosis and increased proliferation through the Ras-Shc-Grb2 pathway, causing uncontrolled cell growth [14]. PTEN is a tumor suppressor, and when mutated, the loss of function causes activation of the P13K/Akt/mTOR pathway, leading to proliferation, growth, and migration [15]. Disrupted regulation of this pathway has been shown to contribute to tumorigenesis and resistance in various cancers [16]. Deletion of CDKN2A, coding for tumor suppressor p16INK4A, was also significantly associated with the classical subtype.

Proneural GBM often presents with increased expression of platelet-derived growth factor receptor alpha (PDGFRA), as well as mutated tumor suppressor p53, isocitrate dehydrogenase 1 (IDH-1), and retinoblastoma genes [11,13,14]. PDGFRA is mitosis-promoting, and overexpression of this mitogen promotes tumor cell proliferation [14]. IDH-1 point mutations were found to occur in about 30% of proneural cases [14]. Mutation of IDH alters DNA and histone methylation and is often found in the early development of diffuse gliomas [17,18]. The p53 gene normally functions as a switch to turn on G1 cell cycle arrest or apoptosis, regulating cell growth [14]. Overexpression of p53 has been shown to negatively regulate MGMT transcription, suggesting that repair of wild-type p53 may increase therapeutic efficacy in GBM therapy [19]. Interestingly, p53 mutations have been found in 54% of proneural GBM but are almost never observed in classical GBM [13]. More recently, interferon-β (IFN-β) has been found to sensitize T98G GBM cells to TMZ, which was also thought to be a function of induced p53 overexpression [20].

Mesenchymal GBM presents most prominently with deletion or mutations of the tumor suppressor gene, neurofibromin 1 (NF-1). Similar to the proneural subtype, p53 and mutations occur in about 32% of mesenchymal cases [13]. Genes in the tumor necrosis factor superfamily are also overexpressed, correlating with the high degree of necrosis observed [13]. Mesenchymal GBM also shows characteristics of epithelial-to-mesenchymal transition (EMT) with high expression of mesenchymal and astrocytic markers, such as CD44 and MERTK [13]. EMT in GBM may be induced by hypoxia [21] or upstream regulators of EMT, including TGF-β [21] and S100A4 [22].

A fourth subtype of GBM, the neural subgroup, has been classified due to its similarity in gene expression to normal neurons or nerve cells. Neural GBM presents with mutations similar to the other subgroups with no outstanding genetic amplification or mutation rates that would differentiate the neural subgroup from the other subgroups [13]. In advanced strategies for treating GBM patients, evaluating the expression of key genes may allow for selection of more personalized and effective therapies. Gene targets in GBM are summarized in Table 1.

Table 1. Gene expression in glioblastoma multiforme (GBM). Common gene targets that are mutated or upregulated in glioblastoma.

Gene Target	Effect	GBM Clinical Subtype	References
EGFR (epidermal growth factor receptor)	Reduction in apoptosis and increased uncontrolled cell proliferation	Classical	[13,14]
PTEN (phosphate and tensin homologue)	Activation of the P13K/Akt/mTOR pathway, leading to cell proliferation, migration and growth	Classical	[1,13,15,16]
PDGFRA (platelet derived growth factor receptor—alpha)	Increased tumor cell proliferation	Proneural	[11,13,14]
IDH-1 (isocitrate dehydrogenase 1)	Alters DNA and histone methylation	Proneural	[17,18]
Tumor suppressor p53	Uncontrolled cell growth	Proneural, mesenchymal	[13,14,20]
NF-1 (neurofibromin 1)	Uncontrolled cell growth	Mesenchymal	[13]

3. Barriers to Drug and Gene Delivery

Various drugs including TMZ, BCNU, and cisplatin have been used in patients with GBM, but several barriers limit effective treatment, including inefficient delivery across the blood-brain barrier (BBB) and chemotherapeutic resistance. The BBB is a cellular barrier that regulates ionic concentrations to allow for synaptic signaling in the brain, while also preventing entry of cells and large molecules via tight junctions between endothelial cells (Figure 1) [5,23]. It exists to regulate the transport of essential nutrients to the brain and to protect the brain from neurotoxins. It is estimated that less than 2% of small molecule drugs and no large molecule drugs or genes are able to cross the BBB [24,25]. It has been found that drugs greater than 400 Da are not often able to cross the BBB. However, this is not a finite cutoff, as peptides greater than 600 Da are known to cross the BBB with relative ease [26], and a 7800 Da molecule, cytokine-induced neutrophil chemoattractant-1 (CNC1), is able to cross the BBB through transmembrane diffusion [27]. Crossing the BBB is highly dependent on many other factors, such as charge, molecular weight, and hydrophobicity, creating a non-linear relationship between size and ability to traverse the BBB [28,29]. Molecules can cross the BBB through various active transport mechanisms, including carrier mediated influx or efflux, and through passive transport mechanisms, including transmembrane diffusion, and paracellular transport (Figure 1) [28,30]. Most molecules that are able to traverse the BBB do so through transmembrane diffusion or active transport. Low molecular weight lipid-soluble molecules are favorable for passive diffusion, whereas water-soluble molecules tend to traverse using active transport processes, including adsorptive- or receptor-mediated transport [26]. The BBB has been a consistent challenge in creating effective delivery systems for therapeutics. Many nanoparticle (NP) delivery systems are designed to rely on diffusion and passive targeting of tumor tissue via the enhanced permeability and retention (EPR) effect [30,31]. Fenestrated capillaries or a dysfunctional endothelium exist in areas of rapidly grown and poorly developed vessels due to increased vascular endothelial growth factor (VEGF) expression and angiogenesis [32,33]. The leaky nature of the vasculature creates an interrupted blood-brain tumor barrier, which may allow for increased therapeutic concentrations in the glioma tissue. However, the EPR effect may be inefficient in therapeutic delivery to brain tumors due to the density of the tumor matrix and the increased interstitial fluid pressure inhibiting both diffusion and convective transport [34]. Furthermore, glioma cells tend to easily travel outside the tumor to other normal regions of the brain. This metastasis not only makes the glioma more difficult to treat, but also reduces the quantity of therapeutic reaching tumor cells in the intact regions of the brain with perfectly functioning BBB [5].

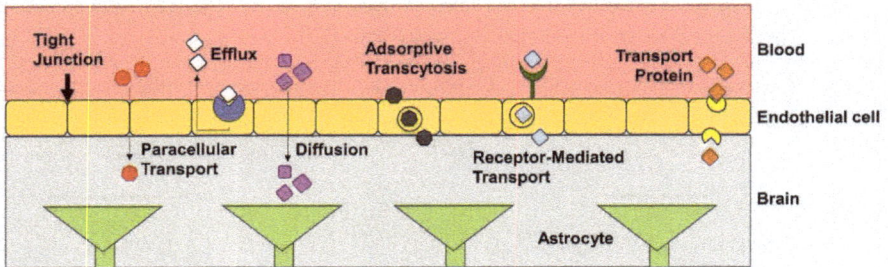

Figure 1. The blood brain barrier (BBB). The BBB regulates entry of nutrients to the brain and prevents entry of cells and large molecules via tight junctions. There are several mechanisms for transporting molecules across the BBB, including paracellular transport, diffusion, protein transporters, receptor-mediated transport, and adsorptive transcytosis.

Drug resistance is another major barrier in the treatment of GBM due to overexpression of drug resistance genes. Additional protection of the BBB exists in the form of various efflux transport systems which remove unwanted substances that are able to traverse the BBB. A largely studied efflux pump, P-glycoprotein or multidrug resistance protein 1 (MDR1), encoded by the ATP-binding cassette sub-family B member 1 (ABCB1) gene, has been a persistent challenge in therapeutic delivery due to its efficacy in removing small molecules from the brain [26]. A wide variety of ATP-dependent substrates are recognized by ABCB1, allowing for resistance to occur when therapeutics are recognized and pumped out of the cell through the efflux pump, reducing cytotoxicity and drug efficacy [35,36]. An MDR1a knockout study demonstrated that P-glycoprotein is a major impediment for drug passage through the BBB, after finding significantly increased drug concentrations in the brains of P-glycoprotein knockout mice [37]. Drug resistance in GBM patients has also been attributed in part to overexpression of the (O)6-methylguanine-DNA- methyltransferase (MGMT) gene. The MGMT gene codes for a protein that removes alkyl adducts at the O(6) position of guanine as a natural repair mechanism to prevent apoptosis due to DNA methylation [38]. Although it is a natural process for DNA repair, when MGMT is upregulated in tumor cells, this mechanism allows for drug resistance when treating GBM with temozolomide (TMZ). TMZ alkylates DNA at the O(6) position of guanine in order to cause DNA damage and programmed cell death [39]. In this case, the damage done by TMZ is possibly reversed due to epigenetic or drug-induced upregulation of MGMT in GBM cells (Figure 2). MGMT methylation status was the first predictive biomarker identified in glioma patients [40]; additionally, Hegi et al. found that epigenetic silencing of MGMT was correlated with longer patient survival when treated with alkylating agents and radiotherapy [39]. In attempts to overcome current barriers to effective treatment, delivery systems for gene and drug therapies have been researched and tested in vivo and/or in the clinic.

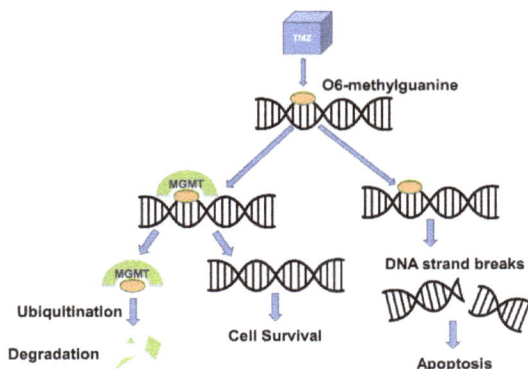

Figure 2. Mechanisms of TMZ and MGMT in DNA damage and repair. TMZ, a DNA alkylating agent, methylates DNA at the O6 position of guanine, resulting in DNA damage and apoptosis of tumor cells. MGMT, a DNA repair protein, removes alkyl adducts from the O6 position of guanine, inhibiting the potentially therapeutic effect of TMZ.

4. Vectors for Glioblastoma Gene Therapy

Gene therapy for cancer treatment conventionally includes the introduction of growth regulating or tumor suppressing genes. More recently, RNA interference (RNAi) has been introduced to inhibit the activity of oncogenes causing tumorigenesis or proliferation. Suicide gene therapy is another approach that is commonly used in viral gene therapy to convert non-toxic prodrugs into lethal active compounds. Other approaches include oncolytic and immunomodulatory gene therapy [41]. Gene or nucleic acid carriers must be used to deliver these therapies successfully into tumor tissue and have been extensively studied. Delivery vectors such as viral vectors, non-polymeric NPs, and polymeric NPs that have been used in GBM gene therapy are discussed in detail in the following sections (Figure 3).

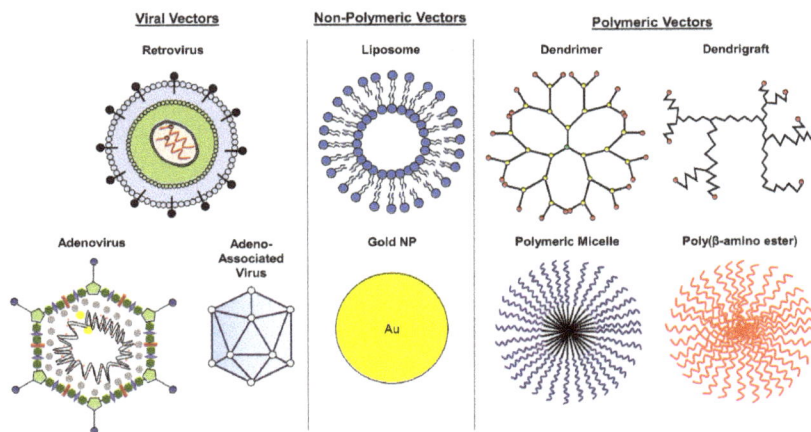

Figure 3. Vectors for glioblastoma gene therapy. Various viral, non-polymeric, and polymeric vectors are used to deliver nucleic acids for GBM gene therapy.

4.1. Viral Vectors

Viral vectors were the first delivery vehicles used for gene therapy in glioma clinical trials and have been studied for glioma gene therapy over the past 25 years. Viral vectors commonly used for GBM gene therapy in clinical trials include neurotropic retroviruses [42] and adenoviruses [43] that are able to infect neurons and glial cells, such as herpes simplex virus-1 (HSV-1) [44]. Adeno-associated viruses have recently shown promise for gene therapy in treating gliomas in preclinical trials, but have not yet been evaluated in clinical trials [45–47]. Current and completed clinical trials using various vectors for gene therapy in glioblastoma treatment have been outlined in Table 2.

Retroviral vectors were the first delivery systems evaluated in clinical trials for glioma gene therapy. The initial trial evaluated the combination of modified murine cells containing retroviral herpes simplex virus-thymidine kinase (HSV-tk) with ganciclovir (Cytovene) and began in 1992 (NCT00001328). HSV-tk functions as a suicide gene and converts the prodrug, ganciclovir, into its active form, ganciclovir-triphosphate, which inhibits DNA replication and cell division in HSV-tk-transfected cells [48]. The results of the study demonstrated intratumorally implanted retroviral vector-producing cells mediated HSV-tk transfection and antitumor activity only in the smaller treated tumors [49], reflecting the limited transfection efficiency of the retroviral vector. Another retroviral vector, Toca 511 delivers suicide gene, cytosine deaminase (CD), and in combination with oral prodrug, Toca FC, the CD enzyme mediates conversion of 5-fluorocytosine into the active antineoplastic drug, 5-fluorouracil [50,51]. Phase I clinical trials demonstrated that Toca 511 and Toca FC were well tolerated and mediated tumor regression in the infusion site in patients with recurrent high-grade glioma [52]. Toca 511 and Toca FC currently make up a regimen in phase 2/3 clinical trials for the treatment of GBM and anaplastic astrocytoma.

Adenoviral vectors have also been widely evaluated in clinical trials. A phase 1 trial of an adenoviral vector carrying the wild-type p53 gene (Ad-p53) demonstrated that Ad-p53 successfully transfected astrocytic tumor cells with minimal toxicity when intratumorally injected pre- and post-resection of the glioma tumor; however, transfected cells were only detected on average within 5 mm of the injection site [53], demonstrating the limited ability of the therapeutic to penetrate the tumor tissue. Another study compared combination therapy using ganciclovir and intratumoral injection of HSV-tk delivered either by retrovirus-packaging cells or adenoviruses in patients with malignant glioma. The results revealed stable tumor in 3/7 patients treated with the adenovirus compared to tumor progression in all patients treated with the retrovirus three months post-treatment [54]. Additionally, survival time nearly doubled in patients treated with adenovirus compared to retrovirus, with averages of 15 months and 7.4 months, respectively [54]. Sandmair et al. concluded that ineffectiveness of retroviruses may be due to low transfection and brain tumor penetration [54]. Several clinical studies have also evaluated the delivery of an adenoviral vector containing HSV-tk (AdV-tk) combined with valacyclovir, an antiherpetic prodrug. Using gene-mediated cytotoxic immunotherapy, thymidine kinase mediates conversion of the prodrug into toxic nucleotide analogs, inducing tumor cell death and activation of antitumor immune cells [55]. A Phase 1B study of AdV-tk with concurrent valacyclovir and radiation therapy followed by TMZ was conducted in patients with recently diagnosed malignant glioma [55]. AdV-tk injected into the tumor bed post-resection followed by radiation and chemotherapy resulted in two and three-year survival rates of 33% and 25% [55], respectively, a small increase over the current standard of care. Of note, CD3$^+$ T-cells were found in tumors analyzed post-treatment [55], indicative of immune activation. A phase 2 trial showed that median survival time significantly increased from 13.5 months for patients that received the standard of care treatment to 17.1 months for patients treated with AdV-tk combined with valacyclovir and standard of care [56]. Additionally, a Phase I trial of AdV-tk with combination valacyclovir and radiation therapy was recently conducted in pediatric malignant glioma and recurrent ependymoma [57]. Half of the patients survived at least 16 months post-treatment with no dose-limited toxicities, though grade 3 lymphopenia was common [57].

Although viral vectors have been studied extensively, they have only resulted in marginal increases in overall survival and have yet to achieve clinical translation through FDA approval to treat patients with GBM after decades of study. Efficient tumor penetration of viral vectors has proven to be a challenge limiting overall efficacy in treating gliomas. However, there has been some clinical success with viral vectors in other cancers. Talimogene laherparepvec is an oncolytic virotherapy consisting of genetically modified HSV-1 containing the human granulocyte-macrophage colony-stimulating factor (GM-CSF) gene and has been FDA approved for the treatment of melanoma [58].

Table 2. Vectors for glioma gene therapy. Vectors that have been evaluated in clinical trials for glioma gene therapy.

Vector	Gene Therapy Agent	Mechanism	Combination Therapy	Clinical Trial Phase	Clinical Trial Number
Retrovirus	HSV-tk	Suicide gene therapy, HSV-tk converts ganciclovir to antiviral drug ganciclovir triphosphate	Ganciclovir	Phase I	NCT00001328
Retrovirus	Toca 511	Suicide gene therapy, CD converts prodrug 5-FC to anti-neoplastic 5-FU	Oral 5-FC	Phase II/III	NCT02414165
Adenovirus	SCH-58500	Tumor suppressor gene therapy, transfects p53 gene	N/A	Phase I	NCT00004080
Adenovirus	Ad-p53	Tumor suppressor gene therapy, transfects p53 gene	N/A	Phase I	NCT00004041
Retro or adenovirus	HSV-tk	Suicide gene therapy, HSV-tk converts ganciclovir to antiviral drug ganciclovir triphosphate	Ganciclovir	Phase I	Sandmair et. al.
Adenovirus	AdV-tk	Gene-mediated cytotoxic immunotherapy, HSV-tk converts valacyclovir to antiviral drug acyclovir	Valacyclovir	Phase I	NCT00751270
Adenovirus	AdV-tk	Gene-mediated cytotoxic immunotherapy, HSV-tk converts valacyclovir to antiviral drug acyclovir	Valacyclovir and radiation therapy	Phase IIa	NCT00589875
Liposome	SGT-53	Tumor suppressor gene therapy, transfects p53 gene	TMZ	Phase II	NCT02340156
Spherical Nucleic Acid Gold NP	NU-0129	RNAi gene therapy, transfects siRNAs targeting oncogene Bcl2L12	N/A	Early Phase I	NCT03020017

4.2. Non-Viral Vectors

In addition to viral vectors, non-viral vectors, including both non-polymeric and polymeric delivery systems (Figure 3), have been explored for delivery of glioma gene therapy and have shown promise as gene vectors for glioma treatment in preclinical and clinical studies. Though these vectors have yet to achieve FDA approval for treatment of GBM, the 2018 approval of the first RNAi therapeutic, Patisiran, a lipid nanoparticle containing siRNAs for the treatment of transthyretin-mediated amyloidosis [59], a neurodegenerative disease, provides a promising outlook for non-viral vector-based nucleic acid therapies. A few non-polymeric vectors have been evaluated clinically for GBM gene therapy, including liposomes, gold nanoparticles, and RNA nanoparticles. SGT-53, a transferrin receptor-targeted liposomal vector encapsulating wild-type p53 plasmid DNA is able to cross the BBB and target GBM cells, resulting in a reduction of MGMT and apoptosis in GBM xenografts in mice [60]. Combination therapy with systemically administered SGT-53 and TMZ enhanced antitumor efficacy compared to TMZ alone [60,61], demonstrating the ability of SGT-53 to improve chemosensitivity. SGT-53 is currently in phase II clinical trials for combination therapy with TMZ in treating recurrent glioblastoma (see Table 2). Additionally, NU-0129, a spherical nucleic acid gold nanoparticle containing siRNAs targeting Bcl-2-like protein 12 (Bcl2L12), which is involved in tumor progression and resistance to apoptosis [62], is in early phase 1 clinical trials for patients with recurrent glioblastoma or gliosarcoma. NU-0129 has previously demonstrated its ability to cross the BBB in xenograft models of GBM in mice after systemic administration, resulting in increased apoptosis of tumor cells and reduced tumor progression [62]. In addition to clinical studies, novel polymeric vectors are being explored in research, such as RNA nanoparticles. RNA nanoparticles completely composed of RNA have been used in preclinical studies for glioma gene therapy. Croce et al. reported using RNA nanoparticles to deliver anti-miR-21 locked nucleic acid sequences to inhibit oncogenic miR-21 in xenograft GBM models in mice, resulting in tumor regression and increased survival compared to untreated mice [63]. Though still in their preclinical stages of development, RNA nanoparticles have shown promise for gene delivery in cancer treatment [64].

4.3. Polymeric Delivery Systems

Polymeric delivery of gene therapy is an emerging approach for cancer treatment to improve therapeutic outcomes. Current research has been focused on micro- and nanoparticles for the systemic or local delivery of genes and/or drugs. These NPs are advantageous for gene therapy because they are highly tailorable, allowing for conjugation of nucleic acids, homing peptides, or targeting ligands. Though they have not yet reached clinical trials specifically for glioblastoma treatment, several polymeric delivery systems have been studied for use in gene therapy for glioma treatment, and are discussed in detail as follows, including their advantages and limitations. The advantages and disadvantages of various vectors that have been studied for glioblastoma gene therapy are summarized in Table 3.

4.3.1. Dendrimers

Dendrimers are highly branched 3D polymers that have been explored for a variety of applications in drug and gene delivery. Cationic dendrimers, such as poly(amidoamine) (PAMAM), are particularly useful for gene therapy in glioma treatment due to their ability to form complexes with negatively charged nucleic acids, penetrate cellular and endosomal membranes, and cross the BBB. Dendrimers have been used to deliver several types of nucleic acids, including antisense oligonucleotides [65], microRNAs [66], siRNAs [67], and genes [68–71] into glioma cells. Functionalized dendrimers have demonstrated a capacity for enhanced transfection and targeted delivery in glioma cells and tissues. Specifically, peptide functionalized dendrimers have been used to increase gene transfection in patient-derived primary glioma cells. Bae et al. showed that PAMAM dendrimers grafted with histidine and arginine residues enhanced transfection efficiency in glioma cells compared to PAMAM alone [68]. This result is likely due to the increased proton buffering capacity provided by the peptides, resulting in enhanced endosomal escape and gene transfection. Additionally, several groups have used PAMAM for targeted delivery of gene therapy by functionalizing the polymer with polyethylene glycol (PEG) to attach a targeting moiety. PAMAM-PEG conjugated with transferrin [69], chlorotoxin [70], or Angiopep-2 [71] have allowed increased distribution of therapeutics in glioma tissue after systemic delivery in mice or rats in comparison to treatment with PAMAM-PEG alone, demonstrating the clinical potential of ligand-conjugated dendrimers for intravenous delivery of gene therapy for glioma treatment. However, one of the critical limitations of PAMAM dendrimers for clinical translation is cytotoxicity due to their high positive surface charge. Studies have shown that PAMAM dendrimers exhibit neurotoxicity by inducing autophagy in glioma cells, resulting in cell death [72]. Strategies to mitigate this effect include reducing the surface charge through acetylation or functionalization using PEG.

Table 3. Comparison of gene delivery vectors. Advantages and disadvantages of various vectors for glioblastoma gene therapy.

Vector	Advantages	Disadvantages
Viral		
Adenovirus	• Deliver large DNA	• Transient gene expression • Elicit immune response
Retrovirus	• Transfer to dividing cells • Sustained expression of vector	• Elicit immune response • Unable to transfect non-dividing cells • Low transfection rate in vivo • Risk of insertion
Adeno-associated virus	• Transfer to dividing and non-dividing cells	• Difficult to produce vectors • Limited transgene capacity • Elicit immune response
Non-Viral		
Liposome	• Non-immunogenic • Ability to co-deliver gene therapy and chemotherapy • Ability to functionalize for targeting	• Short shelf- and half-life • Transient gene expression • Low transfection efficiency • Increased cytotoxicity for cationic lipids
Gold nanoparticles	• Multimodal use for tumor imaging and therapy • Ability to functionalize for targeting	• Non-biodegradable
Dendrimer & Dendrigraft	• Self-assemble with nucleic acids • Ability to functionalize for targeting • Non-immunogenic	• Increased cytotoxicity for cationic dendrimers • Limited release of therapeutics
Polymeric micelles	• Self-assemble with nucleic acids • Ability to functionalize for targeting	• Increased cytotoxicity for PEI and other cationic polymers • Low loading efficiency
Poly(β-amino ester)	• Biodegradable • Lower cytotoxicity than other cationic polymers • High transfection efficiency	• Limited control over release of therapeutics

4.3.2. Dendrigraft

Similar to dendrimers, dendrigrafts are also dendritic structures that can be used to deliver therapeutics. Dendrigraft poly-L-lysine (DGL) was recently discovered as a newer class of dendritic polymers and has shown potential for delivery of nucleic acids. One major advantage of DGL over dendrimers is that DGL is composed entirely of naturally occurring lysine residues and is therefore completely biodegradable. Dendrigrafts are rich in external amino groups, which enables self-assembly with nucleic acids. DGL is also non-immunogenic and has been demonstrated to cross the BBB [73]. To mediate glioma targeting, transferrin- [74] or laminin-targeted [75] peptides have been conjugated to DGL for gene therapy with pORF-hTRAIL or survivin, respectively. The results of both studies revealed that DGL conjugated with targeting peptides exhibits enhanced tumor targeting and long-term survival in xenograft mouse models of U87 human glioblastoma in comparison to non-targeted DGL [74,75]. Transferrin-targeted DGL has also been used successfully for RNAi therapy. Kuang et al. demonstrated that transferrin-targeted DGL mediates increased gene silencing in mouse glioma tissue in comparison to non-targeted DGL [76]. In another study, a cell-penetrating peptide conjugated to DGL for delivery of pcDNA3.1-ING4, a plasmid encoding tumor suppressor gene inhibitor of growth 4 (ING4), demonstrated enhanced apoptosis of U87 tumor cells and resulted in increased survival of mice in comparison to treatment with DGL/pDNA [77]. DGL has also been used for combination delivery of a drug and gene. Li et al. demonstrated that choline-targeted DGL delivers pORF-hTRAIL and doxorubicin to glioma tissue in mice [78]. Similar to dendrimers, the cytotoxicity of DGL is a major disadvantage due to its excessive cationic charge. Studies have shown that the cytotoxicity of DGL/nucleic acid complexes increases in a dose-dependent manner and also results in hemotoxicity [79,80]. To overcome this limitation, toxicity of DGL can be reduced by including anionic polymers [80] or through PEGylation [81]. Though studies with dendrigrafts are relatively new and still evolving, data thus far has shown their potential clinical applicability for nucleic acid delivery in the treatment of gliomas.

4.3.3. Polymeric Micelles

Polymeric micelles are amphiphilic copolymers that have a core/shell structure. They have been widely used for cancer drug delivery [82], but have recently been explored for delivery of nucleic acids and shown promise for treatment of gliomas. Cationic polymers, such as polyethyleneimine (PEI), are commonly combined with hydrophobic polymers for combination delivery of negatively charged nucleic acids and hydrophobic cancer drugs. Cheng et al. demonstrated that a folate (FA)-targeted micelle consisting of PEI and polycaprolactone (PCL) mediated co-delivery of BCL-2 siRNA and doxorubicin in C6 glioma tumors in rats, resulting in increased apoptosis and inhibition of tumor growth following intratumoral injection [83]. In another study, to reduce the cytotoxic effect of PEI as well as enable active targeting, FA- was conjugated to hyperbranched PEI (FA-PEG-PEI) using PEG as spacer for combination gene therapy with CD/5-FC and TRAIL [84]. The results showed increased anticancer activity in C6 glioma tissue in rats after intratumoral delivery compared to treatment with a single therapeutic [84]. An RGD-conjugated PEI-PEG micelle used to co-deliver pORF-hTRAIL and paclitaxel in mice with orthotopic glioblastoma significantly enhanced survival in comparison to mice treated with gene therapy alone [85]. Additionally, intravenous delivery of PEI-PEG conjugated to a tumor homing peptide targeting neuropilin-1, retro-inverso C-end rule (CendR) peptide D(RPPREGR), enhanced gene transfection efficiency in mice with U87 glioma over non-targeted PEG-PEI [86]. Although PEGylation of PEI reduces cytotoxic effects, it can also reduce transfection efficiency by hindering proton buffering capacity. To overcome this limitation, PEI can be reversibly shielded using degradable disulfide (SS) linkages. Lei et al. conjugated RGD peptide to PEI through PEG using a reversible disulfide linkage (RGD-PEG-SS-PEI) for treatment of mice with U87 glioblastoma [87]. Results showed that the reversibly shielded PEI increased gene expression in the mouse brain in comparison to irreversibly shielded RGD-PEG-PEI [87], demonstrating the potential of reversible shielding for reducing cytotoxicity of PEI while maintaining efficient transfection.

4.3.4. Poly(β-amino ester)

Poly(β-amino esters) (PBAEs) are another class of cationic polymer that were designed to meet specific criteria for gene delivery, including DNA condensation, biodegradability, and minimal cytotoxicity [88], PBAEs may contain different types of amines and can be synthesized to create large libraries of polymers using combinatorial chemistry, allowing high-throughput screening of hundreds of polymers and identification of optimal vectors for gene delivery [89]. Research has shown that optimal PBAE vectors can transfect up to 90% of primary GBM cells and mediate up to 85% gene silencing with minimal cytotoxicity [90]. Further, PBAEs have been used as vectors for local injection of therapeutics. PBAE nanoparticles have been proven to penetrate glioma tissue for gene delivery using various strategies. For example, intratumoral injection was used to deliver DNA-containing PBAE nanoparticles in rat models of 9L gliosarcoma [91]. Convection-enhanced delivery (CED) is another local delivery strategy that allows administration of a therapeutic into glioma tumor tissue through catheters placed directly in the tissue with infusion occurring over the course of several hours. CED has been used for PBAE-mediated DNA delivery combined with intraperitoneal administration of ganciclovir in mouse xenograft models of primary brain tumor-initiating cells [92]. To enhance brain penetration, modification of PBAE nanoparticles with PEG has been explored. Mastorakos et al. demonstrated that PEGylated PBAEs could penetrate brain tissue with 20-fold greater volume distribution following CED in comparison to non-PEGylated particles [93]. One of the disadvantages of PBAEs involves their mechanism for cargo release. PBAEs release their cargo through hydrolysis of ester bonds, which can occur over many hours to a couple days [94], resulting in lack of controlled release of therapeutics. To overcome this limitation, bioreducible PBAEs have been synthesized containing disulfide bonds with the ability to trigger release of siRNAs into the cytoplasm [95]. Thus far, PBAEs have shown promise in overcoming limited tissue distribution, a common barrier in clinical applications of local gene therapy. Further research demonstrating enhanced tissue-penetration using PEGylated PBAEs in a glioma model will allow further assessment of the clinical potential of PEGylated PBAEs for local gene therapy of gliomas.

5. Conclusions

Glioblastoma multiforme is a common and currently incurable brain cancer that desperately needs new treatment modalities to improve patient outcomes. The current standard of care including surgical resection, adjuvant chemotherapy, and radiation does not result in remission for the majority of patients. Barriers limiting efficacy include inefficient delivery across the BBB and therapeutic resistance. Gene therapy represents an approach to specifically target and regulate oncogenes and tumor suppressor gene in gliomas. Further, gene therapy can be used to overcome barriers such as chemotherapy resistance by downregulating resistance genes or using approaches such as suicide gene therapy. Viral vectors, including retroviruses and adenoviruses, have been evaluated in clinical trials of GBM for the past few decades for delivery of therapeutic genes or nucleic acids in combination with other therapeutics. However, viral vectors have not reached clinical approval due to immunogenicity, limited tumor penetration and marginal improvement in patient outcomes. Non-viral delivery is an evolving alternative approach that may be used to overcome the barriers of gene delivery. Many non-viral vectors, including polymeric and non-polymeric vectors, are non-immunogenic and can be functionalized with targeting moieties to increase receptor-mediated uptake of vectors into tumor tissue. Multifunctional and multimodal non-polymeric vectors, such as liposomes and gold NPs, respectively, have the ability to co-deliver multiple therapies or be used for tumor imaging as well as therapy. Cationic polymeric vectors have the ability to self-assemble with nucleic acids, enhancing their ease of use for gene therapy over other vectors. Moreover, polymeric vectors such as PBAE, have demonstrated potential for improving tissue penetration, one of the largest barriers to increasing efficacy of gene therapy vectors in glioblastoma. To date, only a few non-viral vectors have been evaluated in clinical trials for GBM; however, further evaluation of non-viral vectors in clinical trials in the future may provide advanced treatment strategies for gene therapy in glioblastoma.

Author Contributions: Conceptualization, A.A.A.-B., B.C. and J.S.L.; formal analysis, A.A.A.-B. and B.C.; investigation, A.A.A.-B. and B.C.; writing—original draft preparation, A.A.A.-B. and B.C.; writing—review and editing, A.A.A.-B. and J.S.L.; supervision, J.S.L.; project administration, J.S.L.; funding acquisition, A.A.A.-B. and J.S.L.

Funding: This work was supported by an Institutional Development Award (IDeA) from the National Institute of General Medical Sciences of the National Institutes of Health under grant number P20GM103444 (J.S.L) and supported in part by the Tiger Talent Postdoctoral Fellowship Program (A.A.A.-B.) Clemson University.

Conflicts of Interest: The authors declare no conflict of interest.

References

1. Alifieris, C.; Trafalis, D.T. Glioblastoma multiforme: Pathogenesis and treatment. *Pharmacol. Ther.* **2015**, *152*, 63–82. [CrossRef] [PubMed]
2. Louis, D.N.; Ohgaki, H.; Wiestler, O.D.; Cavenee, W.K.; Burger, P.C.; Jouvet, A.; Scheithauer, B.W.; Kleihues, P. The 2007 WHO classification of tumours of the central nervous system. *Acta Neuropathol.* **2007**, *114*, 97–109. [CrossRef] [PubMed]
3. Holland, E.C. Glioblastoma multiforme: the terminator. *Proc. Natl. Acad. Sci. USA* **2000**, *97*, 6242–6244. [CrossRef] [PubMed]
4. Ostrom, Q.T.; Gittleman, H.; Farah, P.; Ondracek, A.; Chen, Y.; Wolinsky, Y.; Stroup, N.E.; Kruchko, C.; Barnholtz-sloan, J.S. CBTRUS statistical report: Primary brain and central nervous system tumors diagnosed in the United States in 2006–2010. *J. Neurooncol.* **2013**, *15*, 788–796. [CrossRef]
5. Van Tellingen, O.; Yetkin-Arik, B.; De Gooijer, M.C.; Wesseling, P.; Wurdinger, T.; De Vries, H.E. Overcoming the blood-brain tumor barrier for effective glioblastoma treatment. *Drug Resist. Updat.* **2015**, *19*, 1–12. [CrossRef] [PubMed]
6. Hottinger, A.F.; Stupp, R.; Homicsko, K. Standards of care and novel approaches in the management of glioblastoma multiforme. *Chin. J. Cancer* **2014**, *33*, 32–39. [CrossRef]
7. Bota, D.A.; Desjardins, A.; Quinn, J.A.; Affronti, M.L.; Friedman, H.S. Interstitial chemotherapy with biodegradable BCNU (Gliadel®) wafers in the treatment of malignant gliomas. *Ther. Clin. Risk Manag.* **2007**, *3*, 707–715. [PubMed]
8. Chowdhary, S.A.; Ryken, T.; Newton, H.B. Survival outcomes and safety of carmustine wafers in the treatment of high-grade gliomas: a meta-analysis. *J. Neurooncol.* **2015**, *122*, 367–382. [CrossRef] [PubMed]
9. Cohen, M.H.; Shen, Y.L.; Keegan, P.; Pazdur, R. FDA drug approval summary: Bevacizumab (Avastin) as treatment of recurrent glioblastoma multiforme. *Oncologist* **2009**, *14*, 1131–1138. [CrossRef]
10. Stupp, R.; Hegi, M.E.; Mason, W.P. Effects of radiotherapy with concomitant and adjuvant temozolomide versus radiotherapy alone on survival in glioblastoma in a randomised phase III study: 5-year analysis of the EORTC-NCIC trial. *Lancet Oncol.* **2009**, *10*, 459–466. [CrossRef]
11. Jhanwar-Uniyal, M.; Labagnara, M.; Friedman, M.; Kwasnicki, A.; Murali, R. Glioblastoma: Molecular pathways, stem cells and therapeutic targets. *Cancers (Basel)* **2015**, *7*, 538–555. [CrossRef] [PubMed]
12. Vogelstein, B.; Kinzler, K.W. The multistep nature of cancer. *Trends Genet.* **1993**, *9*, 138–141. [CrossRef]
13. Verhaak, R.; Hoadley, K.; Purdon, E.; Getz, G. An integrated genomic analysis identifies clinically relevant subtypes of glioblastoma characterized by abnormalities in PDGFRA, IDH1, EGFR and NF1. *Cancer Cell* **2010**, *19*, 38–46. [CrossRef]
14. Kleihues, P.; Ohgaki, H. Primary and secondary glioblastomas: from concept to clinical diagnosis. *Neurol. Oncol.* **1999**, *1*, 44–51. [CrossRef] [PubMed]
15. Hay, N.; Sonenberg, N. Upstream and downstream of mTOR. *Genes Dev.* **2004**, *18*, 1926–1945. [CrossRef]
16. Jhanwar-Uniyal, M.; Albert, L.; McKenna, E.; Karsy, M.; Rajdev, P.; Braun, A.; Murali, R. Deciphering the signaling pathways of cancer stem cells of glioblastoma multiforme: Role of Akt/mTOR and MAPK pathways. *Adv. Enzyme Regul.* **2011**, *51*, 164–170. [CrossRef] [PubMed]
17. Liu, X.; Ling, Z.-Q. Role of isocitrate dehydrogenase 1/2 (IDH 1/2) gene mutations in human tumors. *Histol. Histopathol.* **2015**, *30*, 1155–1160. [CrossRef] [PubMed]
18. Watanabe, T.; Nobusawa, S.; Kleihues, P.; Ohgaki, H. IDH1 mutations are early events in the development of astrocytomas and oligodendrogliomas. *Am. J. Pathol.* **2009**, *174*, 1149–1153. [CrossRef]
19. Erasimus, H.; Gobin, M.; Niclou, S.; Van Dyck, E. DNA repair mechanisms and their clinical impact in glioblastoma. *Mutat. Res.-Rev. Mutat. Res.* **2016**, *769*, 19–35. [CrossRef]

20. Yoshino, A.; Ogino, A.; Yachi, K.; Ohta, T.; Fukushima, T.; Watanabe, T.; Kaatayama, Y. Effect of IFN-beta on human glioma cell lines with temozolomide resistance. *Int. J. Oncol.* **2009**, *151*, 414–420. [CrossRef]

21. Joseph, J.V.; Conroy, S.; Tomar, T.; Eggens-Meijer, E.; Bhat, K.; Copray, S.; Walenkamp, A.M.E.; Boddeke, E.; Balasubramanyian, V.; Wagemakers, M.; et al. TGF-beta is an inducer of ZEB1-dependent mesenchymal transdifferentiation in glioblastoma that is associated with tumor invasion. *Cell Death Dis.* **2014**, *5*, e1443. [CrossRef] [PubMed]

22. Chow, K.-H.; Park, H.J.; George, J.; Yamamoto, K.; Gallup, A.D.; Graber, J.H.; Chen, Y.; Jiang, W.; Steindler, D.; Neilson, E.G.; et al. S100A4 is a biomarker and regulator of glioma stem cells that is critical for mesenchymal transition in glioblastoma. *Cancer Res.* **2017**, *77*, 5360–5373. [CrossRef] [PubMed]

23. Abbott, N.J.; Patabendige, A.A.K.; Dolman, D.E.M.; Yusof, S.R.; Begley, D.J. Structure and function of the blood-brain barrier. *Neurobiol. Dis.* **2010**, *37*, 13–25. [CrossRef]

24. Zhou, J.; Patel, T.R.; Sirianni, R.W.; Strohbehn, G.; Zheng, M.-Q.; Duong, N.; Schafbauer, T.; Huttner, A.J.; Huang, Y.; Carson, R.E.; et al. Highly penetrative, drug-loaded nanocarriers improve treatment of glioblastoma. *Proc. Natl. Acad. Sci. USA* **2013**, *110*, 11751–11756. [CrossRef] [PubMed]

25. Pardridge, W.M. The blood-brain barrier and neurotherapeutics. *NeuroRx* **2005**, *2*, 1–2. [CrossRef] [PubMed]

26. Banks, W.A. Characteristics of compounds that cross the blood-brain barrier. *BMC Neurol.* **2009**, *9*, S3. [CrossRef] [PubMed]

27. Pan, W.; Kastin, A.J. Changing the chemokine gradient: CINC1 crosses the blood–brain barrier. *J. Neuroimmunol.* **2001**, *115*, 64–70. [CrossRef]

28. Karanth, H.; Rayasa, M. Nanotechnology in Brain Targeting. *Int. J. Pharm. Sci. Nanotechnol.* **2008**, *1*, 924.

29. Masserini, M. Nanoparticles for brain drug delivery. *ISRN Biochem.* **2013**, *2013*, 238428. [CrossRef]

30. Kamaly, N.; Xiao, Z.; Valencia, P.M.; Radovic-Moreno, A.F.; Farokhzad, O.C. Targeted polymeric therapeutic nanoparticles: design, development and clinical translation. *Chem. Soc. Rev.* **2012**, *41*, 2971–3010. [CrossRef]

31. Peer, D.; Karp, J.M.; Hong, S.; Farokhzad, O.C.; Margalit, R.; Langer, R. Nanocarriers as an emerging platform for cancer therapy. *Nat. Nanotechnol.* **2007**, *2*, 751–760. [CrossRef] [PubMed]

32. Hardee, M.E.; Zagzag, D. Mechanisms of glioma-associated neovascularization. *Am. J. Pathol.* **2012**, *181*, 1126–1141. [CrossRef]

33. Kim, S.S.; Harford, J.B.; Pirollo, K.F.; Chang, E.H. Effective treatment of glioblastoma requires crossing the blood-brain barrier and targeting tumors including cancer stem cells: The promise of nanomedicine. *Biochem. Biophys. Res. Commun.* **2015**, *468*, 485–489. [CrossRef] [PubMed]

34. Séhédic, D.; Cikankowitz, A.; Hindré, F.; Davodeau, F.; Garcion, E. Nanomedicine to overcome radioresistance in glioblastoma stem-like cells and surviving clones. *Trends Pharmacol. Sci.* **2015**, *36*, 236–252. [CrossRef]

35. Juliano, R.L.; Ling, V. A surface glycoprotein modulating drug permeability in Chinese hamster ovary cell mutants. *BBA-Biomembr.* **1976**, *455*, 152–162. [CrossRef]

36. Horio, M.; Gottesman, M.M.; Pastan, I. ATP-dependent transport of vinblastine in vesicles from human multidrug-resistant cells. *Proc. Natl. Acad. Sci. USA* **1988**, *85*, 3580–3584. [CrossRef]

37. Schinkel, A.H.; Wagenaar, E.; Mol, C.A.A.M.; Van Deemter, L. P-glycoprotein in the blood-brain barrier of mice influences the brain penetration and pharmacological activity of many drugs. *J. Clin. Investig.* **1996**, *97*, 2517–2524. [CrossRef]

38. Zhang, J.; Stevens, M.F.G.; Bradshaw, T.D. Temozolomide: Mechanisms of Action, Repair and Resistance. *Curr. Mol. Pharmacol.* **2012**, *5*, 102–114. [CrossRef]

39. Hegi, M.E.; Diserens, A.-C.; Gorlia, T.; Hamou, M.-F.; de Tribolet, N.; Weller, M.; Kros, J.M.; Hainfellner, J.A.; Mason, W.; Mariani, L.; et al. MGMT gene silencing and benefit from temozolomide in glioblastoma. *N. Engl. J. Med.* **2005**, *352*, 997–1003. [CrossRef] [PubMed]

40. Esteller, M.; Garcia-Foncillas, J.; Andion, E.; Goodman, S.N.; Hidalgo, O.F.; Vanaclocha, V.; Baylin, S.B.; Herman, J.G. Inactivation of the DNA-repair gene MGMT and the clinical response of gliomas to alkylating agents. *N. Engl. J. Med.* **2000**, *343*, 1350–1354. [CrossRef]

41. Okura, H.; Smith, C.A.; Rutka, J.T. Gene therapy for malignant glioma. *Mol. Cell. Ther.* **2014**, *2*, 21. [CrossRef] [PubMed]

42. Watanabe, R.; Takase-Yoden, S. Gene expression of neurotropic retrovirus in the CNS. *Prog. Brain Res.* **1995**, *105*, 255–262. [PubMed]

43. Peltékian, E.; Garcia, L.; Danos, O. Neurotropism and Retrograde Axonal Transport of a Canine Adenoviral Vector: A Tool for Targeting Key Structures Undergoing Neurodegenerative Processes. *Mol. Ther.* **2002**, *5*, 25–32. [CrossRef]

44. Braun, E. Neurotropism of herpes simplex virus type 1 in brain organ cultures. *J. Gen. Virol.* **2006**, *87*, 2827–2837. [CrossRef] [PubMed]

45. Crommentuijn, M.H.W.; Maguire, C.A.; Niers, J.M.; Vandertop, W.P.; Badr, C.E.; Wurdinger, T.; Tannous, B.A. Intracranial AAV-sTRAIL combined with lanatoside C prolongs survival in an orthotopic xenograft mouse model of invasive glioblastoma. *Mol. Oncol.* **2016**, *10*, 625–634. [CrossRef] [PubMed]

46. GuhaSarkar, D.; Su, Q.; Gao, G.; Sena-Esteves, M. Systemic AAV9-IFNbeta gene delivery treats highly invasive glioblastoma. *Neurol. Oncol.* **2016**, *18*, 1508–1518. [CrossRef]

47. Meijer, D.H.; Maguire, C.A.; LeRoy, S.G.; Sena-Esteves, M. Controlling brain tumor growth by intraventricular administration of an AAV vector encoding IFN-beta. *Cancer Gene Ther.* **2009**, *16*, 664–671. [CrossRef]

48. Rainov, N. A phase III clinical evaluation of herpes simplex virus type 1 thymidine kinase and ganciclovir gene therapy as an adjuvant to surgical resection and radiation in adults with previously untreated glioblastoma multiforme. *Hum. Gene Ther.* **2000**, *11*, 2389–2401. [CrossRef]

49. Ram, Z.; Culver, K.W.; Oshiro, E.M.; Viola, J.J.; DeVroom, H.L.; Otto, E.; Long, Z.; Chiang, Y.; McGarrity, G.J.; Muul, L.M.; et al. Therapy of malignant brain tumors by intratumoral implantation of retroviral vector-producing cells. *Nat. Med.* **1997**, *3*, 1354–1361. [CrossRef]

50. Huang, T.T.; Hlavaty, J.; Ostertag, D.; Espinoza, F.L.; Martin, B.; Petznek, H.; Rodriguez-Aguirre, M.; Ibanez, C.E.; Kasahara, N.; Gunzburg, W.; et al. Toca 511 gene transfer and 5-fluorocytosine in combination with temozolomide demonstrates synergistic therapeutic efficacy in a temozolomide-sensitive glioblastoma model. *Cancer Gene Ther.* **2013**, *20*, 544–551. [CrossRef]

51. Takahashi, M.; Valdes, G.; Hiraoka, K.; Inagaki, A.; Kamijima, S.; Micewicz, E.; Gruber, H.E.; Robbins, J.M.; Jolly, D.J.; McBride, W.H.; et al. Radiosensitization of gliomas by intracellular generation of 5-fluorouracil potentiates prodrug activator gene therapy with a retroviral replicating vector. *Cancer Gene Ther.* **2014**, *21*, 405–410. [CrossRef]

52. Aghi, M.; Vogelbaum, M.A.; Kesari, S.; Chen, C.C.; Liau, L.M.; Piccioni, D.; Portnow, J.; Chang, S.; Robbins, J.M.; Boyce, T.; et al. AT-02 Intratumoral delivery of the retroviral replicating vector (RRV) TOCA 511 in subjects with recurrent high grade glioma: Interim report of phase I study (NCT 01156584). *Neurol. Oncol.* **2014**, *16*. [CrossRef]

53. Lang, F.F.; Bruner, J.M.; Fuller, G.N.; Aldape, K.; Prados, M.D.; Chang, S.; Berger, M.S.; McDermott, M.W.; Kunwar, S.M.; Junck, L.R.; et al. Phase I Trial of Adenovirus-Mediated p53 Gene Therapy for Recurrent Glioma: Biological and Clinical Results. *J. Clin. Oncol.* **2003**, *21*, 2508–2518. [CrossRef]

54. Sandmair, A.M.; Loimas, S.; Puranen, P.; Immonen, A.; Kossila, M.; Puranen, M.; Hurskainen, H.; Tyynela, K.; Turunen, M.; Vanninen, R.; et al. Thymidine kinase gene therapy for human malignant glioma, using replication-deficient retroviruses or adenoviruses. *Hum. Gene Ther.* **2000**, *11*, 2197–2205. [CrossRef]

55. Chiocca, E.A.; Aguilar, L.K.; Bell, S.D.; Kaur, B.; Hardcastle, J.; Cavaliere, R.; McGregor, J.; Lo, S.; Ray-Chaudhuri, A.; et al. Phase IB Study of Gene-Mediated Cytotoxic Immunotherapy Adjuvant to Up-Front Surgery and Intensive Timing Radiation for Malignant Glioma. *J. Clin. Oncol.* **2011**, *29*, 3611–3619. [CrossRef] [PubMed]

56. Wheeler, L.A.; Manzanera, A.G.; Bell, S.D.; Cavaliere, R.; McGregor, J.M.; Grecula, J.C.; Newton, H.B.; Lo, S.S.; Badie, B.; Portnow, J.; et al. Phase II multicenter study of gene-mediated cytotoxic immunotherapy as adjuvant to surgical resection for newly diagnosed malignant glioma. *Neurol. Oncol.* **2016**, *18*, 1137–1145. [CrossRef]

57. Kieran, M.W.; Goumnerova, L.; Manley, P.; Chi, S.N.; Marcus, K.; Manzanera, A.G.; Aguilar-Cordova, E.; DiPatri, A.J.; Tomita, T.; Lulla, R.; et al. EPT-14 Phase I study of gene mediated cytotoxic immunotherapy with AdV-tk as adjuvant to surgery and radiation therapy for pediatric malignant glioma and recurrent ependymoma. *Neurol. Oncol.* **2016**, *18*, iii26–iii27. [CrossRef]

58. Conry, R.M.; Westbrook, B.; McKee, S.; Norwood, T.G. Talimogene laherparepvec: First in class oncolytic virotherapy. *Hum. Vaccin. Immunother.* **2018**, *14*, 839–846. [CrossRef]

59. Adams, D.; Gonzalez-Duarte, A.; O'Riordan, W.D.; Yang, C.-C.; Ueda, M.; Kristen, A.V.; Tournev, I.; Schmidt, H.H.; Coelho, T.; Berk, J.L.; et al. Patisiran, an RNAi Therapeutic, for Hereditary Transthyretin Amyloidosis. *N. Engl. J. Med.* **2018**, *379*, 11–21. [CrossRef]

60. Kim, S.-S.; Rait, A.; Kim, E.; Pirollo, K.F.; Nishida, M.; Farkas, N.; Dagata, J.A.; Chang, E.H. A Nanoparticle Carrying the p53 Gene Targets Tumors Including Cancer Stem Cells, Sensitizes Glioblastoma to Chemotherapy and Improves Survival. *ACS Nano* **2014**, *8*, 5494–5514. [CrossRef] [PubMed]

61. Kim, S.-S.; Rait, A.; Kim, E.; Pirollo, K.F.; Chang, E.H. A Tumor-targeting p53 Nanodelivery System Limits Chemoresistance to Temozolomide Prolonging Survival in a Mouse Model of Glioblastoma Multiforme. *Nanomedicine* **2015**, *11*, 301–311. [CrossRef] [PubMed]

62. Jensen, S.A.; Day, E.S.; Ko, C.H.; Hurley, L.A.; Luciano, J.P.; Kouri, F.M.; Merkel, T.J.; Luthi, A.J.; Patel, P.C.; Cutler, J.I.; et al. Spherical Nucleic Acid Nanoparticle Conjugates as an RNAi-Based Therapy for Glioblastoma. *Sci. Transl. Med.* **2013**, *5*, 209ra152. [CrossRef] [PubMed]

63. Lee, T.J.; Yoo, J.Y.; Shu, D.; Li, H.; Zhang, J.; Yu, J.-G.; Jaime-Ramirez, A.C.; Acunzo, M.; Romano, G.; Cui, R.; et al. RNA Nanoparticle-Based Targeted Therapy for Glioblastoma through Inhibition of Oncogenic miR-21. *Mol. Ther.* **2017**, *25*, 1544–1555. [CrossRef] [PubMed]

64. Shu, Y.; Pi, F.; Sharma, A.; Rajabi, M.; Haque, F.; Shu, D.; Leggas, M.; Evers, B.M.; Guo, P. Stable RNA nanoparticles as potential new generation drugs for cancer therapy. *Adv. Drug Deliv. Rev.* **2014**, *66*, 74–89. [CrossRef]

65. Kang, C.; Yuan, X.; Li, F.; Pu, P.; Yu, S.; Shen, C.; Zhang, Z.; Zhang, Y. Evaluation of folate-PAMAM for the delivery of antisense oligonucleotides to rat C6 glioma cells in vitro and in vivo. *J. Biomed. Mater. Res. A* **2010**, *93*, 585–594. [CrossRef] [PubMed]

66. Liu, X.; Li, G.; Su, Z.; Jiang, Z.; Chen, L.; Wang, J.; Yu, S.; Liu, Z. Poly(amido amine) is an ideal carrier of miR-7 for enhancing gene silencing effects on the EGFR pathway in U251 glioma cells. *Oncol. Rep.* **2013**, *29*, 1387–1394. [CrossRef] [PubMed]

67. Waite, C.L.; Roth, C.M. PAMAM-RGD Conjugates Enhance siRNA Delivery Through a Multicellular Spheroid Model of Malignant Glioma. *Bioconjug. Chem.* **2009**, *20*, 1908–1916. [CrossRef]

68. Bae, Y.; Green, E.S.; Kim, G.Y.; Song, S.J.; Mun, J.Y.; Lee, S.; Park, J.I.; Park, J.S.; Ko, K.S.; Han, J.; et al. Dipeptide-functionalized Polyamidoamine dendrimer-mediated apoptin gene delivery facilitates apoptosis of human primary glioma cells. *Int. J. Pharm.* **2016**, *515*, 186–200. [CrossRef]

69. Gao, S.; Li, J.; Jiang, C.; Hong, B.; Hao, B. Plasmid pORF-hTRAIL targeting to glioma using transferrin-modified polyamidoamine dendrimer. *Drug Des. Dev. Ther.* **2016**, *10*, 1–11. [CrossRef]

70. Huang, R.; Ke, W.; Han, L.; Li, J.; Liu, S.; Jiang, C. Targeted delivery of chlorotoxin-modified DNA-loaded nanoparticles to glioma via intravenous administration. *Biomaterials* **2011**, *32*, 2399–2406. [CrossRef]

71. Huang, S.; Li, J.; Han, L.; Liu, S.; Ma, H.; Huang, R.; Jiang, C. Dual targeting effect of Angiopep-2-modified, DNA-loaded nanoparticles for glioma. *Biomaterials* **2011**, *32*, 6832–6838. [CrossRef] [PubMed]

72. Wang, S.; Li, Y.; Fan, J.; Wang, Z.; Zeng, X.; Sun, Y.; Song, P.; Ju, D. The role of autophagy in the neurotoxicity of cationic PAMAM dendrimers. *Biomaterials* **2014**, *35*, 7588–7597. [CrossRef] [PubMed]

73. Liu, Y.; Li, J.; Shao, K.; Huang, R.; Ye, L.; Lou, J.; Jiang, C. A leptin derived 30-amino-acid peptide modified pegylated poly-L-lysine dendrigraft for brain targeted gene delivery. *Biomaterials* **2010**, *31*, 5246–5257. [CrossRef]

74. Liu, S.; Guo, Y.; Huang, R.; Li, J.; Huang, S.; Kuang, Y.; Han, L.; Jiang, C. Gene and doxorubicin co-delivery system for targeting therapy of glioma. *Biomaterials* **2012**, *33*, 4907–4916. [CrossRef]

75. Liu, Y.; He, X.; Kuang, Y.; An, S.; Wang, C.; Guo, Y.; Ma, H.; Lou, J.; Jiang, C. A bacteria deriving peptide modified dendrigraft poly-l-lysines (DGL) self-assembling nanoplatform for targeted gene delivery. *Mol. Pharm.* **2014**, *11*, 3330–3341. [CrossRef] [PubMed]

76. Kuang, Y.; An, S.; Guo, Y.; Huang, S.; Shao, K.; Liu, Y.; Li, J.; Ma, H.; Jiang, C. T7 peptide-functionalized nanoparticles utilizing RNA interference for glioma dual targeting. *Int. J. Pharm.* **2013**, *454*, 11–20. [CrossRef] [PubMed]

77. Yao, H.; Wang, K.; Wang, Y.; Wang, S.; Li, J.; Lou, J.; Ye, L.; Yan, X.; Lu, W.; Huang, R. Enhanced blood-brain barrier penetration and glioma therapy mediated by a new peptide modified gene delivery system. *Biomaterials* **2015**, *37*, 345–352. [CrossRef]

78. Li, J.; Guo, Y.; Kuang, Y.; An, S.; Ma, H.; Jiang, C. Choline transporter-targeting and co-delivery system for glioma therapy. *Biomaterials* **2013**, *34*, 9142–9148. [CrossRef]

79. Kodama, Y.; Nakamura, T.; Kurosaki, T.; Egashira, K.; Mine, T.; Nakagawa, H.; Muro, T.; Kitahara, T.; Higuchi, N.; Sasaki, H. Biodegradable nanoparticles composed of dendrigraft poly-L-lysine for gene delivery. *Eur. J. Pharm. Biopharm.* **2014**, *87*, 472–479. [CrossRef]

80. Kodama, Y.; Kuramoto, H.; Mieda, Y.; Muro, T.; Nakagawa, H.; Kurosaki, T.; Sakaguchi, M.; Nakamura, T.; Kitahara, T.; Sasaki, H. Application of biodegradable dendrigraft poly-l-lysine to a small interfering RNA delivery system. *J. Drug Target.* **2017**, *25*, 49–57. [CrossRef]

81. Tang, M.; Dong, H.; Li, Y.; Ren, T. Harnessing the PEG-cleavable strategy to balance cytotoxicity, intracellular release and the therapeutic effect of dendrigraft poly-l-lysine for cancer gene therapy. *J. Mater. Chem. B* **2016**, *4*, 1284–1295. [CrossRef]

82. Oerlemans, C.; Bult, W.; Bos, M.; Storm, G.; Nijsen, J.F.W.; Hennink, W.E. Polymeric Micelles in Anticancer Therapy: Targeting, Imaging and Triggered Release. *Pharm. Res.* **2010**, *27*, 2569–2589. [CrossRef] [PubMed]

83. Cheng, D.; Cao, N.; Chen, J.; Yu, X.; Shuai, X. Multifunctional nanocarrier mediated co-delivery of doxorubicin and siRNA for synergistic enhancement of glioma apoptosis in rat. *Biomaterials* **2012**, *33*, 1170–1179. [CrossRef] [PubMed]

84. Liang, B.; He, M.L.; Chan, C.Y.; Chen, Y.C.; Li, X.P.; Li, Y.; Zheng, D.; Lin, M.C.; Kung, H.F.; Shuai, X.T.; et al. The use of folate-PEG-grafted-hybranched-PEI nonviral vector for the inhibition of glioma growth in the rat. *Biomaterials* **2009**, *30*, 4014–4020. [CrossRef] [PubMed]

85. Zhan, C.; Wei, X.; Qian, J.; Feng, L.; Zhu, J.; Lu, W. Co-delivery of TRAIL gene enhances the anti-glioblastoma effect of paclitaxel in vitro and in vivo. *J. Control. Release* **2012**, *160*, 630–636. [CrossRef]

86. Wang, J.; Lei, Y.; Xie, C.; Lu, W.; Wagner, E.; Xie, Z.; Gao, J.; Zhang, X.; Yan, Z.; Liu, M. Retro-inverso CendR peptide-mediated polyethyleneimine for intracranial glioblastoma-targeting gene therapy. *Bioconjug. Chem.* **2014**, *25*, 414–423. [CrossRef] [PubMed]

87. Lei, Y.; Wang, J.; Xie, C.; Wagner, E.; Lu, W.; Li, Y.; Wei, X.; Dong, J.; Liu, M. Glutathione-sensitive RGD-poly(ethylene glycol)-SS-polyethylenimine for intracranial glioblastoma targeted gene delivery. *J. Gene Med.* **2013**, *15*, 291–305. [CrossRef]

88. Green, J.J.; Zugates, G.T.; Langer, R.; Anderson, D.G. Poly (β-amino esters): Procedures for Synthesis and Gene Delivery. *Methods Mol. Biol.* **2009**, *480*, 53–63. [CrossRef]

89. Anderson, D.G.; Lynn, D.M.; Langer, R. Semi-automated synthesis and screening of a large library of degradable cationic polymers for gene delivery. *Angew Chem. Int. Ed. Engl.* **2003**, *42*, 3153–3158. [CrossRef] [PubMed]

90. Tzeng, S.Y.; Green, J.J. Subtle changes to polymer structure and degradation mechanism enable highly effective nanoparticles for siRNA and DNA delivery to human brain cancer. *Adv Heal. Mater* **2013**, *2*, 468–480. [CrossRef]

91. Guerrero-Cazares, H.; Tzeng, S.Y.; Young, N.P.; Abutaleb, A.O.; Quinones-Hinojosa, A.; Green, J.J. Biodegradable polymeric nanoparticles show high efficacy and specificity at DNA delivery to human glioblastoma in vitro and in vivo. *ACS Nano* **2014**, *8*, 5141–5153. [CrossRef] [PubMed]

92. Mangraviti, A.; Tzeng, S.Y.; Kozielski, K.L.; Wang, Y.; Jin, Y.; Gullotti, D.; Pedone, M.; Buaron, N.; Liu, A.; Wilson, D.R.; et al. Polymeric Nanoparticles for Nonviral Gene Therapy Extend Brain Tumor Survival in Vivo. *ACS Nano* **2015**, *9*, 1236–1249. [CrossRef] [PubMed]

93. Mastorakos, P.; Song, E.; Zhang, C.; Berry, S.; Park, H.W.; Kim, Y.E.; Park, J.S.; Lee, S.; Suk, J.S.; Hanes, J. Biodegradable DNA Nanoparticles that Provide Widespread Gene Delivery in the Brain. *Small* **2016**, *12*, 678–685. [CrossRef] [PubMed]

94. Sunshine, J.C.; Peng, D.Y.; Green, J.J. Uptake and transfection with polymeric nanoparticles are dependent on polymer end-group structure, but largely independent of nanoparticle physical and chemical properties. *Mol. Pharm.* **2012**, *9*, 3375–3383. [CrossRef]

95. Kozielski, K.L.; Tzeng, S.Y.; Green, J.J. A bioreducible linear poly(β-amino ester) for siRNA delivery. *Chem. Commun. (Camb)* **2013**, *49*, 5319–5321. [CrossRef] [PubMed]

![nanomaterials logo] *nanomaterials*

MDPI

Review

Intracellular Imaging with Genetically Encoded RNA-Based Molecular Sensors

Zhining Sun, Tony Nguyen, Kathleen McAuliffe and Mingxu You *

Department of Chemistry, University of Massachusetts Amherst, Amherst, MA 01003, USA;
zhiningsun@umass.edu (Z.S.); tonnguyen@umass.edu (T.N.); kmmcauliffe@umass.edu (K.M.)
* Correspondence: mingxuyou@umass.edu

Received: 22 January 2019; Accepted: 5 February 2019; Published: 8 February 2019

Abstract: Genetically encodable sensors have been widely used in the detection of intracellular molecules ranging from metal ions and metabolites to nucleic acids and proteins. These biosensors are capable of monitoring in real-time the cellular levels, locations, and cell-to-cell variations of the target compounds in living systems. Traditionally, the majority of these sensors have been developed based on fluorescent proteins. As an exciting alternative, genetically encoded RNA-based molecular sensors (GERMS) have emerged over the past few years for the intracellular imaging and detection of various biological targets. In view of their ability for the general detection of a wide range of target analytes, and the modular and simple design principle, GERMS are becoming a popular choice for intracellular analysis. In this review, we summarize different design principles of GERMS based on various RNA recognition modules, transducer modules, and reporting systems. Some recent advances in the application of GERMS for intracellular imaging are also discussed. With further improvement in biostability, sensitivity, and robustness, GERMS can potentially be widely used in cell biology and biotechnology.

Keywords: RNA aptamers; biosensors; live-cell imaging; fluorogenic RNA; riboswitch; ribozyme

---

1. Introduction

The detection and quantification of cellular proteins, nucleic acids, and metabolites is critical in understanding cellular signaling pathways and many other physiological processes. These cellular molecules are tightly regulated in living systems. Both their cellular levels and distributions play essential roles for their biological functions. As a result, the development of sensors to characterize the spatial and temporal distributions of cellular targets and to accurately quantify their cellular levels has been a major focus in current biochemical studies [1–3].

Although the expression levels of many biomolecules can be measured using traditional methods such as gel electrophoresis, mass spectrometry, liquid chromatography, and NMR spectroscopy [4], most of these techniques require complex pre- and post-treatments on cells and can only deal with cell lysates. These in vitro assays provide limited information on the cellular distributions, live-cell dynamics, or cell-to-cell variations of the target analytes.

Fluorescence imaging, on the other hand, overcomes most of these challenges [5–7]. Synthetic fluorescent compounds, such as fluorescein, rhodamine, BODIPY, and cyanine, have been widely used as reporters in developing small-molecule sensors for cellular imaging [8–16]. However, the limited biocompatibilities, cellular interferences, and cellular distributions of these non-natural compounds remain major issues that limit their actual biological applications [17–19].

Sensors based on naturally occurring proteins or RNA molecules could potentially address these issues in cellular analysis. For example, fluorescent protein (FP)-based sensors were developed soon after the isolation of green fluorescent protein (GFP) from the luminous organ of the jellyfish

Aequorea victoria [20,21]. FP-based Förster resonance energy transfer (FRET) sensors have advanced the field of bioimaging by quantitatively detecting various classes of targets in living systems [22–26]. However, many critical cellular targets cannot be feasibly detected using these protein-based sensors. This fact is largely due to the limited choice of protein domains that can selectively bind to the target molecules, which should also induce conformational changes that lead to significant FRET changes. Furthermore, the detection range and the signal-to-noise ratio of many FP-based sensors are not ideal for the cellular imaging and detection of target biomolecules [27,28].

Recently, an alternative class of RNA-based fluorescent biosensors has been developed for intracellular applications [29–32]. In general, these Genetically Encoded RNA-based Molecular Sensors (GERMS) consist of three components: a recognition module, a reporting system, and a transducer module. The recognition module, such as an RNA aptamer (RNA aptamers will be described in more depth in Section 4.1), is an RNA sequence that can specifically recognize target molecules and bind to them with a high affinity [33,34]. The reporting system is normally a fluorescent protein or a fluorogenic RNA aptamer that can bind and induce the fluorescence of its cognate small-molecule dye [35,36]. The transducer module is used to connect the recognition module and the reporting system. These transducers act as switches that can convert target binding events into detectable signals [37].

These novel RNA-based sensors can be genetically encoded and transcribed by cells on their own for long-term studies. GERMS can be easily and rationally modified for the detection of a wide range of target molecules with good selectivity and sensitivity. These genetically encodable sensors have shown promising potential in detecting intracellular RNAs, proteins, metabolites, signaling molecules, and metal ions [29,30,32,38–41]. GERMS have started to be used to monitor cellular signaling pathways as well as other biological processes [41,42]. There are several great reviews and articles about the design and application of RNA-based nanodevices [43–49]. In this review, we will focus on a specific emerging group of RNA devices that can be genetically encoded for the intracellular detection of biological analytes. We will first illustrate how to design and engineer the three components of GERMS: the recognition module, transducer module, and reporting system. Recent examples will be further provided to demonstrate the intracellular applications of these novel RNA-based sensors.

2. Transducer Modules in GERMS

Because GERMS are used to sense essential biomolecules in live cells, a fundamental question arises: How do GERMS recognize the target molecules and then provide a corresponding signal? The transducer module couples the recognition module with the reporting system in order to realize the entire sensing process. These RNA-based transducers provide an additional layer of modulation to permit an efficient signal transmission. In this section, we will discuss existing transducer modules in the design of GERMS.

2.1. RNA Duplex Formation or Helix Slipping

In the general design of GERMS, target binding to the recognition module triggers a conformational change in the transducer module, adjusting the activity of the reporting system. One of the most straightforward conformational changes in RNA devices is the folding and unfolding of a duplex structure (Figure 1A). A duplex formation based on the Watson-Crick or wobble base pairs can be rationally designed as the bridge between the recognition module and the reporting system. Indeed, as demonstrated in the crystal structures of several naturally occurring riboswitches, the most common target binding-induced RNA structural changes are the formation of new duplex regions or the disruption of existing duplexes [50]. In addition, the folding and activation of many reporting systems in GERMS, such as the fluorogenic RNAs, ribosomal binding sites, and transcriptional activators, can also be tuned merely by the formation of a duplex. As a result, duplex formation is one of the most popular and powerful transducer modules in developing allosterically controlled RNA devices, including GERMS.

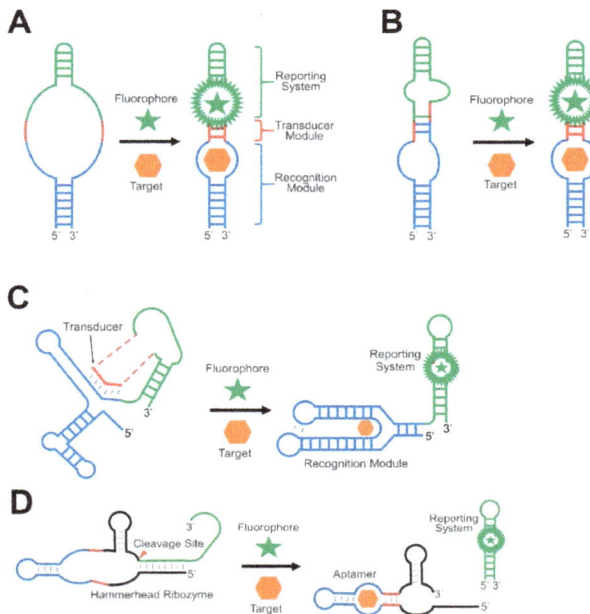

Figure 1. Schematics of different types of transducer module in GERMS. (**A**) The target binding-induced proper folding of the recognition module (blue) can trigger the formation of RNA duplexes (red), which further activate the reporting system (green). (**B**) The target binding induces helix slipping in the transducer module (red) to generate the signal. (**C**) Similar to natural riboswitches, the target binding induces the strand displacement of the transducer sequence (red) from the recognition module (blue) to the reporting system (green). (**D**) The target binding induces the folding (red) and activation of a hammerhead ribozyme to induce the catalytic cleavage and activation of the reporting system (green).

Helix slipping is another strategy to regulate the formation of the duplex. Helix slipping is a local nucleotide shift in the transducer region. Here, target binding induces a structural change in the recognition module, leading normally to shifts in only one or two nucleotides in the transducer helix, which further activates the reporting system, such as a ribosomal binding site (Figure 1B). The rationale behind this helix slipping principle is that even in the absence of a target, the transducer module should preferably still form a structure, instead of a free form, to better inhibit the activity of the reporting system. As a result, a large signal-to-noise ratio will be realized after the slipping.

2.2. RNA Strand Displacement

The structural rearrangement of the transducer module can also be realized through a strand displacement reaction. Strand displacement-based RNA signal transductions have been widely used in natural riboswitches. Riboswitches are regions in mRNAs that contain a specific evolutionarily conserved target-binding aptamer domain and an expression platform that enables the regulation of the downstream transcription or translation. The competitive binding of a transducer sequence to either a switching sequence in the aptamer domain or the expression platform is critical for the function of riboswitches (Figure 1C). For example, in a naturally abundant thiamine pyrophosphate (TPP) riboswitch, the addition of TPP allows for the formation of a TPP-binding pocket in the aptamer domain, which displaces the transducer sequence, further allowing the formation of an expression platform duplex to inhibit the translation [51].

In general, target binding with riboswitches will alter the relative stability or accessibility of the RNA duplex involved in the displacement reaction. As a result, new thermodynamically more stable duplexes will replace the previously favorable conformations. If the newly formed structure can induce the activation of a reporting system, such strand displacement reactions can be used to engineer RNA-based sensors.

Inspired by the mechanism of these naturally evolved riboswitches, synthetic riboswitches have been engineered into biosensors to detect different biological targets. Here, artificial aptamer domains and synthetic expression platforms are conjugated based on a strand displacement reaction. Computational predictions of the RNA folding and energy landscapes are often used in the generation of these synthetic biosensors. For example, an automated design model has been engineered to generate synthetic riboswitches from aptamers that can activate the translation initiation by up to 383-fold [52]. Statistical thermodynamics models have been made to measure the sequence-structure-function relationships to convert synthetic RNA aptamers into translational regulating riboswitches [53]. There are several factors determining the efficiency of such synthetic riboswitches, including their target-binding affinities, overall induced conformational changes, target and RNA expression levels, interactions with ribosomes and other protein/RNA complexes, as well as the macromolecular crowding effect. Due to the existence of these complex factors, the intracellular and in vivo behaviors of many synthetic riboswitches are still not easily predictable. In situ experimental optimizations are often necessary. It is expected that the further development of advanced computational tools and simplified high-throughput in vivo screening approaches will dramatically improve the performance of these synthetic riboswitch tools.

2.3. Ribozyme-Based Transducers

The transducer modules of GERMS can also stem from catalytic cleavage functions, as shown in naturally occurring RNA ribozymes. For example, the hammerhead ribozyme is the most widely studied natural catalytic RNA for this purpose [54–56]. The minimal catalytic domain of a hammerhead ribozyme comprises three duplexed stem regions. The proper folding of all these three regions is required for the catalytic self-cleavage of the hammerhead ribozyme. By fusing a target-binding recognition module and a reporting system into two of the three stem regions, the hammerhead ribozyme can function as a transducer in developing RNA-based sensors (Figure 1D). Here, a target-bound recognition module activates the ribozyme so that it self-cleaves and releases the reporting system from the original connection. As a result, biological analytes can allosterically regulate the reporting system in a highly modular pattern. The structure and function of hammerhead ribozymes have been well characterized with rapid kinetics, simple design, and small sizes [57]. Ribozyme-based transducers have been engineered for the in vitro and intracellular measurement of many metabolites [54,58], as well as for intracellular gene regulation [55,56].

In addition to hammerhead ribozymes, several other ribozymes have been identified as potential platforms for engineering the transducer modules. Most of these ribozymes, including twister ribozymes, twister sister ribozymes, pistol ribozymes, Varkud satellite ribozymes, and hairpin ribozymes [59–62], are known as "small self-cleaving ribozymes" ranging between 50 and 150 nucleotides in length [63]. Having been evolved directly in the living system, these ribozyme scaffolds will likely still function properly after incorporation into genetically encoded RNA devices. The diverse choice and advantageous small sizes of these ribozyme units can be potentially useful for the generation of versatile GERMS, and in the detection of a wide range of cellular targets.

3. Reporting Systems in GERMS

3.1. Protein-Based Reporters

As mentioned above, fluorescent protein-based sensors have revolutionized cellular imaging. Fluorescent proteins like GFP have been widely used as genetically encodable tags that can be fused to virtually any protein molecules. Various fluorescent proteins with optimized physical and optical properties have been evolved, providing a rich toolbox to study cell biology. Fluorescent protein-based reporting systems are straightforward choices in engineering RNA-based genetic devices. Similar to that shown in synthetic riboswitches, the target recognition module and transducer module can be inserted into transcripts encoding fluorescent proteins. As a result, variations in the cellular target levels will lead to changes in the cellular fluorescence.

Luciferase-induced luminescence signals have also been used to report the efficiency of RNA-based devices. Luciferase is a class of enzymes that can emit light by oxidizing their small-molecule luciferin substrates. Without the light excitement that induces cellular auto-fluorescence, the luciferase-based reporting system can provide a better signal-to-noise ratio than that of fluorescent protein reporters. Furthermore, luciferase signals can be easily quantified [64]. Being widely used for in vitro analysis, the intracellular functions of these bioluminescent systems can be hindered by their overall dim signals, limited choice of wavelengths, and due to the limited availability of the luciferin substrates [65].

To realize such an RNA-based regulation of the protein expression, the RNA sensors normally function at the cotranscriptional level (by alternating RNA splicing or intron synthesis) [66–68], the post-transcriptional level (by regulating mRNA stability or availability) [69–73], or the translational level (by controlling the initiation, termination, and specificity of translation) [74–77]. In addition to the relatively low signal-to-noise ratio, one major challenge is the limited temporal resolution of such fluorescent protein or luciferase-based reporters. This is mainly due to the time required for the translation and for nascent fluorescent proteins or enzymes to mature into their activated forms.

3.2. Fluorogenic RNA Complexes

Fluorogenic RNA complexes are composed of a fluorogenic aptamer and a small-molecule chromophore that exhibit fluorescence when bound together [78]. A fluorogenic RNA aptamer is a short nucleic acid strand that can specifically bind to and activate the fluorescence of its corresponding chromophore. For example, Spinach (Figure 2A,C), an RNA mimic of GFP, is one of the most popular fluorogenic aptamers in developing GERMS [35]. Spinach binds to a DFHBI chromophore and turns on its fluorescence. DFHBI (Figure 2F) is cell membrane permeable and has a low cellular background signal. After genetically conjugating Spinach to the target RNA molecules, DFHBI can be added externally to track the cellular locations and concentrations of the RNA targets.

Spinach can also be engineered as the reporter for the detection of metabolites and proteins. There is a sequence-independent stem region in Spinach that plays an important structural role in the activation of the DFHBI fluorescence [29]. By fusing a target-binding aptamer and a transducer module into this stem region, the binding of the target will fold the aptamer and subsequently stabilize the stem of Spinach to exhibit fluorescence [29]. It is critical that the target-binding aptamers should be unstructured until they are bound to the target. These Spinach-based RNA sensors can be used to detect concentration variations of the targets in real time both in vitro and in living cells [79].

To improve the folding of Spinach, the systematic mutagenesis of the original Spinach RNA has led to the development of Spinach2 [80]. Spinach2 exhibits a brighter fluorescence and increased thermal stability than Spinach in living cells [80]. Another notable Spinach derivative is named Baby Spinach [81]. The shortened sequence of Baby Spinach reduces the overall size of the Spinach tag, which may allow for the incorporation of multiple fluorogenic RNAs for cellular tracking, and which may increase the cellular biostability of these fluorogenic RNA complexes [81].

To improve the intracellular folding and brightness of these fluorogenic RNA complexes, fluorescence-activated cell sorting (FACS) has been used to identify Broccoli (Figure 2B) [79]. Broccoli is a short sequence, which shows increased folding and fluorescence in cells, even at low magnesium levels, making it a suitable option for live-cell imaging. A particularly useful version of Broccoli in engineering biosensors is called Split-Broccoli, where the Broccoli fluorescence is activated only upon the reassembly of two split pieces of Broccoli RNA [82]. For example, Split-Broccoli can be used to visualize intracellular RNA-RNA hybridizations with faster kinetics than fluorescent protein-based reporters [82].

Another recent advance in these fluorogenic RNA complexes is the Corn/DFHO system [83]. This complex is unique in its high photostability and red-shifted fluorescence emission. For example, in imaging the activities of RNA Polymerase III in live mammalian cells, a better performance has been demonstrated in Corn than in Broccoli [83].

Similarly, several other fluorogenic RNA complexes have been developed as potential reporting systems in engineering GERMS. For example, RNA Mango (Figure 2D) is a class of fluorogenic RNA aptamers that exhibits a bright fluorescence when bound to TO1-biotin (Figure 2E) or TO3-biotin dyes [84]. Mutagenesis of the original Mango I aptamer has resulted in the generation of Mango II, Mango III, and Mango IV aptamers with optimized fluorescence intensities, chromophore-binding affinities, and a salt dependency [85]. In another example, the binding of a DNB or SRB-2 aptamer can separate and activate the fluorescence of a sulforhodamine-dinitroaniline (SR-DN) dye-quencher pair, exhibiting a bright orange/red fluorescence for intracellular imaging [86,87].

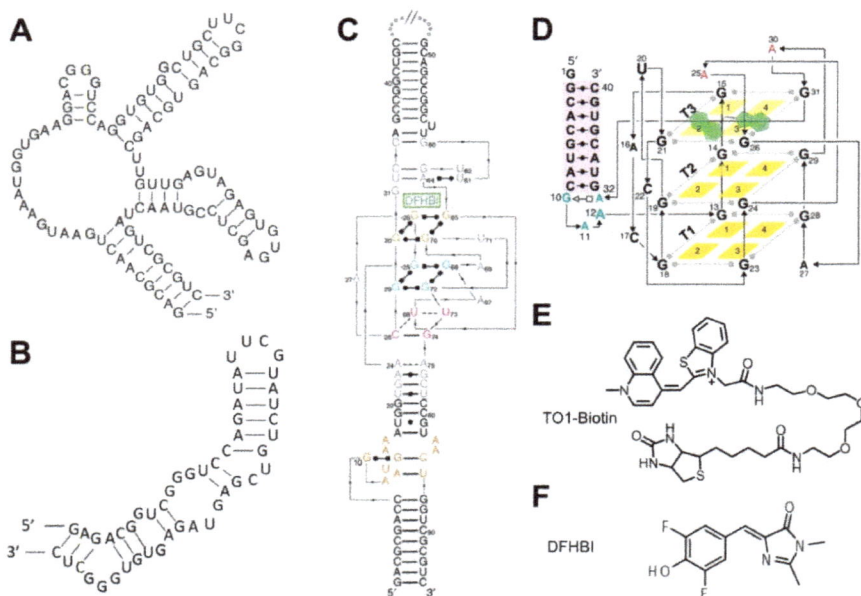

Figure 2. 2D/3D structures of fluorogenic aptamers and the chemical structures of their ligands. (**A**) Spinach aptamer 2D structure. (**B**) Broccoli aptamer 2D structure. (**C**) Spinach aptamer 3D structure [88]. (**D**) Mango aptamer 3D structure [85]. (**E**) Mango aptamer's cognate dye TO1-Biotin. (**F**) Spinach and Broccoli aptamers' cognate dye DFHBI.

Table 1 summarizes the commonly used fluorogenic RNA complexes that can be used as potential reporters for the sensor development [89]. Right now, these fluorogenic RNA complexes are still far less versatile than the existing fluorescent protein toolbox. However, with the rapid development of brighter, more photostable, and multi-color fluorogenic RNA complexes, we expect that more sensitive, multiplexed, and quantitative imaging can be achieved by these direct RNA-based reporting systems.

Table 1. Commonly used fluorogenic RNA complexes and their spectral properties.

Aptamer	Fluorophore	K_D (nM)	Ex./Em. (nm)	ε (M^{-1}cm^{-1}) [a]	Φ [b]	Brightness [c]
Spinach [35]	DFHBI	540	469/501	24,300	0.72	100
Spinach2 [31]	DFHBI	530	447/501	22,000	0.72	91
Spinach2 [31]	DFHBI-1T	560	482/505	31,000	0.94	167
Spinach2 [31]	DFHBI-2T	1300	500/523	29,000	0.12	20
Broccoli [79]	DFHBI-1T	360	472/507	29,600	0.94	159
Corn [83]	DFHO	70	505/545	29,000	0.25	41
Mango [84]	TO1-Biotin	3.2	510/535	77,500	0.14	62
Mango [84]	TO3-Biotin	5.1	637/658	9300	N/A [d]	N/A
Mango II [85]	TO1-Biotin	0.7	510/535	77,000	0.2	88
Mango III [85]	TO1-Biotin	5.6	510/535	77,000	0.56	247
Mango IV [85]	TO1-Biotin	11.1	510/535	77,000	0.42	185
DNB [87]	SR-DN	800	572/591	50,250	0.98	282
SRB-2 [86]	SR-DN	1400	579/596	85,200	0.65	317

[a] ε—Extinction Coefficient. [b] Φ—Quantum Yield. [c] Brightness—Extinction Coefficient × Quantum Yield relative to Spinach-DFHBI. [d] N/A—Not Available.

4. Recognition Modules in GERMS

The target-specific recognition module is another critical unit in GERMS. In general, to detect cellular nucleic acid targets, RNA strands with complementary sequences can be directly used as highly specific recognition modules. On the other hand, for most non-nucleic acid targets, RNA aptamers can be engineered as the recognition modules in GERMS.

4.1. Aptamers and Conventional SELEX

Aptamers, first reported in 1990 [90,91], are oligonucleotide strands that have a high binding affinity and specificity toward their targets. Aptamers can be comparable with antibodies in many ways. Aptamers can be either selected from a large random library pool using Systematic Evolution of Ligands by EXponential enrichment (SELEX) or directly adapted from naturally existing riboswitches [90–95]. Depending on the sequence, RNA aptamers can form diverse and intricate three-dimensional structures, allowing them to tightly and specifically bind with various biological targets.

SELEX has been widely used in aptamer selection. In general, SELEX begins with a chemically synthesized DNA library. The library contains numerous (normally 10^{14}–10^{15}) oligonucleotides with a random sequence in the same region, which is flanked by known fixed sequences. After the PCR and in vitro transcription of the synthetic DNA library into an RNA library, several selection steps are introduced to remove unwanted unbound oligonucleotides. The RNA sequences that are bound to the target are then released and reverse transcribed into DNA, before being further amplified by PCR. Such multiplied DNA molecules are then transcribed, in vitro, back into RNA, and a new selection round begins. Up to 20 selection rounds are usually performed in conventional SELEX to enrich aptamers with a high target binding affinity. Negative and counter SELEX are often processed at the same time to ensure a selective binding toward the target [96].

Using SELEX, RNA aptamers have been identified toward various targets, ranging from metal ions (e.g., Co^{2+}) [97], small organic molecules (e.g., amino acids [98], ATP [99], antibiotics [100], vitamins [101], and organic dyes [102]), to proteins (e.g., thrombin [103], transcription factors [104], and HIV-associated proteins [105]), and even to entire cells or microorganisms (e.g., virus and bacteria [106]). Through the SELEX procedure, RNA aptamers can be generated toward essentially almost any type of biomolecule.

4.2. Advanced SELEX Approaches for GERMS

In addition to the conventional SELEX procedure, several other advanced SELEX methods have been developed that are particularly suitable for engineering GERMS. Among others, three notable methods are Capture-SELEX, ribozyme-based SELEX, and graphene oxide-based SELEX.

Capture-SELEX is different from conventional SELEX in that it does not require the immobilization of the target compounds to beads or surfaces [107]. In a regular Capture-SELEX method (Figure 3A), short capture DNA strands are first attached to the surface of magnetic beads, and then an oligonucleotide library is immobilized to the beads by binding to the capture strands through the fixed sequence region in each oligonucleotide. By adding a solution of the solvated target, aptamers that can bind to the target and undergo conformational changes to displace the capture strands are then eluted for further enrichment. This method opens opportunities for RNA aptamer selection against target molecules that cannot be easily immobilized or chemically modified, such as several small metabolites and signaling molecules. In addition, similar to riboswitches, the identified aptamers in the Capture-SELEX have been already optimized to respond to target binding by changing the RNA conformation, which is important for sensor development. Instead of merely screening for the recognition module, Capture-SELEX allows the direct identification of both the recognition module and the transducer module for the development of GERMS.

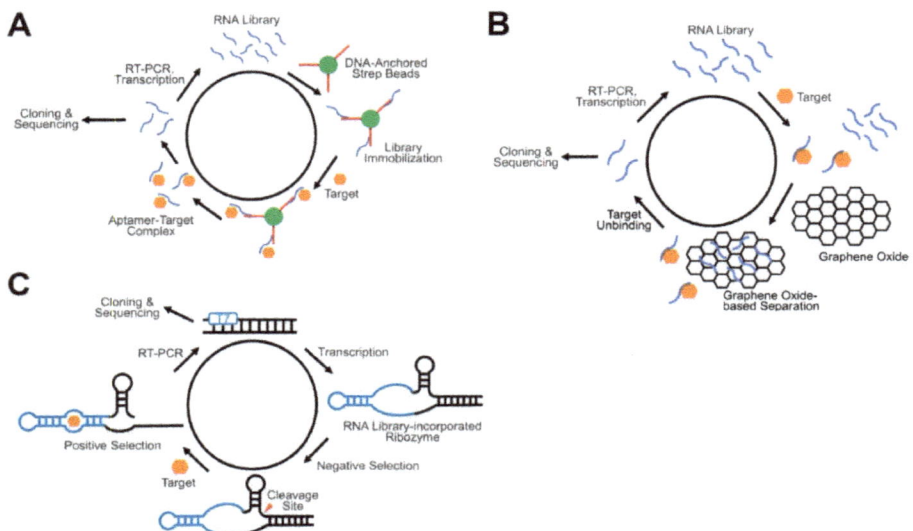

Figure 3. Schematics of advanced SELEX approaches for GERMS. (**A**) In a Capture-SELEX, an RNA library is immobilized on the surface of beads via the attached short capture DNA. RNA aptamers that can bind with the target and undergo conformational changes will be eluted and further amplified. (**B**) In a graphene oxide SELEX, target unbound RNA strands adsorb onto the surface of graphene oxide and separate from target-binding aptamers. (**C**) In a ribozyme-based SELEX, target-binding aptamers induce the catalytic self-cleavage of the ribozyme. Based on the band shift in gel electrophoresis, aptamer-containing constructs can be isolated from the RNA pool.

Ribozyme-based SELEX is another powerful method for developing RNA-based sensors. As mentioned above, ribozymes are naturally occurring RNA strands that can catalytically trans- or cis-cleave at a particular position or sequence [108]. One potential challenge in performing small molecule-targeting by SELEX is that these small targets normally cannot lead to significant conformational or property changes between the bound complexes and the free aptamers. Ribozyme-based SELEX, however, utilizes the self-cleaving properties of ribozymes to realize massive target-induced size changes in the RNA strands. To design an oligonucleotide library for such a purpose, a random RNA region is inserted into a structurally critical domain of the ribozymes, such as one of the three stem regions in a hammerhead ribozyme [109] (Figure 3C). Ribozyme-based SELEX starts with a negative selection, in the absence of the targets, to remove autonomously self-cleaved RNA strands. Uncleaved aptamers are then PAGE gel-purified and incubated with targets in the positive selection. During this step, the cleaved RNA strands are isolated via gel-purification, further reversibly transcribed, amplified by PCR, and transcribed back into full-length RNA strands for the next round of selection. The hammerhead ribozyme is the most widely used ribozyme in this type of SELEX. The identified aptamers can selectively bind with the target and further induce the folding of a stem region (i.e., the transducer) of the hammerhead ribozyme. Again, both the recognition module and the transducer module can be directly identified for the development of GERMS. In addition, with the diverse choices of different classes of naturally occurring ribozymes, various target molecules can potentially be recognized with different signal transduction mechanisms.

Graphene oxide (GO) is another platform which has recently become popular in screening for aptamers. GO-SELEX is based on the non-specific adsorption of the oligonucleotide library by graphene oxide [110]. The library is normally pre-incubated with the target, after which GO is added. Single-stranded oligonucleotides can be adsorbed by GO due to π–π stacking, while target-bound complexes remain free in the solution. After removing sequences not bound to GO through centrifugation, the target-bound oligonucleotides are then separated and amplified by reverse transcription, PCR and transcription (Figure 3B). GO-SELEX also does not require a target immobilization. The selected aptamers have also been optimized in order to obtain the property of target-induced conformational changes. GO-SELEX is a simple, high-speed, high-throughput aptamer screening method that can be applied to various target molecules [111].

4.3. Riboswitch-Based Recognition Modules

Riboswitches are naturally occurring recognition modules for many critical cellular metabolites and signaling molecules [112,113]. Another way of developing GERMS is by directly adopting these riboswitches as recognition modules. As mentioned previously in this manuscript, a riboswitch consists of an aptamer domain, a switching sequence, and an expression platform. The aptamer domains in the riboswitches have been naturally evolved to selectively bind with various cellular targets including enzyme cofactors, nucleotide precursors, amino acids and atomic ions [114]. For example, the *metH* S-adenosylhomocysteine (SAH) riboswitch can selectively recognize SAH in preference to S-adenosylmethionine (SAM) by 1000-fold [115], while SAM and SAH differ only by a methyl group. As a result, these SAH riboswitches have been used to develop sensors to measure SAH levels as well as the methyltransferase activities in vitro, which further facilitates the screening of novel MTase inhibitors [116].

During the conventional in vitro SELEX process, it is difficult to perform negative or counter SELEX against all the diverse and structurally related molecules in the cell. In addition, obtaining aptamers that have a suitable target-binding affinity is in many cases still a challenge. Most in vitro identified aptamers should be further tested and optimized in the real cellular environment. The major advantage of riboswitches over SELEX-generated aptamers is that riboswitches have been evolved to have the type of in vivo selectivity and binding affinity needed to recognize cellular targets.

4.4. Specific Base Pair Formation

RNA-based recognition modules can also be designed based on sequence-specific base pairings. In addition to the traditional Watson-Crick (A to U and C to G) base pairs, wobble base pairs (e.g., G to U or I to C), G-quadruplexes, and metallo-base pairs can also be engineered as specific recognition modules for the development of RNA-based sensors. For example, we have recently developed a C–Ag^+–C metallo-base pair-based fluorogenic RNA sensor for the intracellular imaging of Ag^+ ions [117]. In this study, these metallo base pairs can function as both the recognition module and the transducer module. The signal transduction mechanism is similar to the one discussed above in Section 2.2.

5. Recent Examples of GERMS

GERMS have been successfully applied in multiple intracellular studies. For example, the Jaffrey lab developed a type of allosteric Spinach sensor. Similar to the one shown in Figure 1A, the allosteric Spinach sensor comprises a target-binding aptamer (recognition module), a transducer duplex (transducer module), and a Spinach aptamer (reporting system). This type of sensor has been engineered to detect diverse metabolites and proteins, such as adenosine diphosphate, SAM, guanosine triphosphate, thrombin, and MCP coat protein [29,30]. In Table 2, we have shown some of the existing GERMS for intracellular applications. The optimal sensors normally exhibited 10- to 40-fold increases in fluorescence upon binding their cognate ligands. Notably, a SAM-targeting allosteric Spinach sensor has been used to reveal cell-to-cell variations in the SAM metabolism, which cannot be observed via conventional methods [29].

We previously engineered Spinach riboswitches, nature-inspired GERMS for detecting metabolites in the cytosol of cells with high target selectivity. For example, by engineering the Spinach aptamer into the expression platform in a natural *thiM* TPP riboswitch, we developed TPP-targeting GERMS. Similar to that shown in Figure 1C, the TPP-dependent natural switching mechanism of the riboswitch enables the proper folding of the Spinach aptamer, which then activates the fluorescence of DFHBI [32]. Compared to aptamers selected by in vitro SELEX, naturally occurring riboswitches have inherent advantages in their high affinity and selectivity for cellular targets. Currently, many naturally occurring riboswitches have been discovered, and the Spinach riboswitch strategy enables the direct conversion of riboswitches into functional GERMS.

Table 2. Properties and design of existing GERMS for intracellular applications. [a]

Target	Recognition Module Source	Transducer Module Type	Reporting System Type	EC$_{50}$ (μM)	ON/OFF [b]	Cell System
ADP [29]	SELEX	Duplex Formation	Spinach	270	20	Bacteria
5-HTP [118]	SELEX	Duplex Formation	Broccoli	N/A [c]	>5	Bacteria
L-DOPA [118]	SELEX	Duplex Formation	Broccoli	N/A	>5	Bacteria
MS2 coat protein [30]	SELEX	Duplex Formation	Spinach	~0.6	41.7	Bacteria
MS2 coat protein [119]	SELEX	Ribozyme	BFP	N/A	1.8	Mammalian
Neomycin [120]	SELEX	Ribozyme	β-galactosidase	N/A	25	Yeast
Streptavidin [30]	SELEX	Duplex Formation	Spinach	<0.2	10.3	Bacteria
Tetracycline/Theophylline [121]	SELEX	Ribozyme	EGFP	N/A	N/A	Yeast
Tetracycline [122]	SELEX	Ribozyme	Luciferase/EGFP	35.4	4.8	Mammalian
Theophylline [123]	SELEX	Ribozyme	EGFP	~200	10	Bacteria
c-AMP-GMP [38]	Riboswitch	Duplex Formation	Spinach	4.2	~8 (37 °C)	Bacteria
c-di-AMP [39]	Riboswitch	Strand Displacement	Spinach2	3.4 & 29	2.4 & 9.1	Bacteria
c-di-GMP [38]	Riboswitch	Duplex Formation	Spinach	0.23	~6 (37 °C)	Bacteria
c-di-GMP [124]	Riboswitch	Duplex Formation	Spinach2	0.005–0.4	~6 (37 °C)	Bacteria
c-di-GMP [125]	Riboswitch	Strand Displacement	TurboRFP	N/A	38	Bacteria
Guanine [126]	Riboswitch	Ribozyme	EGFP	N/A	9.6	Mammalian
SAM [29]	Riboswitch	Duplex Formation	Spinach	120	25	Bacteria
TPP [32]	Riboswitch	Strand Displacement	Spinach	9	15.9	Bacteria
TPP [127]	Riboswitch	Strand Displacement	EGFP	N/A	~5	Plant
TPP [128]	Riboswitch	Ribozyme	tRNA	N/A	43	Bacteria
N-peptide [129]	Ribonucleoprotein complexes	Ribozyme	EGFP/SEAP	N/A	~12	Mammalian
RNA [130]	Base Pairing	Ribozyme	EGFP	N/A	~10	Bacteria
RNA [131]	Base Pairing	Strand Displacement	Split Broccoli	~0.001	2.2	Bacteria

[a] Only one example is given when the same design principle has been used to detect the same target molecules.
[b] ON/OFF indicates the number of fold enhancements in the *in vitro* fluorescence of GERMS after adding the target (measured at 25 °C unless otherwise stated). [c] N/A—Not Available.

We recently engineered another class of RNA-based fluorescent sensors, termed RNA integrators, for the intracellular detection of low-abundance metabolites [132]. In this design, the self-cleaving property of hammerhead ribozymes is used to activate the Broccoli aptamer upon binding to target molecules. Similar to that shown in Figure 1D, in the presence of target molecules, the recognition module rearranges to form the binding pocket, which leads to the formation of the catalytic pocket in the hammerhead ribozyme. As a result, target binding induces the activation of the self-cleavage of the ribozyme and releases the downstream Broccoli aptamer sequence, which then binds DFHBI in order to emit fluorescence. Here, each target molecule can induce the cleavage of multiple copies of the RNA integrator, resulting in an amplified signal.

In addition to these nature-inspired designs, GERMS can also be engineered based on recent advancements in DNA and RNA nanotechnology. For example, our lab has recently developed the first GERMS based on an RNA logic circuit, termed the Catalytic Hairpin Assembly RNA circuit, that is Genetically Encoded (CHARGE) [131]. In our CHARGE sensor design, two complementary RNA hairpins stay separate from each other in the absence of a target (Figure 4A). After adding the target, one hairpin opens based on a toehold-mediated strand displacement reaction, and then induces the subsequent hybridization of both hairpins [133]. The target can then be recycled to trigger the hybridization of multiple copies of hairpins. By coupling with split-Broccoli, we were able to image cellular RNA targets with a high sensitivity [131].

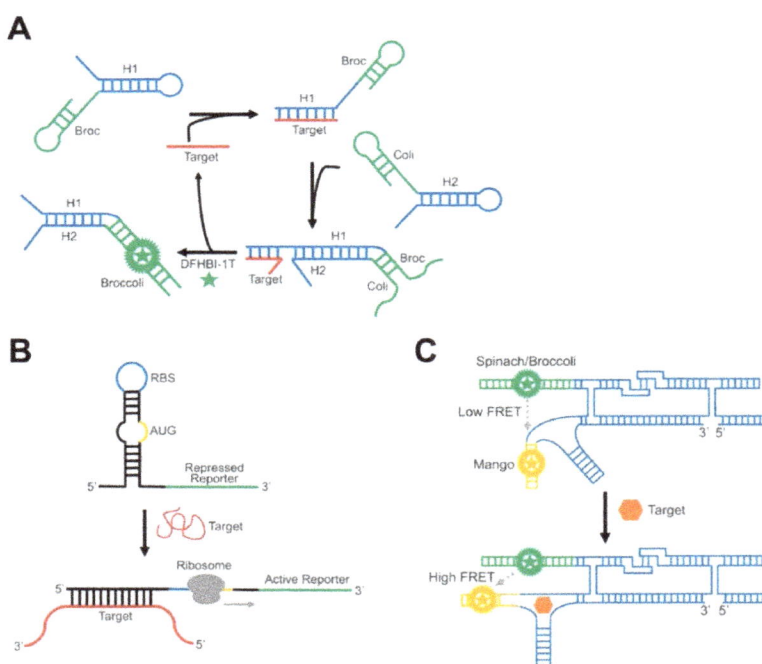

Figure 4. Schematics of RNA nanotechnology-inspired GERMS. (**A**) In a CHARGE circuit, target binding (red) induces the catalytic hybridization of multiple hairpin assemblies (blue), further activating an amplified signal from reassembled Broccoli RNA (green). (**B**) In a toehold switch sensor, target binding releases the ribosome binding site (RBS) and a start codon (AUG), which activates the expression of the reporting system (green). (**C**) In an RNA origami construct, target-induced structure change can regulate the distance and FRET efficiency between two fluorogenic RNA complexes.

In another example showing that dynamic RNA nanotechnology can contribute to the design of GERMS, the Yin group has developed toehold switches to detect target RNAs with an average ON/OFF ratio of over 400 [134]. The toehold riboswitch functions by the target-induced post-transcriptional activation of the gene expression (Figure 4B). Taking advantage of toehold-mediated linear-linear interactions [133], target RNA can bind with the sequences around the ribosome binding sites (RBS) and a start codon (AUG), triggering a branch migration process to expose the RBS and the start codon. As a result, the presence of the target RNA strand initiates the translation of the downstream fluorescent protein and emits a corresponding fluorescence signal. The orthogonality and programmability of toehold switches can even allow for the independent regulation of 12 genes and is used to construct complex genetic circuits [134].

Synthetic RNA nanotechnology, i.e., the design and construction of artificial RNA nanostructures, can also provide a useful scaffold to improve the performance of GERMS. For example, the Andersen group has recently reported a single-stranded RNA origami FRET system [46]. In their nanoconstruct, two fluorogenic RNA aptamers, Spinach and Mango, were placed in close proximity following a designed pattern (Figure 4C). In the absence of target molecules, the Spinach and Mango pair produced a limited FRET signal. Upon target binding, the origami rearranged the structure, bringing the two aptamers closer to each other and producing a large FRET signal. This construct has been successfully genetically encoded in *E. coli* cells, demonstrating its potential for intracellular imaging.

Other than the examples described above, GERMS can also function, in a way, as logic gates. Alam et al. showed that Split Broccoli aptamers can be converted into an AND gate for monitoring the assembly of RNA–RNA hybrids [82]. The Khisamutdinov group has recently demonstrated a new generation of smart RNA nanodevices based on RNA aptamers [129]. In their approach, the Malachite Green aptamer and the Broccoli aptamer were engineered into four types of oligonucleotide-responsive RNA logic gates (AND, OR, NAND and NOR), which offer a new route to engineer "label-free" ligand-sensing regulatory circuits and nucleic acid detection systems.

6. Conclusions and Outlook

Over the past few years, GERMS have emerged for live-cell imaging and the detection of various RNAs, proteins, metabolites, synthetic compounds, and ions. The high versatility of these RNA nanostructures has provided GERMS with a wide choice regarding the recognition modules, transducer modules, and the reporting systems. GERMS can be developed toward various targets with both a high binding affinity and selectivity. The sensitivity, modularity, and dynamic range of these RNA-based sensors have been dramatically improved.

One critical challenge in the rational design of GERMS is to understand how the recognition module changes its conformation after binding to its target. Indeed, it can be difficult to design transducer modules if the structures of both the apo- and holo- forms of the recognition module remain unknown. The crystal structures for most existing riboswitches have been solved. However, for many SELEX-generated aptamers, we still have limited knowledge about their tertiary structures. On the other hand, computational simulations have been used to assist the design and engineering of GERMS. Unfortunately, it is still challenging to accurately simulate many complex intramolecular interactions among different modules within these functional RNA structures, without mentioning the challenge in predicting how target binding can thermodynamically and kinetically change the conformation of GERMS.

Currently, it is still taking a long time and many trials to develop a functional RNA-based sensor. The number of selection rounds will be greatly reduced if there are guidelines for the pairing of different modules in GERMS. In other words, if we could design the transducer module simply by looking at the sequence of the recognition module and its binding pocket, this would greatly improve the design efficiency. Potential milestones in engineering GERMS will likely depend on revolutionary algorithms in computational simulations and a more comprehensive understanding of the relationships between RNA sequences and their corresponding tertiary structures.

Another limitation in applying GERMS for mammalian cells or in vivo imaging is RNA degradation and low-level expression. Short RNA constructs, like those in most GERMS, can be rapidly degraded in eukaryotic cells. One potential solution for improving the expression level of GERMS is based on circular RNA constructs. Circular RNAs have been identified in vivo as naturally evolved stable RNA molecules. Circular RNAs do not have either a free 5′- or 3′-end, which makes them invulnerable to most cellular exonucleases. Recent studies have shown that these circular RNAs can be stable for days-to-weeks and that they accumulate at high levels in diverse eukaryotic organisms [134–138]. The potential incorporation of the circular RNA strategy in GERMS may open a new window for the in vivo imaging and detection of targets that have not been successfully studied with available RNA- or protein-based sensors.

In conclusion, we have summarized in this review the basic design principles and recent applications of GERMS for bioimaging and the detection of cellular targets. The versatility of these RNA-based sensors makes GERMS highly useful for studying essentially any molecule in living cells. GERMS have shown great potential for future live-cell imaging. After improving their biostability, sensitivity, target selectivity, and kinetics, the next steps will likely be the engineering of GERMS into working sensors in eukaryotic cells, as well as the generation of universal protocols for developing GERMS toward any target of interest.

Author Contributions: Conceptualization, M.Y. and Z.S.; investigation, Z.S., T.N. and K.M.; writing—original draft preparation, Z.S., T.N. and K.M.; writing—review and editing, Z.S. and M.Y.; supervision, M.Y.; funding acquisition, M.Y.

Funding: This work was supported by the start-up grant from UMass Amherst, a National Institutes of Health grant R01AI136789, and a National Science Foundation CAREER award #1846152.

Acknowledgments: The authors are grateful to Kathryn R. Williams for help with manuscript preparation. The authors also thank other members in the You Lab for useful discussion and valuable comments.

Conflicts of Interest: The authors declare no conflict of interest.

Abbreviations

GERMS	Genetically Encoded RNA-based Molecular Sensors
FRET	Förster Resonance Energy Transfer
FACS	Fluorescence-Activated Cell Sorting
SELEX	Systematic Evolution of Ligands by EXponential enrichment
GO	Graphene Oxide
PCR	Polymerase Chain Reaction
PAGE	Polyacrylamide Gel Electrophoresis
CHARGE	Catalytic Hairpin Assembly RNA circuit that is Genetically Encoded
BODIPY	Boron Dipyrromethene
FP, GFP, EGFP, BFP, RFP	Fluorescent Protein, Green Fluorescent Protein, Enhanced Green Fluorescent Protein, Blue Fluorescent Protein, Red Fluorescent Protein
DFHBI	3,5-difluoro-4-hydroxybenzylidene Imidazolinone
DFHO	3,5-difluoro-4-hydroxybenzylidene Imidazolinone-2-oxime
DNB, SRB, SR-DN	Dinitroaniline-Binding aptamer, Sulforhodamine B, Sulforhodamine-Dinitroaniline
TO-Biotin	Thizole Orange-Biotin
Co^{2+}, Ag^+	Cobalt ion, Silver ion
ATP, ADP, AMP, GMP	Adenosine Triphosphate, Adenosine Diphosphate, Adenosine Monophosphate, Guanosine Monophosphate
HIV	Human Immunodeficiency Virus
SAH, SAM	S-adenosylhomocysteine, S-adenosylmethionine
MTase	Methyltransferase
TPP	Thiamine 5′-pyrophosphate
MCP	MS2 Coat Protein
5-HTP	5-hydroxytryptophan
L-DOPA	L-3,4-dihydroxyphenylalanine

References

1. Spiller, D.G.; Wood, C.D.; Rand, D.A.; White, M.R.H. Measurement of Single-Cell Dynamics. *Nature* **2010**, *465*, 736–745. [CrossRef] [PubMed]

2. Carlo, D.D.; Lee, L.P. Dynamic Single-Cell Analysis for Quantitative Biology. *Anal. Chem.* **2006**, *78*, 7918–7925. [CrossRef]

3. Jares-Erijman, E.A.; Jovin, T.M. FRET Imaging. *Nat. Biotechnol.* **2003**, *21*, 1387–1395. [CrossRef]

4. Manz, A.; Dittrich, P.S.; Pamme, N.; Iossifidis, D. *Bioanalytical Chemistry*; IMPERIAL COLLEGE PRESS: London, UK, 2015.

5. Zhang, J.; Campbell, R.E.; Ting, A.Y.; Tsien, R.Y. Creating New Fluorescent Probes for Cell Biology. *Nat. Rev. Mol. Cell Biol.* **2002**, *3*, 906–918. [CrossRef] [PubMed]

6. Specht, E.A.; Braselmann, E.; Palmer, A.E. A Critical and Comparative Review of Fluorescent Tools for Live-Cell Imaging. *Annu. Rev. Physiol.* **2017**, *79*, 93–117. [CrossRef]

7. Wysocki, L.M.; Lavis, L.D. Advances in the Chemistry of Small Molecule Fluorescent Probes. *Curr. Opin. Chem. Biol.* **2011**, *15*, 753–759. [CrossRef]

8. Yuan, L.; Lin, W.; Zheng, K.; Zhu, S. FRET-Based Small-Molecule Fluorescent Probes: Rational Design and Bioimaging Applications. *Acc. Chem. Res.* **2013**, *46*, 1462–1473. [CrossRef]

9. Fernández-Suárez, M.; Ting, A.Y. Fluorescent Probes for Super-Resolution Imaging in Living Cells. *Nat. Rev. Mol. Cell Biol.* **2008**, *9*, 929–943. [CrossRef] [PubMed]

10. Oliveira, E.; Bértolo, E.; Núñez, C.; Pilla, V.; Santos, H.M.; Fernández-Lodeiro, J.; Fernández-Lodeiro, A.; Djafari, J.; Capelo, J.L.; Lodeiro, C. Green and Red Fluorescent Dyes for Translational Applications in Imaging and Sensing Analytes: A Dual-Color Flag. *ChemistryOpen* **2018**, *7*, 3. [CrossRef] [PubMed]

11. Chen, H.; Zhou, B.; Ye, R.; Zhu, J.; Bao, X. Synthesis and Evaluation of a New Fluorescein and Rhodamine B-Based Chemosensor for Highly Sensitive and Selective Detection of Cysteine over Other Amino Acids and Its Application in Living Cell Imaging. *Sens. Actuat. B-Chem.* **2017**, *251*, 481–489. [CrossRef]

12. Loudet, A.; Burgess, K. BODIPY Dyes and Their Derivatives: Syntheses and Spectroscopic Properties. *Chem. Rev.* **2007**, *107*, 4891–4932. [CrossRef]

13. Luo, S.; Zhang, E.; Su, Y.; Cheng, T.; Shi, C. A Review of NIR Dyes in Cancer Targeting and Imaging. *Biomaterials* **2011**, *32*, 7127–7138. [CrossRef]

14. Ueno, T.; Nagano, T. Fluorescent Probes for Sensing and Imaging. *Nat. Methods* **2011**, *8*, 642–645. [CrossRef]

15. Van de Linde, S.; Aufmkolk, S.; Franke, C.; Holm, T.; Klein, T.; Löschberger, A.; Proppert, S.; Wolter, S.; Sauer, M. Investigating Cellular Structures at the Nanoscale with Organic Fluorophores. *Chem. Biol.* **2013**, *20*, 8–18. [CrossRef]

16. Zhu, H.; Fan, J.; Du, J.; Peng, X. Fluorescent Probes for Sensing and Imaging within Specific Cellular Organelles. *Acc. Chem. Res.* **2016**, *49*, 2115–2126. [CrossRef]

17. Alford, R.; Simpson, H.M.; Duberman, J.; Hill, G.C.; Ogawa, M.; Regino, C.; Kobayashi, H.; Choyke, P.L. Toxicity of Organic Fluorophores Used in Molecular Imaging: Literature Review. *Mol. Imaging* **2009**, *8*, 7290.2009.00031. [CrossRef]

18. Crawford, J.M.; Braunwald, N.S. Toxicity in Vital Fluorescence Microscopy: Effect of Dimethylsulfoxide, Rhodamine-123, and DiI-Low Density Lipoprotein on Fibroblast Growth in Vitro. *Vitr. Cell. Dev. Biol.-Anim.* **1991**, *27*, 633–638. [CrossRef]

19. Fei, X.; Gu, Y. Progress in Modifications and Applications of Fluorescent Dye Probe. *Prog. Nat. Sci.* **2009**, *19*, 1–7. [CrossRef]

20. Heim, R.; Prasher, D.C.; Tsien, R.Y. Wavelength Mutations and Posttranslational Autoxidation of Green Fluorescent Protein. *Proc. Natl. Acad. Sci. USA* **1994**, *91*, 12501–12504. [CrossRef]

21. Morise, H.; Shimomura, O.; Johnson, F.H.; Winant, J. Intermolecular Energy Transfer in the Bioluminescent System of Aequorea. *Biochemistry* **1974**, *13*, 2656–2662. [CrossRef]

22. Chudakov, D.M.; Matz, M.V.; Lukyanov, S.; Lukyanov, K.A. Fluorescent Proteins and Their Applications in Imaging Living Cells and Tissues. *Physiol. Rev.* **2010**, *90*, 1103–1163. [CrossRef]

23. Day, R.N.; Davidson, M.W. The Fluorescent Protein Palette: Tools for Cellular Imaging. *Chem. Soc. Rev.* **2009**, *38*, 2887. [CrossRef]

24. Kremers, G.-J.; Gilbert, S.G.; Cranfill, P.J.; Davidson, M.W.; Piston, D.W. Fluorescent Proteins at a Glance. *J. Cell Sci.* **2011**, *124*, 157–160. [CrossRef]

25. Davidson, M.W.; Campbell, R.E. Engineered Fluorescent Proteins: Innovations and Applications. *Nat. Methods* **2009**, *6*, 713–717. [CrossRef]

26. Shaner, N.C.; Patterson, G.H.; Davidson, M.W. Advances in Fluorescent Protein Technology. *J. Cell Sci.* **2007**, *120*, 4247–4260. [CrossRef]

27. Okumoto, S.; Jones, A.; Frommer, W.B. Quantitative Imaging with Fluorescent Biosensors. *Annu. Rev. Plant Biol.* **2012**, *63*, 663–706. [CrossRef]

28. Piston, D.W.; Kremers, G.-J. Fluorescent Protein FRET: The Good, the Bad and the Ugly. *Trends Biochem. Sci.* **2007**, *32*, 407–414. [CrossRef]

29. Paige, J.S.; Nguyen-Duc, T.; Song, W.; Jaffrey, S.R. Fluorescence Imaging of Cellular Metabolites with RNA. *Science* **2012**, *335*, 1194. [CrossRef]

30. Song, W.; Strack, R.L.; Jaffrey, S.R. Imaging Bacterial Protein Expression Using Genetically Encoded RNA Sensors. *Nat. Methods* **2013**, *10*, 873–875. [CrossRef]

31. Song, W.; Strack, R.L.; Svensen, N.; Jaffrey, S.R. Plug-and-Play Fluorophores Extend the Spectral Properties of Spinach. *J. Am. Chem. Soc.* **2014**, *136*, 1198–1201. [CrossRef]

32. You, M.; Litke, J.L.; Jaffrey, S.R. Imaging Metabolite Dynamics in Living Cells Using a Spinach-Based Riboswitch. *Proc. Natl. Acad. Sci. USA* **2015**, *112*, E2756–E2765. [CrossRef]

33. Song, K.-M.; Lee, S.; Ban, C. Aptamers and Their Biological Applications. *Sensors* **2012**, *12*, 612–631. [CrossRef]

34. Lakhin, A.V.; Tarantul, V.Z.; Gening, L.V. Aptamers: Problems, Solutions and Prospects. *Acta Nat.* **2013**, *5*, 34–43.

35. Paige, J.S.; Wu, K.Y.; Jaffrey, S.R. RNA Mimics of Green Fluorescent Protein. *Science* **2011**, *333*, 642–646. [CrossRef]

36. Iii, R.J.T.; Truong, L.; Ferré-D'amaré, A.R. Structural Principles of Fluorescent RNA Aptamers. *Trends Pharmacol. Sci.* **2017**, *38*, 928–939. [CrossRef]

37. Ouellet, J. RNA Fluorescence with Light-Up Aptamers. *Front. Chem.* **2016**, *4*, 29. [CrossRef]

38. Kellenberger, C.A.; Wilson, S.C.; Sales-Lee, J.; Hammond, M.C. RNA-Based Fluorescent Biosensors for Live Cell Imaging of Second Messengers Cyclic Di-GMP and Cyclic AMP-GMP. *J. Am. Chem. Soc.* **2013**, *135*, 4906–4909. [CrossRef]

39. Kellenberger, C.A.; Chen, C.; Whiteley, A.T.; Portnoy, D.A.; Hammond, M.C. RNA-Based Fluorescent Biosensors for Live Cell Imaging of Second Messenger Cyclic Di-AMP. *J. Am. Chem. Soc.* **2015**, *137*, 6432–6435. [CrossRef]

40. Cromie, M.J.; Shi, Y.; Latifi, T.; Groisman, E.A. An RNA Sensor for Intracellular Mg^{2+}. *Cell* **2006**, *125*, 71–84. [CrossRef]

41. Bose, D.; Su, Y.; Marcus, A.; Raulet, D.H.; Hammond, M.C. An RNA-Based Fluorescent Biosensor for High-Throughput Analysis of the CGAS-CGAMP-STING Pathway. *Cell Chem. Biol.* **2016**, *23*, 1539–1549. [CrossRef]

42. Win, M.N.; Smolke, C.D. A Modular and Extensible RNA-Based Gene-Regulatory Platform for Engineering Cellular Function. *Proc. Natl. Acad. Sci. USA* **2007**, *104*, 14283–14288. [CrossRef]

43. Wang, R.E.; Zhang, Y.; Cai, J.; Cai, W.; Gao, T. Aptamer-Based Fluorescent Biosensors. *Curr. Med. Chem.* **2011**, *18*, 4175–4184. [CrossRef]

44. Strack, R.L.; Jaffrey, S.R. New Approaches for Sensing Metabolites and Proteins in Live Cells Using RNA. *Curr. Opin. Chem. Biol.* **2013**, *17*, 651–655. [CrossRef]

45. Kellenberger, C.A.; Hammond, M.C. In Vitro Analysis of Riboswitch–Spinach Aptamer Fusions as Metabolite-Sensing Fluorescent Biosensors. *Methods Enzymol.* **2015**, *550*, 147–172. [CrossRef]

46. Jepsen, M.D.E.E.; Sparvath, S.M.; Nielsen, T.B.; Langvad, A.H.; Grossi, G.; Gothelf, K.V.; Andersen, E.S. Development of a Genetically Encodable FRET System Using Fluorescent RNA Aptamers. *Nat. Commun.* **2018**, *9*, 1–10. [CrossRef]

47. Zhang, C.; Wei, Z.H.; Ye, B.C. Imaging and Tracing of Intracellular Metabolites Utilizing Genetically Encoded Fluorescent Biosensors. *Biotechnol. J.* **2013**, *8*, 1280–1291. [CrossRef]

48. Chakraborty, K.; Veetil, A.T.; Jaffrey, S.R.; Krishnan, Y. Nucleic Acid-Based Nanodevices in Biological Imaging. *Annu. Rev. Biochem.* **2016**, *85*, 349–373. [CrossRef]

49. Feagin, T.A.; Maganzini, N.; Soh, H.T. Strategies for Creating Structure-Switching Aptamers. *ACS Sens.* **2018**, *3*, 1611–1615. [CrossRef]

50. Hall, B.; Hesselberth, J.R.; Ellington, A.D. Computational Selection of Nucleic Acid Biosensors via a Slip Structure Model. *Biosens. Bioelectron.* **2007**, *22*, 1939–1947. [CrossRef]

51. Winkler, W.; Nahvi, A.; Breaker, R.R. Thiamine Derivatives Bind Messenger RNAs Directly to Regulate Bacterial Gene Expression. *Nature* **2002**, *419*, 952–956. [CrossRef]

52. Espah Borujeni, A.; Mishler, D.M.; Wang, J.; Huso, W.; Salis, H.M. Automated Physics-Based Design of Synthetic Riboswitches from Diverse RNA Aptamers. *Nucleic Acids Res.* **2016**, *44*, 1–13. [CrossRef] [PubMed]

53. Chappell, J.; Westbrook, A.; Verosloff, M.; Lucks, J.B. Computational Design of Small Transcription Activating RNAs for Versatile and Dynamic Gene Regulation. *Nat. Commun.* **2017**, *8*, 1051. [CrossRef] [PubMed]

54. Ausländer, S.; Fuchs, D.; Hürlemann, S.; Ausländer, D.; Fussenegger, M. Engineering a Ribozyme Cleavage-Induced Split Fluorescent Aptamer Complementation Assay. *Nucleic Acids Res.* **2016**, *44*, e94. [CrossRef] [PubMed]

55. Felletti, M.; Stifel, J.; Wurmthaler, L.A.; Geiger, S.; Hartig, J.S. Twister Ribozymes as Highly Versatile Expression Platforms for Artificial Riboswitches. *Nat. Commun.* **2016**, *7*, 12834. [CrossRef] [PubMed]

56. Furukawa, K.; Gu, H.; Breaker, R.R. In Vitro Selection of Allosteric Ribozymes That Sense the Bacterial Second Messenger C-Di-GMP. *Methods Mol. Biol.* **2014**, *1111*, 209–220. [CrossRef] [PubMed]

57. Murray, J.B.; Terwey, D.P.; Maloney, L.; Karpeisky, A.; Usman, N.; Beigelman, L.; Scott, W.G. The Structural Basis of Hammerhead Ribozyme Self-Cleavage. *Cell* **1998**, *92*, 665–673. [CrossRef]

58. Lee, E.R.; Baker, J.L.; Weinberg, Z.; Sudarsan, N.; Breaker, R.R. An Allosteric Self-Splicing Ribozyme Triggered by a Bacterial Second Messenger. *Science* **2010**, *329*, 845–848. [CrossRef]

59. Roth, A.; Weinberg, Z.; Chen, A.G.Y.; Kim, P.B.; Ames, T.D.; Breaker, R.R. A Widespread Self-Cleaving Ribozyme Class Is Revealed by Bioinformatics. *Nat. Chem. Biol.* **2014**, *10*, 56–60. [CrossRef]

60. Weinberg, Z.; Kim, P.B.; Chen, T.H.; Li, S.; Harris, K.A.; Lünse, C.E.; Breaker, R.R. New Classes of Self-Cleaving Ribozymes Revealed by Comparative Genomics Analysis. *Nat. Chem. Biol.* **2015**, *11*, 606–610. [CrossRef]

61. Lilley, D.M.J. The Varkud Satellite Ribozyme. *RNA* **2004**, *10*, 151–158. [CrossRef]

62. Fedor, M.J. Structure and Function of the Hairpin Ribozyme. *J. Mol. Biol.* **2000**, *297*, 269–291. [CrossRef] [PubMed]

63. Ferré-D'Amaré, A.R.; Scott, W.G. Small Self-Cleaving Ribozymes. *Cold Spring Harb. Perspect. Biol.* **2010**, *2*, a003574. [CrossRef] [PubMed]

64. Vesuna, F.; Winnard, P.; Raman, V. Enhanced Green Fluorescent Protein as an Alternative Control Reporter to Renilla Luciferase. *Anal. Biochem.* **2005**, *342*, 345–347. [CrossRef] [PubMed]

65. Zhang, B.S.; Jones, K.A.; McCutcheon, D.C.; Prescher, J.A. Pyridone Luciferins and Mutant Luciferases for Bioluminescence Imaging. *ChemBioChem* **2018**, *19*, 470–477. [CrossRef] [PubMed]

66. Culler, S.J.; Hoff, K.G.; Smolke, C.D. Reprogramming Cellular Behavior with RNA Controllers Responsive to Endogenous Proteins. *Science* **2010**, *330*, 1251–1255. [CrossRef] [PubMed]

67. Yoshimatsu, T.; Nagawa, F. Control of Gene Expression by Artificial Introns in Saccharomyces Cerevisiae. *Science* **1989**, *244*, 1346–1348. [CrossRef] [PubMed]

68. Swinburne, I.A.; Miguez, D.G.; Landgraf, D.; Silver, P.A. Intron Length Increases Oscillatory Periods of Gene Expression in Animal Cells. *Genes Dev.* **2008**, *22*, 2342–2346. [CrossRef]

69. Carothers, J.M.; Goler, J.A.; Juminaga, D.; Keasling, J.D. Model-Driven Engineering of RNA Devices to Quantitatively Program Gene Expression. *Science* **2011**, *334*, 1716–1719. [CrossRef]

70. Purnick, P.E.M.; Weiss, R. The second wave of synthetic biology: From modules to systems. *Nat. Rev. Mol. Cell Biol.* **2009**, *10*, 410–422. [CrossRef]

71. Ausländer, S.; Ketzer, P.; Hartig, J.S. A Ligand-Dependent Hammerhead Ribozyme Switch for Controlling Mammalian Gene Expression. *Mol. Biosyst.* **2010**, *6*, 807. [CrossRef]

72. Qi, L.; Haurwitz, R.E.; Shao, W.; Doudna, J.A.; Arkin, A.P. RNA Processing Enables Predictable Programming of Gene Expression. *Nat. Biotechnol.* **2012**, *30*, 1002–1006. [CrossRef] [PubMed]

73. Salmena, L.; Poliseno, L.; Tay, Y.; Kats, L.; Pandolfi, P.P. A CeRNA Hypothesis: The Rosetta Stone of a Hidden RNA Language? *Cell* **2011**, *146*, 353–358. [CrossRef] [PubMed]

74. Rackham, O.; Chin, J.W. A Network of Orthogonal Ribosome·mRNA Pairs. *Nat. Chem. Biol.* **2005**, *1*, 159–166. [CrossRef] [PubMed]

75. Anderson, J.C.; Voigt, C.A.; Arkin, A.P. Environmental Signal Integration by a Modular AND Gate. *Mol. Syst. Biol.* **2007**, *3*, 133. [CrossRef] [PubMed]

76. Venkatesan, A.; Dasgupta, A. Novel Fluorescence-Based Screen to Identify Small Synthetic Internal Ribosome Entry Site Elements. *Mol. Cell Biol.* **2001**, *21*, 2826–2837. [CrossRef] [PubMed]

77. Salis, H.M.; Mirsky, E.A.; Voigt, C.A. Automated Design of Synthetic Ribosome Binding Sites to Control Protein Expression. *Nat. Biotechnol.* **2009**, *27*, 946–950. [CrossRef] [PubMed]

78. Ketterer, S.; Fuchs, D.; Weber, W.; Meier, M. Systematic Reconstruction of Binding and Stability Landscapes of the Fluorogenic Aptamer Spinach. *Nucleic Acids Res.* **2015**, *43*, 9564–9572. [CrossRef]

79. Filonov, G.S.; Moon, J.D.; Svensen, N.; Jaffrey, S.R. Broccoli: Rapid Selection of an RNA Mimic of Green Fluorescent Protein by Fluorescence-Based Selection and Directed Evolution. *J. Am. Chem. Soc.* **2014**, *136*, 16299–16308. [CrossRef]

80. Strack, R.L.; Disney, M.D.; Jaffrey, S.R. A Superfolding Spinach2 Reveals the Dynamic Nature of Trinucleotide Repeat–containing RNA. *Nat. Methods* **2013**, *10*, 1219–1224. [CrossRef]

81. Okuda, M.; Fourmy, D.; Yoshizawa, S. Use of Baby Spinach and Broccoli for Imaging of Structured Cellular RNAs. *Nucleic Acids Res.* **2016**, *45*, gkw794. [CrossRef]

82. Alam, K.K.; Tawiah, K.D.; Lichte, M.F.; Porciani, D.; Burke, D.H. A Fluorescent Split Aptamer for Visualizing RNA–RNA Assembly *In Vivo*. *ACS Synth. Biol.* **2017**, *6*, 1710–1721. [CrossRef] [PubMed]

83. Song, W.; Filonov, G.S.; Kim, H.; Hirsch, M.; Li, X.; Moon, J.D.; Jaffrey, S.R. Imaging RNA Polymerase III Transcription Using a Photostable RNA-Fluorophore Complex. *Nat. Chem. Biol.* **2017**, *13*, 1187–1194. [CrossRef] [PubMed]

84. Dolgosheina, E.V.; Jeng, S.C.Y.; Panchapakesan, S.S.S.; Cojocaru, R.; Chen, P.S.K.; Wilson, P.D.; Hawkins, N.; Wiggins, P.A.; Unrau, P.J. RNA Mango Aptamer-Fluorophore: A Bright, High-Affinity Complex for RNA Labeling and Tracking. *ACS Chem. Biol.* **2014**, *9*, 2412–2420. [CrossRef] [PubMed]

85. Autour, A.; Jeng, S.C.Y.; Cawte, A.D.; Abdolahzadeh, A.; Galli, A.; Panchapakesan, S.S.S.; Rueda, D.; Ryckelynck, M.; Unrau, P.J.; Jeng, S.C.Y.; et al. Fluorogenic RNA Mango Aptamers for Imaging Small Non-Coding RNAs in Mammalian Cells. *Nat. Commun.* **2018**, *9*, 656. [CrossRef]

86. Sunbul, M.; Jäschke, A. Contact-Mediated Quenching for RNA Imaging in Bacteria with a Fluorophore-Binding Aptamer. *Angew. Chem. Int. Ed.* **2013**, *52*, 13401–13404. [CrossRef] [PubMed]

87. Arora, A.; Sunbul, M.; Jäschke, A. Dual-Colour Imaging of RNAs Using Quencher- and Fluorophore-Binding Aptamers. *Nucleic Acids Res.* **2015**, *43*, gkv718. [CrossRef] [PubMed]

88. Warner, K.D.; Chen, M.C.; Song, W.; Strack, R.L.; Thorn, A.; Jaffrey, S.R.; Ferré-D'Amaré, A.R. Structural Basis for Activity of Highly Efficient RNA Mimics of Green Fluorescent Protein. *Nat. Struct. Mol. Biol.* **2014**, *21*, 658–663. [CrossRef]

89. Karunanayake Mudiyanselage, A.P.K.K.; Wu, R.; Leon-Duque, M.A.; Ren, K.; You, M. "Second-Generation" Fluorogenic RNA-Based Sensors. *Methods* **2019**, in press. [CrossRef]

90. Ellington, A.D.; Szostak, J.W. In Vitro Selection of RNA Molecules that Bind Specific Ligands. *Nature* **1990**, *346*, 818–822. [CrossRef]

91. Tuerk, C.; Gold, L. Systematic Evolution of Ligands by Exponential Enrichment: RNA Ligands to Bacteriophage T4 DNA Polymerase. *Science* **1990**, *249*, 505–510. [CrossRef]

92. Stoltenburg, R.; Reinemann, C.; Strehlitz, B. SELEX—A (r)Evolutionary Method to Generate High-Affinity Nucleic Acid Ligands. *Biomol. Eng.* **2007**, *24*, 381–403. [CrossRef] [PubMed]

93. Mandal, M.; Breaker, R.R. Gene Regulation by Riboswitches. *Nat. Rev. Mol. Cell Biol.* **2004**, *5*, 451–463. [CrossRef] [PubMed]

94. Wieland, M.; Benz, A.; Klauser, B.; Hartig, J.S. Artificial Ribozyme Switches Containing Natural Riboswiteh Aptamer Domains. *Angew. Chem. Int. Ed.* **2009**, *48*, 2715–2718. [CrossRef] [PubMed]

95. Breaker, R.R.; McCown, P.; Corbino, K.; Stav, S.; Sherlock, M. Riboswitch Diversity and Distribution. *RNA* **2017**, rna-061234. [CrossRef]

96. Aquino-Jarquin, G.; Toscano-Garibay, J.D. RNA Aptamer Evolution: Two Decades of SELEction. *Int. J. Mol. Sci.* **2011**, *12*, 9155–9171. [CrossRef] [PubMed]

97. Wrzesinski, J.; Ciesiolka, J. Characterization of Structure and Metal Ions Specificity of Co^{2+}-Binding RNA Aptamers. *Biochemistry* **2005**, *44*, 6257–6268. [CrossRef]

98. Yarus, M. A Specific Amino Acid Binding Site Composed of RNA. *Science* **1988**, *240*, 1751–1758. [CrossRef] [PubMed]

99. Sassanfar, M.; Szostak, J.W. An RNA Motif that Binds ATP. *Nature* **1993**, *364*, 550–553. [CrossRef] [PubMed]

100. Mehlhorn, A.; Rahimi, P.; Joseph, Y. Aptamer-Based Biosensors for Antibiotic Detection: A Review. *Biosensors* **2018**, *8*, 54. [CrossRef] [PubMed]

101. Lorsch, J.R.; Szostak, J.W. In Vitro Selection of RNA Aptamers Specific for Cyanocobalamin. *Biochemistry* **1994**, *33*, 973–982. [CrossRef]

102. Bouhedda, F.; Autour, A.; Ryckelynck, M. Light-up RNA Aptamers and Their Cognate Fluorogens: From Their Development to Their Applications. *Int. J. Mol. Sci.* **2018**, *19*, 44. [CrossRef] [PubMed]

103. White, R.; Rusconi, C.; Scardino, E.; Wolberg, A.; Lawson, J.; Hoffman, M.; Sullenger, B. Generation of Species Cross-Reactive Aptamers Using "Toggle" SELEX. *Mol. Ther.* **2001**, *4*, 567–573. [CrossRef]

104. Mondragón, E.; Maher, L.J., III. Anti-Transcription Factor RNA Aptamers as Potential Therapeutics. *Nucleic Acid Ther.* **2016**, *26*, 29–43. [CrossRef] [PubMed]

105. Tuerk, C.; MacDougal-Waugh, S. In Vitro Evolution of Functional Nucleic Acids: High-Affinity RNA Ligands of HIV-1 Proteins. *Gene* **1993**, *137*, 33–39. [CrossRef]

106. Davydova, A.; Vorobjeva, M.; Pyshnyi, D.; Altman, S.; Vlassov, V.; Venyaminova, A. Aptamers against Pathogenic Microorganisms. *Crit. Rev. Microbiol.* **2016**, *42*, 847–865. [CrossRef] [PubMed]

107. Stoltenburg, R.; Nikolaus, N.; Strehlitz, B. Capture-SELEX: Selection of DNA Aptamers for Aminoglycoside Antibiotics. *J. Anal. Methods Chem.* **2012**, *2012*, 415697. [CrossRef] [PubMed]

108. Kruger, K.; Grabowski, P.J.; Zaug, A.J.; Sands, J.; Gottschling, D.E.; Cech, T.R. Self-Splicing RNA: Autoexcision and Autocyclization of the Ribosomal RNA Intervening Sequence of Tetrahymena. *Cell* **1982**, *31*, 147–157. [CrossRef]

109. Usman, N.; Beigelman, L.; McSwiggen, J.A. Hammerhead ribozyme engineering. *Curr. Opin. Struct. Biol.* **1996**, *6*, 527–533. [CrossRef]

110. Park, J.-W.; Tatavarty, R.; Kim, D.W.; Jung, H.-T.; Gu, M.B. Immobilization-Free Screening of Aptamers Assisted by Graphene Oxide. *Chem. Commun.* **2012**, *48*, 2071–2073. [CrossRef]

111. Nguyen, V.-T.; Kwon, Y.S.; Kim, J.H.; Gu, M.B. Multiple GO-SELEX for Efficient Screening of Flexible Aptamers. *Chem. Commun.* **2014**, *50*, 10513–10516. [CrossRef] [PubMed]

112. Garst, A.; Edwards, A.L.; Batey, R.T. Riboswitches: Structures and Mechanisms. *Cold Spring Harb. Perspect. Biol.* **2011**, *3*, a003533. [CrossRef] [PubMed]

113. Serganov, A.; Patel, D.J. Metabolite Recognition Principles and Molecular Mechanisms Underlying Riboswitch Function. *Annu. Rev. Biophys.* **2012**, *41*, 343–370. [CrossRef] [PubMed]

114. Hammann, C.; Westhof, E. Searching Genomes for Ribozymes and Riboswitches. *Genome Biol.* **2007**, *8*, 210. [CrossRef] [PubMed]

115. Wang, J.X.; Lee, E.R.; Morales, D.R.; Lim, J.; Breaker, R.R. Riboswitches That Sense S-Adenosylhomocysteine and Activate Genes Involved in Coenzyme Recycling. *Mol. Cell* **2008**, *29*, 691–702. [CrossRef] [PubMed]

116. Su, Y.; Hickey, S.F.; Keyser, S.G.L.; Hammond, M.C. *In Vitro* and *In Vivo* Enzyme Activity Screening via RNA-Based Fluorescent Biosensors for *S*-Adenosyl-L-Homocysteine (SAH). *J. Am. Chem. Soc.* **2016**, *138*, 7040–7047. [CrossRef] [PubMed]

117. Yu, Q.; Shi, J.; Mudiyanselage, A.P.K.K.K.; Wu, R.; Zhao, B.; Zhou, M.; You, M. Genetically Encoded RNA-Based Sensors for Intracellular Imaging of Silver Ions. *Chem. Commun.* **2019**, *55*, 707–710. [CrossRef] [PubMed]

118. Porter, E.B.; Polaski, J.T.; Morck, M.M.; Batey, R.T. Recurrent RNA Motifs as Scaffolds for Genetically Encodable Small-Molecule Biosensors. *Nat. Chem. Biol.* **2017**, *13*, 295–301. [CrossRef]

119. Kennedy, A.B.; Vowles, J.V.; d'Espaux, L.; Smolke, C.D. Protein-Responsive Ribozyme Switches in Eukaryotic Cells. *Nucleic Acids Res.* **2014**, *42*, 12306–12321. [CrossRef]

120. Klauser, B.; Atanasov, J.; Siewert, L.K.; Hartig, J.S. Ribozyme-Based Aminoglycoside Switches of Gene Expression Engineered by Genetic Selection in *S. Cerevisiae*. *ACS Synth. Biol.* **2015**, *4*, 516–525. [CrossRef]

121. Win, M.N.; Smolke, C.D. Higher-Order Cellular Information Processing with Synthetic RNA Devices. *Science* **2008**, *322*, 456–460. [CrossRef]

122. Beilstein, K.; Wittmann, A.; Grez, M.; Suess, B. Conditional Control of Mammalian Gene Expression by Tetracycline-Dependent Hammerhead Ribozymes. *ACS Synth. Biol.* **2015**, *4*, 526–534. [CrossRef] [PubMed]

123. Wieland, M.; Hartig, J.S. Improved Aptazyme Design and In Vivo Screening Enable Riboswitching in Bacteria. *Angew. Chem. Int. Ed.* **2008**, *47*, 2604–2607. [CrossRef] [PubMed]

124. Wang, X.C.; Wilson, S.C.; Hammond, M.C. Next-Generation RNA-Based Fluorescent Biosensors Enable Anaerobic Detection of Cyclic Di-GMP. *Nucleic Acids Res.* **2016**, *44*, e139. [CrossRef] [PubMed]

125. Zhou, H.; Zheng, C.; Su, J.; Chen, B.; Fu, Y.; Xie, Y.; Tang, Q.; Chou, S.-H.; He, J. Characterization of a Natural Triple-Tandem c-Di-GMP Riboswitch and Application of the Riboswitch-Based Dual-Fluorescence Reporter. *Sci. Rep.* **2016**, *6*, 20871. [CrossRef] [PubMed]

126. Nomura, Y.; Kumar, D.; Yokobayashi, Y. Synthetic Mammalian Riboswitches Based on Guanine Aptazyme. *Chem. Commun.* **2012**, *48*, 7215. [CrossRef] [PubMed]

127. Wachter, A.; Tunc-Ozdemir, M.; Grove, B.C.; Green, P.J.; Shintani, D.K.; Breaker, R.R. Riboswitch Control of Gene Expression in Plants by Splicing and Alternative 3′ End Processing of MRNAs. *Plant Cell* **2007**, *19*, 3437–3450. [CrossRef] [PubMed]

128. Saragliadis, A.; Hartig, J.S. Ribozyme-Based Transfer RNA Switches for Post-Transcriptional Control of Amino Acid Identity in Protein Synthesis. *J. Am. Chem. Soc.* **2013**, *135*, 8222–8226. [CrossRef]

129. Goldsworthy, V.; LaForce, G.; Abels, S.; Khisamutdinov, E. Fluorogenic RNA Aptamers: A Nano-Platform for Fabrication of Simple and Combinatorial Logic Gates. *Nanomaterials* **2018**, *8*, 984. [CrossRef]

130. Hansen, T.B.; Jensen, T.I.; Clausen, B.H.; Bramsen, J.B.; Finsen, B.; Damgaard, C.K.; Kjems, J. Natural RNA Circles Function as Efficient MicroRNA Sponges. *Nature* **2013**, *495*, 384–388. [CrossRef]

131. Karunanayake Mudiyanselage, A.P.K.K.; Yu, Q.; Leon-Duque, M.A.; Zhao, B.; Wu, R.; You, M. Genetically Encoded Catalytic Hairpin Assembly for Sensitive RNA Imaging in Live Cells. *J. Am. Chem. Soc.* **2018**, *140*, 8739–8745. [CrossRef]

132. You, M.; Litke, J.L.; Wu, R.; Jaffrey, S.R. Detection of Low-Abundance Metabolites in Live Cells Using an RNA Integrator. *Cell Chem. Biol.* **2019**, *26*. in press. [CrossRef]

133. Yin, P.; Choi, H.M.T.; Calvert, C.R.; Pierce, N.A. Programming Biomolecular Self-Assembly Pathways. *Nature* **2008**, *451*, 318–322. [CrossRef] [PubMed]

134. Green, A.A.; Silver, P.A.; Collins, J.J.; Yin, P. Toehold Switches: De-Novo-Designed Regulators of Gene Expression. *Cell* **2014**, *159*, 925–939. [CrossRef] [PubMed]

135. Memczak, S.; Jens, M.; Elefsinioti, A.; Torti, F.; Krueger, J.; Rybak, A.; Maier, L.; Mackowiak, S.D.; Gregersen, L.H.; Munschauer, M.; et al. Circular RNAs Are a Large Class of Animal RNAs with Regulatory Potency. *Nature* **2013**, *495*, 333–338. [CrossRef]

136. Barrett, S.P.; Salzman, J. Circular RNAs: Analysis, Expression and Potential Functions. *Development* **2016**, *143*, 1838–1847. [CrossRef] [PubMed]

137. Holdt, L.M.; Kohlmaier, A.; Teupser, D. Molecular Roles and Function of Circular RNAs in Eukaryotic Cells. *Cell. Mol. Life Sci.* **2018**, *75*, 1071–1098. [CrossRef] [PubMed]

138. Wesselhoeft, R.A.; Kowalski, P.S.; Anderson, D.G. Engineering Circular RNA for Potent and Stable Translation in Eukaryotic Cells. *Nat. Commun.* **2018**, *9*, 2629. [CrossRef] [PubMed]

MDPI

St. Alban-Anlage 66

4052 Basel

Switzerland

Tel. +41 61 683 77 34

Fax +41 61 302 89 18

www.mdpi.com

Nanomaterials Editorial Office

E-mail: nanomaterials@mdpi.com

www.mdpi.com/journal/nanomaterials

www.ingramcontent.com/pod-product-compliance
Lightning Source LLC
Chambersburg PA
CBHW051856210326
41597CB00033B/5914